P. K. Varshney · M. K. Arora

Advanced Image Processing Techniques
for Remotely Sensed Hyperspectral Data

P. K. Varshney · M. K. Arora

Advanced Image Processing Techniques for Remotely Sensed Hyperspectral Data

With 128 Figures and 30 Tables

Professor Dr. Pramod K. Varshney
Syracuse University
Department of Electrical Engineering
and Computer Science
Syracuse, NY 13244
U.S.A.

Dr. Manoj K. Arora
Indian Institute of Technology Roorkee
Department of Civil Engineering
Roorkee 247 667
India

Cover: HyMap hyperspectral image of Ningaloo Reef, Western Australia.
Source: Merton, UNSW CRSGIS.

ISBN 3-540-21668-5   Springer Berlin Heidelberg New York

Library of Congress Control Number: 2004104167

This work is subject to copyright. All rights are reserved, whether the whole or part of this material is concerned, specifically the rights of translation, reprinting, reuse of illustrations, recitations, broadcasting, reproduction on microfilm or in any other way, and storage in data banks. Duplication of this publication or parts thereof is permitted only under the provisions of the German Copyright Law of September 9, 1965, in its current version, and permission for use must always be obtained from Springer-Verlag. Violations are liable to prosecution under the German Copyright Law.

Springer is a part of Springer Science+Business Media

springeronline.com

© Springer-Verlag Berlin Heidelberg 2004
Printed in Germany

The use of general descriptive names, registered names, trademarks, etc. in this publication does not imply, even in the absence of a specific statement, that such names are exempt from the relevant protective laws and regulations and therefore free for general use.

Typesetting: LE-TeX Jelonek, Schmidt & Vöckler GbR, Leipzig
Cover Design: Erich Kirchner, Heidelberg
Production: Almas Schimmel
Printing: Mercedes-Druck, Berlin
Binding: Stein+Lehmann, Berlin

Printed on acid free paper   30/3141/as   5 4 3 2 1 0

Preface

Over the last fifty years, a large number of spaceborne and airborne sensors have been employed to gather information regarding the earth's surface and environment. As sensor technology continues to advance, remote sensing data with improved temporal, spectral, and spatial resolution is becoming more readily available. This widespread availability of enormous amounts of data has necessitated the development of efficient data processing techniques for a wide variety of applications. In particular, great strides have been made in the development of digital image processing techniques for remote sensing data. The goal has been efficient handling of vast amounts of data, fusion of data from diverse sensors, classification for image interpretation, and development of user-friendly products that allow rich visualization.

This book presents some new algorithms that have been developed for high-dimensional datasets, such as multispectral and hyperspectral imagery. The contents of the book are based primarily on research carried out by some members and alumni of the Sensor Fusion Laboratory at Syracuse University. Early chapters that provide an overview of multispectral and hyperspectral sensing, and digital image processing have been prepared by other leading experts in the field. The intent of this book is to present the material at a level suitable for a diverse audience ranging from beginners to advanced remote sensing researchers and practitioners. This book can be used as a text or a reference in courses on image analysis and remote sensing. It can also be used as a reference book by remote sensing researchers, scientists and users, as well as by resource managers and planners.

We would like to thank all the authors for their enthusiasm and active cooperation during the course of this project. The research conducted at Syracuse University was supported by NASA under grant NAG5-11227. We thank Syracuse University for assistance in securing this grant. We also thank the Indian Institute of Technology, Roorkee for granting a postdoctoral leave that enabled Manoj K. Arora to participate in this endeavor. We are grateful to several organizations for providing and allowing the use of remote sensing data in the illustrative examples presented in this book. They include the Laboratory for Applications of Remote Sensing – Purdue University (AVIRIS data), USGS (Landsat ETM+ data), Eastman Kodak (IRS PAN, Radarsat SAR, and HyMap data as well as digital aerial photographs), NASA (IKONOS data) and DIRS laboratory of the Center for Imaging Science at Rochester Institute of Technology.

April 2004 *Pramod K. Varshney*
Manoj K. Arora

Contents

Introduction   1
- The Challenge 1
- What is Hyperspectral Imaging? 2
- Structure of the Book 4

Part I General

1 Hyperspectral Sensors and Applications   11
- 1.1 Introduction 11
- 1.2 Multi-spectral Scanning Systems (MSS) 11
- 1.3 Hyperspectral Systems 14
  - 1.3.1 Airborne sensors 14
  - 1.3.2 Spaceborne sensors 23
- 1.4 Ground Spectroscopy 27
- 1.5 Software for Hyperspectral Processing 29
- 1.6 Applications 30
  - 1.6.1 Atmosphere and Hydrosphere 30
  - 1.6.2 Vegetation 33
  - 1.6.3 Soils and Geology 38
  - 1.6.4 Environmental Hazards and Anthropogenic Activity 39
- 1.7 Summary 40

2 Overview of Image Processing   51
- 2.1 Introduction 51
- 2.2 Image File Formats 52
- 2.3 Image Distortion and Rectification 53
  - 2.3.1 Radiometric Distortion 53
  - 2.3.2 Geometric Distortion and Rectification 54
- 2.4 Image Registration 56
- 2.5 Image Enhancement 57
  - 2.5.1 Point Operations 57
  - 2.5.2 Geometric Operations 63
- 2.6 Image Classification 66

	2.6.1	Supervised Classification	67
	2.6.2	Unsupervised Classification	69
	2.6.3	Crisp Classification Algorithms	71
	2.6.4	Fuzzy Classification Algorithms	74
	2.6.5	Classification Accuracy Assessment	76
2.7	Image Change Detection		79
2.8	Image Fusion		80
2.9	Automatic Target Recognition		81
2.10	Summary		82

Part II Theory

3 Mutual Information: A Similarity Measure for Intensity Based Image Registration — 89

3.1	Introduction	89
3.2	Mutual Information Similarity Measure	90
3.3	Joint Histogram Estimation Methods	93
	3.3.1 Two-Step Joint Histogram Estimation	93
	3.3.2 One-Step Joint Histogram Estimation	94
3.4	Interpolation Induced Artifacts	95
3.5	Generalized Partial Volume Estimation of Joint Histograms	99
3.6	Optimization Issues in the Maximization of MI	103
3.7	Summary	107

4 Independent Component Analysis — 109

4.1	Introduction	109
4.2	Concept of ICA	109
4.3	ICA Algorithms	113
	4.3.1 Preprocessing using PCA	113
	4.3.2 Information Minimization Solution for ICA	115
	4.3.3 ICA Solution through Non-Gaussianity Maximization	121
4.4	Application of ICA to Hyperspectral Imagery	123
	4.4.1 Feature Extraction Based Model	124
	4.4.2 Linear Mixture Model Based Model	125
	4.4.3 An ICA algorithm for Hyperspectral Image Processing	126
4.5	Summary	129

5 Support Vector Machines — 133

5.1	Introduction	133
5.2	Statistical Learning Theory	135
	5.2.1 Empirical Risk Minimization	136
	5.2.2 Structural Risk Minimization	137
5.3	Design of Support Vector Machines	138
	5.3.1 Linearly Separable Case	139
	5.3.2 Linearly Non-Separable Case	143

Contents

		5.3.3 Non-Linear Support Vector Machines	146
	5.4	SVMs for Multiclass Classification	148
		5.4.1 One Against the Rest Classification	149
		5.4.2 Pairwise Classification	149
		5.4.3 Classification based on Decision Directed Acyclic Graph and Decision Tree Structure	150
		5.4.4 Multiclass Objective Function	152
	5.5	Optimization Methods	152
	5.6	Summary	154

6 Markov Random Field Models — 159

6.1	Introduction	159
6.2	MRF and Gibbs Distribution	161
	6.2.1 Random Field and Neighborhood	161
	6.2.2 Cliques, Potential and Gibbs Distributions	162
6.3	MRF Modeling in Remote Sensing Applications	165
6.4	Optimization Algorithms	167
	6.4.1 Simulated Annealing	168
	6.4.2 Metropolis Algorithm	173
	6.4.3 Iterated Conditional Modes Algorithm	175
6.5	Summary	177

Part III Applications

7 MI Based Registration of Multi-Sensor and Multi-Temporal Images — 181

7.1	Introduction	181
7.2	Registration Consistency	183
7.3	Multi-Sensor Registration	184
	7.3.1 Registration of Images Having a Large Difference in Spatial Resolution	184
	7.3.2 Registration of Images Having Similar Spatial Resolutions	188
7.4	Multi-Temporal Registration	190
7.5	Summary	197

8 Feature Extraction from Hyperspectral Data Using ICA — 199

8.1	Introduction	199
8.2	PCA vs ICA for Feature Extraction	200
8.3	Independent Component Analysis Based Feature Extraction Algorithm (ICA-FE)	202
8.4	Undercomplete Independent Component Analysis Based Feature Extraction Algorithm (UICA-FE)	203
8.5	Experimental Results	210
8.6	Summary	215

9 Hyperspectral Classification Using ICA Based Mixture Model 217
- 9.1 Introduction ... 217
- 9.2 Independent Component Analysis Mixture Model (ICAMM) – Theory ... 219
  - 9.2.1 ICAMM Classification Algorithm ... 220
- 9.3 Experimental Methodology ... 222
  - 9.3.1 Feature Extraction Techniques ... 223
  - 9.3.2 Feature Ranking ... 224
  - 9.3.3 Feature Selection ... 224
  - 9.3.4 Unsupervised Classification ... 225
- 9.4 Experimental Results and Analysis ... 225
- 9.5 Summary ... 233

10 Support Vector Machines for Classification of Multi- and Hyperspectral Data 237
- 10.1 Introduction ... 237
- 10.2 Parameters Affecting SVM Based Classification ... 239
- 10.3 Remote Sensing Images ... 241
  - 10.3.1 Multispectral Image ... 241
  - 10.3.2 Hyperspectral Image ... 242
- 10.4 SVM Based Classification Experiments ... 243
  - 10.4.1 Multiclass Classification ... 243
  - 10.4.2 Choice of Optimizer ... 245
  - 10.4.3 Effect of Kernel Functions ... 248
- 10.5 Summary ... 254

11 An MRF Model Based Approach for Sub-pixel Mapping from Hyperspectral Data 257
- 11.1 Introduction ... 257
- 11.2 MRF Model for Sub-pixel Mapping ... 259
- 11.3 Optimum Sub-pixel Mapping Classifier ... 261
- 11.4 Experimental Results ... 265
  - 11.4.1 Experiment 1: Sub-pixel Mapping from Multispectral Data ... 266
  - 11.4.2 Experiment 2: Sub-pixel Mapping from Hyperspectral Data ... 271
- 11.5 Summary ... 276

12 Image Change Detection and Fusion Using MRF Models 279
- 12.1 Introduction ... 279
- 12.2 Image Change Detection using an MRF model ... 279
  - 12.2.1 Image Change Detection (ICD) Algorithm ... 281
  - 12.2.2 Optimum Detector ... 284
- 12.3 Illustrative Examples of Image Change Detection ... 285
  - 12.3.1 Example 1: Synthetic Data ... 287
  - 12.3.2 Example 2: Multispectral Remote Sensing Data ... 290

12.4	Image Fusion using an MRF model	292
	12.4.1 Image Fusion Algorithm	294
12.5	Illustrative Examples of Image Fusion	299
	12.5.1 Example 1: Multispectral Image Fusion	299
	12.5.2 Example 2: Hyperspectral Image Fusion	303
12.6	Summary	306

Color Plates   309

Index   317

List of Contributors

Manoj K. Arora
Department of Civil Engineering
Indian Institute of Technology Roorkee
Roorkee 247 667, India
Tel: +91-1332-285417; Fax: +91-1332-273560
E-mail: manojfce@iitr.ernet.in

Hua-mei Chen
Department of Computer Science and Engineering
The University of Texas at Arlington
Box 19015, Arlington, TX 76019-0015
Tel: +1-817-272-1394; Fax: +1-817-272-3784
E-mail: hchen@cse.uta.edu

Teerasit Kasetkasem
Electrical Engineering Department
Kasetsart University, 50 Phanonyothin Rd.
Chatuchak, Bangkok 10900, Thailand
E-mail: fengtsk@ku.ac.th

Richard M. Lucas
Institute of Geography and Earth Sciences,
The University of Wales, Aberystwyth,
Aberystwyth, Ceredigion,
SY23 3DB, Wales, UK
Tel: +44-1970-622612; Fax: +44-1970-622659
E-mail: rml@aber.ac.uk

Ray Merton
School of Biological, Earth & Environmental Sciences
The University of New South Wales
Sydney NSW 2052, Australia
Tel: +61-2-9385 8713;
Fax: +61-2-9385 1558
E-mail: r.merton@unsw.edu.au

Olaf Niemann

Department of Geography
University of Victoria
PO BOX 3050 STN CSC
Victoria, B.C., V8W 3P5, Canada
Tel: +1-250-472 4624; Fax: +1-250-721 6216
E-mail: oniemann@office.geog.uvic.ca

Mahesh Pal

Department of Civil Engineering
National Institute of Technology
Kurukshetra, 136119, Haryana, India
Tel: +91-1744-239276; Fax: +91-1744-238050
E-mail: mpce_pal@yahoo.co.uk

Raghuveer M. Rao

Department of Electrical Engineering
Rochester Institute of Technology
79 Lomb Memorial Drive
Rochester NY 14623-5603
Tel.: +1-585-475 2185; Fax: +1-585-475 5845
E-mail: mrreee@rit.edu

Stefan A. Robila

301 Richardson Hall
Department of Computer Science
Montclair State University
1, Normal Ave, Montclair, NJ 07043
E-mail: robilas@mail.montclair.edu

Aled Rowlands

Institute of Geography and Earth Sciences,
The University of Wales, Aberystwyth,
Aberystwyth, Ceredigion, SY23 3DB, Wales, UK
Tel: +44-1970-622598; Fax: +44-1970-622659
E-mail: alr@aber.ac.uk

Chintan A. Shah

Department of Electrical Engineering Computer Science
121 Link Hall
Syracuse University, Syracuse, NY, 13244, USA
E-mail: cashah@ecs.syr.edu

Pramod K. Varshney

Department of Electrical Engineering Computer Science
121 Link Hall
Syracuse University, Syracuse, NY, 13244, USA
Tel: +1-315-443 4013; Fax: +1-315-443 2583
E-mail: varshney@syr.edu

Pakorn Watanachaturaporn

Department of Electrical Engineering Computer Science
121 Link Hall
Syracuse University, Syracuse, NY, 13244, USA
E-mail: pwatanac@syr.edu

Introduction

Pramod K. Varshney, Manoj K. Arora

The Challenge

From time immemorial, man has had the urge to see the unseen, to peer beneath the earth, and to see distant bodies in the heavens. This primordial curiosity embedded deep in the psyche of humankind, led to the birth of satellites and space programs. Satellite images, due to their synoptic view, map like format, and repetitive coverage are a viable source of gathering extensive information. In recent years, the extraordinary developments in satellite remote sensing have transformed this science from an experimental application into a technology for studying many aspects of earth sciences. These sensing systems provide us with data critical to weather prediction, agricultural forecasting, resource exploration, land cover mapping and environmental monitoring, to name a few. In fact, no segment of society has remained untouched by this technology.

Over the last few years, there has been a remarkable increase in the number of remote sensing sensors on-board various satellite and aircraft platforms. Noticeable is the availability of data from hyperspectral sensors such as AVIRIS, HYDICE, HyMap and HYPERION. The hyperspectral data together with geographical information system (GIS) derived ancillary data form an exceptional spatial database for any scientific study related to Earth's environment. Thus, significant advances have been made in remote sensing data acquisition, storage and management capabilities.

The availability of huge spatial databases brings in new challenges for the extraction of quality information. The sheer increase in the volume of data available has created the need for the development of new techniques that can automate extraction of useful information to the greatest degree. Moreover, these techniques need to be objective, reproducible, and feasible to implement within available resources (DeFries and Chan, 2000). A number of image analysis techniques have been developed to process remote sensing data with varied amounts of success. A majority of these techniques have been standardized and implemented in various commercial image processing software systems such as ERDAS Imagine, ENVI and ER Mapper etc. These techniques are suitable for the processing of multispectral data but have limitations when it comes to an efficient processing of the large amount of hyperspectral data available in hundreds of bands. Thus, the conventional techniques may be inappropriate for information extraction from hyperspectral, multi-source, multi-sensor and multi-temporal data sets.

A number of textbooks on conventional digital processing of remote sensing data are available to the academic community (e.g. Jensen 1996; Richards and Jia 1999; Mather 1999; Campbell 2002). These deal with the basic digital image processing techniques together with fundamental principles of remote sensing technology. They are excellent resources for undergraduate and beginning graduate-level teaching. However, the researchers and graduate students working on the development and application of advanced techniques for remote sensing image processing have to depend on texts available in electrical engineering, and computer science literature, which do not focus on remote sensing data. This book contains chapters written by authors with backgrounds in different disciplines and thus attempts to bridge this gap. It addresses the recent developments in the area of image registration, fusion and change detection, feature extraction, classification of remote sensing data and accuracy assessment. The aim of the book is to introduce the reader to a new generation of information extraction techniques for various image processing operations for multispectral and particularly hyperspectral data. We expect that the book will form a useful text for graduate students and researchers taking advanced courses in remote sensing and GIS, image processing and pattern recognition areas. The book will also prove to be of interest to the professional remote sensing data users such as geologists, hydrologists, ecologists, environmental scientists, civil and electrical engineers and computer scientists. In addition to remote sensing, the algorithms presented will have far-reaching applicability in fields such as signal processing and medical imaging.

What is Hyperspectral Imaging?

Since the initial acquisition of satellite images, remote sensing technology has not looked back. A number of earth satellites have been launched to advance our understanding of Earth's environment. The satellite sensors, both active and passive, capture data from visible to microwave regions of the electromagnetic spectrum. The multispectral sensors gather data in a small number of bands (also called features) with broad wavelength intervals. No doubt, multispectral sensors are innovative. However, due to relatively few spectral bands, their spectral resolution is insufficient for many precise earth surface studies.

When spectral measurement is performed using hundreds of narrow contiguous wavelength intervals, the resulting image is called a *hyperspectral image*, which is often represented as a *hyperspectral image cube* (see Fig. 1) (JPL, NASA). In this cube, the x and y axes specify the size of the images, whereas the z axis denotes the number of bands in the hyperspectral data. An almost continuous spectrum can be generated for a pixel and hence hyperspectral imaging is also referred to as imaging spectrometry. The detailed spectral response of a pixel assists in providing accurate and precise extraction of information than is obtained from multispectral imaging. Reduction in the cost of sensors as well as advances in data storage and transmission technolo-

Introduction

Fig. 1. The hyperspectral cube (source: http://aviris.jpl.nasa.gov/html/aviris.cube.html). For a colored version of this figure, see the end of the book

gies have made the hyperspectral imaging technology more readily available. Much like improvements in spectral resolution, spatial resolution has also been dramatically increased by the installation of hyperspectral sensors on aircrafts (airborne imagery), opening the door for a wide array of applications.

Nevertheless, the processing of hyperspectral data remains a challenge since it is very different from multispectral processing. Specialized, cost effective and computationally efficient procedures are required to process hundreds of bands acquiring 12-bit and 16-bit data. The whole process of hyperspectral imaging may be divided into three steps: preprocessing, radiance to reflectance transformation and data analysis.

Preprocessing is required for the conversion of raw radiance into at-sensor radiance. This is generally performed by the data acquisition agencies and the user is supplied with the at-sensor radiance data. The processing steps involve operations like spectral calibration, geometric calibration and geocoding, signal to noise adjustment, de-striping etc. Since, radiometric and geometric accuracy of hyperspectral data vary significantly from one sensor to the other, the users are advised to discuss these issues with the data providing agencies before the purchase of data.

Further, due to topographical and atmospheric effects, many spectral and spatial variations may occur in at-sensor radiance. Therefore, the at-sensor data need to be normalized in the second step for accurate determination of the reflectance values in difference bands. A number of atmospheric models and correction methods have been developed to perform this operation. Since the focus of this book is on data analysis aspects, the reader is referred to (van der Meer 1999) for a more detailed overview of steps 1 and 2.

Data analysis is aimed at extracting meaningful information from the hyperspectral data. A limited number of image analysis algorithms have been developed to exploit the extensive information contained in hyperspectral imagery for many different applications such as mineral mapping, military target detection, pixel and sub-pixel level land cover classification etc. Most of these algorithms have originated from the ones used for analysis of multispectral data, and thus have limitations. A number of new techniques for the processing of hyperspectral data are discussed in this book. Specifically, these include techniques based on independent component analysis (ICA), mutual information (MI), Markov random field (MRF) models and support vector machines (SVM).

Structure of the Book

This book is organized in three parts, each containing several chapters. Although it is expected that the reader is familiar with the basic principles of remote sensing and digital image processing, yet for the benefit of the reader, two overview chapters one on hyperspectral sensors and the other on conventional image processing techniques are provided in the first part. These chapters are introductory in nature and may be skipped by readers having sufficient relevant background. In the second part, we have introduced the theoretical and mathematical background in four chapters. This background is valuable for understanding the algorithms presented in later chapters. The last part is focused on applications and contains six chapters discussing the implementation of these techniques for processing a variety of multi and hyperspectral data for image registration and feature extraction, image fusion, classification and change detection. A short bibliography is included at the end of each chapter. Colored versions of several figures are included at the end of the book and this is indicated in figure captions where appropriate.

Part I: General

Chapter 1 provides a description of a number of hyperspectral sensors onboard various aircraft and space platforms that were, are and will be in operation in the near future. The spectral, spatial, temporal and radiometric characteristics of these sensors have also been discussed, which may provide sufficient guidance on the selection of appropriate hyperspectral data for a particular application. A section on ground based spectroscopy has also been included where the utility of laboratory and field based sensors has been indicated. Laboratory and field measurements of spectral reflectance form an important component of understanding the nature of hyperspectral data. These measurements help in the creation of spectral libraries that may be used for calibration and validation purposes. A list of some commercially available software packages and tools has also been provided. Finally, some application areas have been identified where hyperspectral imaging may be used successfully.

An overview of basic image processing tasks that are necessary for multi and hyperspectral data analysis is given in Chap. 2. Various sections in this chapter discuss the concepts of radiometric and geometric rectification, image registration, feature extraction, image enhancement, classification, fusion and change detection operations. The objective of these sections is to briefly familiarize the reader with the role of each image processing operation in deriving useful information from the remote sensing data. Each operation has been clearly explained with the help of illustrative examples based on real image data, wherever necessary.

Part II: Theory

Having established the background on the basic concepts of data acquisition and analysis systems, Part II of the book imparts theoretical knowledge on the advanced techniques described later in this book. In this part, four chapters are included. Chapter 3 deals with the description of the mutual information (MI) similarity measure that has origins in information theory. MI is used for automatic intensity based registration of multi and hyperspectral data. As the computation of MI depends on the accurate estimation of joint histograms, two methods of joint histogram estimation have been explained. The existing joint histogram estimation methods suffer from the problem of interpolation-induced artifacts, which has been categorically highlighted and a new joint histogram estimation algorithm called generalized partial volume estimation has been proposed to reduce the effect of artifacts. Some optimization issues to maximize the MI have also been addressed.

In Chap. 4, details of a technique called independent component analysis (ICA), originally used in signal processing, have been provided. The theory of ICA discussed in this chapter forms the basis of its utilization for feature extraction and classification of hyperspectral images. ICA is a multivariate data analysis technique that commences with a linear mixture of unknown independent sources and proceeds to recover them from the original data. The merit of ICA lies in the fact that it uses higher order statistics unlike principal component analysis (PCA) that uses only second order statistics to model the data. After introducing the concept of ICA, several algorithms for determining the ICA solution have been presented, which is followed with its implementation in a couple of applications of hyperspectral data processing.

Image classification is perhaps the key image processing operation to retrieve information from remote sensing data for a particular application. The use of non-parametric classifiers has been advocated since these do not depend on data distribution assumptions. Recently, support vector machines (SVM) have been proposed for the classification of hyperspectral data. These are a relatively new generation of techniques for classification and regression problems and are based on *statistical learning theory* having its origins in *machine learning*. Chapter 5 builds the theoretical background of SVM. A section is exclusively devoted to a brief description of statistical learning theory. SVM formulations for three different cases; linearly separable and non-separable cases, and the non-linear case, have been presented. Since SVM is essentially a binary classification technique, a number of multi-class methods have also been described. Optimization, being the key to an efficient implementation of

SVM has also been given due consideration and a separate section is written to discuss the merits and demerits of existing optimization methods that have been used in SVM classification.

Chapter 6 provides the theoretical setting of Markov random field (MRF) models that have been used by statistical physicists to explain various phenomena occurring among neighboring particles because of their ability to describe local interactions between them. The concept of MRF model suits image analysis because many image properties, such as texture, depend highly on the information obtained from the intensity values of neighboring pixels, as these are known to be highly correlated. As a result of this, MRF models have been found useful in image classification, fusion and change detection applications. A section in this chapter is devoted to a detailed discussion of MRF and its equivalent form (i.e. Gibbs fields). Some approaches for the use of MRF modeling are explained. Several widely used optimization methods including simulated annealing are also introduced and discussed.

Part III: Applications

The theoretical background, concepts and knowledge introduced in the second part of the book are exploited to develop processing techniques for different applications of hyperspectral data in this part. In Chap. 7, MI based registration has been applied for automatic registration of a variety of multi and hyperspectral images at different spatial resolutions. Two different cases namely multi-sensor and multi-temporal registration, have been considered. The performance of various interpolation algorithms has been evaluated using registration consistency, which is a measure to assess the quality of registration in the absence of ground control points.

Since, the hyperspectral data are obtained in hundreds of bands, for many applications, it may be inefficient and undesirable to utilize the data from all the bands thereby increasing the computational time and cost of analysis. Hence, it may be essential to choose the most effective features that are sufficient for extracting the desired information and efficient in reducing the computational time. Chapter 8 focuses on the use of ICA for feature extraction from hyperspectral data. After clarifying the distinction between PCA and ICA, the details of two ICA based algorithms for the extraction of features have been given. The aim of these algorithms is to identify those features, which allow us to discriminate different classes that are hidden in the hyperspectral data, with high degree of accuracy. A method called spectral screening has also been proposed to increase the computational speed of the ICA based feature extraction algorithm.

In Chap. 9, the concept of ICA is further advanced to develop an ICA mixture model (ICAMM) for unsupervised classification of hyperspectral data. The advantage of using ICAMM lies in the fact that it can accurately map non-Gaussian classes unlike other conventional statistical unsupervised classification algorithms that require that the classes be Gaussian. The ICAMM finds independent components and the mixing matrix for each class using the *extended infomax learning algorithm* and computes the class membership probability for each pixel. The complete ICAMM algorithm has been explained in a simplified manner. Unsupervised classification of hyperspectral data is performed using ICAMM and its performance evaluated *vis a vis* the most widely used K-means algorithm.

Chapter 10 describes the application of SVM for supervised classification of multi and hyperspectral data. Several issues, which may have a bearing on the performance of SVM, have been considered. The effect of a number of kernel functions, multi class methods and optimization techniques on the accuracy and efficiency of classification has been assessed.

The two classification algorithms, unsupervised ICAMM and supervised SVM, discussed in Chaps. 9 and 10 respectively, are regarded as per pixel classifiers, as they allocate each pixel of an image to one class only. Often the images are dominated by mixed pixels, which contain more than one class. Since a mixed pixel displays a composite spectral response, which may be dissimilar to each of its component classes, the pixel may not be allocated to any of its component classes. Therefore, error is likely to occur in the classification of mixed pixels, if per pixel classification algorithms are used. Hence, sub-pixel classification methods such as fuzzy c-means, linear mixture modeling and artificial neural networks have been proposed in the literature. In Chap. 11, a novel method based on MRF models has been introduced for sub-pixel mapping of hyperspectral data. The method is based on an optimization algorithm whereby raw coarse resolution images are first used to generate an initial sub-pixel classification, which is then iteratively refined to accurately characterize the spatial dependence between the class proportions of the neighboring pixels. Thus, spatial relations within and between pixels are considered throughout the process of generating the sub-pixel map. The implementation of the complete algorithm is discussed and it is illustrated by means of an example

The spatial dependency concept of MRF models is further extended to change detection and image fusion applications in Chap. 12. Image change detection is one of the basic image analysis tools and is frequently used in many remote sensing applications to quantify temporal information, whereas image fusion is primarily intended to improve, enhance and highlight certain features of interest in remote sensing images for extracting useful information. Individual MRF model based algorithms for these two applications have been described and illustrated through experiments on multi and hyperspectral data.

References

Campbell JB (2002) Introduction to remote sensing, 3rd edition. Guilford Press, New York

DeFries RS, Chan JC (2000) Multiple criteria for evaluating machine learning algorithms for land cover classification from satellite data. Remote Sensing of Environment 74: 503–515

Jensen JR (1996) Introductory digital image processing: a remote sensing perspective, 2nd edition. Prentice Hall, Upper Saddle River, N.J.

JPL, NASA. AVIRIS image cube. http://aviris.jpl.nasa.gov/html/aviris.cube.html

Mather PM (1999) Computer processing of remotely sensed images: an introduction. Wiley, Chichester

Richards JA and Jia X (1999) Remote sensing digital image analysis: an introduction, 3rd edition. Springer Verlag Berlin

Van der Meer F (1999) Imaging spectrometry for geologic remote sensing. Geologie en Mijnbouw 77: 137–151

Part I
General

CHAPTER 1

Hyperspectral Sensors and Applications

Richard Lucas, Aled Rowlands, Olaf Niemann, Ray Merton

1.1
Introduction

Since the beginning of remote sensing observation, scientists have created a "toolbox" with which to observe the varying dimensions of the Earth's dynamic surface. Hyperspectral imaging represents one of the later additions to this toolbox, emerging from the fields of aerial photography, ground spectroscopy and multi-spectral imaging. This new tool provides capacity to characterise and quantify, in considerable detail, the Earth's diverse environments.

This chapter has two goals. The first is to chronicle the development of hyperspectral remote sensing. In doing so, an overview of the past, present and future ground, airborne and spaceborne sensors and their unique attributes which render them most suited for specific applications is presented. The second is to provide an overview of the applications where hyperspectral remote sensing has, to date, made the greatest impact. Key areas considered include the atmosphere, snow and ice, marine and freshwater environments, vegetation, soils, geology, environmental hazards (e. g., fire) and anthropogenic activity (e. g., land degradation).

1.2
Multi-spectral Scanning Systems (MSS)

Since the mid 1950s, a diverse range of airborne and spaceborne sensors have recorded the reflectance of the Earth's surface in the spectral wavelength region extending from $\sim$ 400 to $\sim$ 2500 nanometres (nm). Early sensors collected data initially onto black and white and subsequently colour infrared film although later, filters and multi-camera systems were developed for creating four "color" image combinations (blue, green, red and near infrared (NIR)). With the advent and advancement of more readily available computer technology and the declassification of sensing technology developed for military use, a number of civilian airborne (Table 1.1) and, later, spaceborne sensors were developed and deployed. These initial digital sensors acquired data in broad, widely-spaced spectral ranges which were referred to commonly as bands or channels. Such multi-spectral scanning systems (MSS) have been, and continue to be, the backbone of the optical data acquisition systems.

Table 1.1. Basic characteristics of airborne multispectral sensors

Sensor	Availability	Number of Bands	Spectral range (nm)	Band width at FWHM (nm)	Notes	Technical Reference
Multispectral scanner: Airborne Thematic Mapper ATM-Daedalus 1268	1986	11	420–13000	420–450; 450–520; 520–600; 600–620; 630–690; 690–750 760–900; 910–1050 1550–1750; 2080–2350 8500–13000		
AMTIS Multi-angle Thermal/Visible Imaging System	Proposed	3	400–13000	400–700 700–900 7500–13000	Multi-angle system: 45°, 33.75°, 22.5°, 11.25° and 0°	(Zhou et al. 2003)
GER EPS-M series	1994	3	300–12000	300–400 600–700 8000–12000	16 bits	

Since the early 1970s, a large number of spaceborne multispectral sensors have been launched. In Earth observing programs, key sensors have been those on board the LANDSAT, SPOT and Indian Remote Sensing (IRS) series of satellites. From the early 1970s through to the early 1990s, the LANDSAT (1–3) MSS provided global observations at 80 m spatial resolution, although only in four broad wavelength regions; two in the visible (green and red) and two in the NIR respectively. The successors of the MSS, the LANDSAT Thematic Mapper (TM on board LANDSAT 4–5) and Enhanced Thematic Mapper (ETM+ on board LANDSAT 7) provided additional measurements in the shortwave infrared (SWIR) region centred on 1600 nm and 2100 nm and at a finer (30 m) spatial resolution and level of quantization. The provision of a 15 m panchromatic band on the LANDSAT ETM+ also facilitated greater resolving of ground features and opportunities for data fusion. The SPOT-series of sensors, developed in France, were first launched in 1986. Initially, the SPOT 1–3 High Resolution Visible (HRV) recorded reflectance (at 20 m spatial resolution) in three visible and NIR (VNIR) channels, although the subsequent High Resolution Visible InfraRed (HRVIR; SPOT 4) supported a SWIR waveband. The launch of SPOT-5, in 2003, heralded a new era in fine spatial resolution remote sensing, with the panchromatic (PAN) sensor increasing in spatial resolution from 10 m to 5 m, with potential to obtain 2.5 m spatial resolution data through simultaneous observation. The resolution of the SPOT-5 multispectral (XS) sensor also increased to 10 m (maintaining 20 m in the SWIR region). The IRS series of spacecraft, which were first launched by India in 1988, complemented the Landsat and SPOT-series, as several sensors were in orbit, observing in four VNIR bands and at spatial resolutions of approximately 23.5–36.25 m in XS (70 m in SWIR) and 6 m in PAN mode.

With time, these sensors have provided consistent historical data with which to observe and quantify intra-annual, inter-annual or decadal changes occurring on the Earth's surface. The addition of other sensors (e.g., IKONOS and Quickbird) observing at a variety of spatial and temporal resolutions and also band configurations and quantizations has served to extend the time-series and to provide additional information on Earth surface processes and dynamics. The continuation and development of these Earth observing programs is testament to the important role that data from these sensors play in understanding the longer-term dynamics of both natural and anthropogenic change. This also indicates the commitment of national and international organisations to Earth observation for environmental monitoring.

A fundamental limitation of these multispectral sensors, however, is that data are acquired in a few broad (typically 100–200 nm in width) irregularly space spectral bands (van der Meer and de Jong 2001). As such, narrow spectral features may not be readily discriminated and tend to be either averaged across the spectral sampling range or masked by stronger proximal features (Kumar et al. 2001), thereby resulting in a reduction or loss in the environmentally relevant information that can potentially be extracted. This deficiency in the spectral, radiometric, and spatial resolution of many multispectral systems, therefore render them unsuitable for identifying many surface materials and features, particularly minerals (Goetz 1995).

1.3
Hyperspectral Systems

In recognising the limitations of multispectral systems, an improved technology based on spectroscopy was developed whereby the limited number of discrete spectral bands was enhanced by a series (nominally > 50 bands) of narrow (Full-Width-Half-Maximum; FWHM 2–20 nm) contiguous bands (Aspinall et al. 2002) over the VNIR and SWIR wavelength regions. Similar advances were also made in the thermal infrared (TIR) regions. Matrices of spectral samples were then designed to be built up on a line-by-line basis to form a two dimensional image (x- and y-axes), with a third dimension (z-axis) holding the spectral data for each sample (pixel). These new imaging spectrometers were subsequently able to retrieve near-laboratory quality reflectance spectra such that the data associated with each pixel approximated the true spectral signature of a target material, with sufficiently high signal-to-noise ratio (SNR) across the full contiguous wavelength range (nominally 400–2500 nm) to ultimately form a 3-dimensional datacube.

Within these hyperspectral images, molecular absorption and particle scattering signatures of materials could be detected to unambiguously identify and quantify the abundance of surface constituents (Buckingham et al. 2002). Specific reflectance or absorption features could also be associated with different minerals or even biological and chemical processes, thereby providing an opportunity to better characterise surface environments and dynamics. With this advancement, the era of hyperspectral imaging was developed with airborne sensors deployed initially albeit largely for research purposes. Several spaceborne hyperspectral sensors have also been deployed although, to date, the data have been used largely for technology demonstration and research.

The following sections chronicle the development of these airborne and spaceborne hyperspectral sensors and subsequently consider the characteristics of several, in the context of temporal, spatial, spectral and radiometric resolutions, which render them unique for environmental applications.

1.3.1
Airborne sensors

Airborne hyperspectral remote sensing has been available since the early 1980s (Table 1.2). Early developments were characterized by small, purpose-built sensors. Amongst the earliest of the scanning imaging spectrometers was a one-dimensional profiler developed by the Geophysical Environmental Research Company (GER) in 1981. This sensor gathered data in 576 channels over the 400–2500 nm wavelength range. In the same year, the Shuttle Multi-spectral Infrared Radiometer (SMIRR) became operational and Canada's Department of Fisheries and Oceans introduced the Fluorescence Line Imager (FLI). In 1983, the Airborne Imaging Spectrometer (AIS) was first flown following a three year period of development at NASA's Jet Propulsion Laboratory (Goetz 1995). The principal driving force behind many of these initial developments came from geological disciplines.

Table 1.2. Basic characteristics of airborne hyperspectral sensors

Sensor	Availability	Number of Bands	Spectral range (nm)	Band width at FWHM (nm)	Notes	Technical Reference
AAHIS Advanced Airborne Hyperspectral Imaging Sensors	1994	288	432– 832	6		(Hochberg and Atkinson 2003) STI industries www.sti-industries.com/
AHI Airborne Hyperspectral Imager	1994	256	7500–11700	100	12 bits	Hawaii Institute of Geophysics and Planetology www.higp.hawaii.edu/
AIS-1/2	1982–1985 1985–1987	128	900– 2100 800– 2400	9.3 10.6		(Vane et al. 1984)
AISA+ AISA Eagle Airborne Imaging Spectrometer for Different Applications	1997 2002	244	400– 970	2.9	AISA Eagle has a 1000 pixel swath, compared with 500 for AISA+	(Kallio et al. 2001) www.specim.fi/
AISA Hawk	2003	240	1000– 2400	8		www.specim.fi/
ARGUS		400 (300 in VNIR and SWIR)	370– 2500 780– 2500	5–10	Hyperspectral Mineral Mapping	www.fugroairborne.com www.hyvista.com
ASAS The Advanced Solid-state Array Spectro-radiometer	1987 upgraded in 1991/92	62	400– 1060	11.5	signal-to-noise ratio ~1000:1 12 bits	(Irons et al. 1991) asas.gsfc.nasa.gov/

Sensor	Availability	Number of Bands	Spectral range (nm)	Band width at FWHM (nm)	Notes	Technical Reference
APEX Airborne Prism Experiment	2005	Programmable to a max of 300	380–2500	10	312 spectral rows in the VNIR and 195 spectral rows in the SWIR.	(Schaepman et al. 2003)
ASTER Simulator	1992	1 3 20	700–1000 3000–5000 8000–12000	300 600–700 200	16 bits	(Mills et al. 1993) www.cis.rit.edu/class/simg707/Web_Pages/Survey_report.htm#_Toc404106559
AVIRIS Airborne Visible/ Infrared Imaging Spectrometer	1987	224	400–2450	9.4–16	10 bits until 1994, 12 bits from 1995	(Curran and Dungan 1989; Curran and Dungan 1990) makalu.jpl.nasa.gov/aviris.html
AVIS-2, Airborne Visible/ Infrared Imaging Spectrometer	2001	64	400–850	7	14 bits	(Mauser 2003) Ground Truth Center, Germany (www.gtco.de)
CASI Compact Airborne Spectrographic Imager	1989 CASI 3 in 2002	288 at 2.2 nm intervals	400–1050	2.2 @ 650	12-bits 14 bits for CASI 3 Signal to noise of 480:1 CASI3	ITRES (www.itres.com)
DAIS 7915 Digital Airborne Imaging Spectrometer	1994	32 8 32 1 6	498–1010 1000–1800 1970–2450 3000–5000 8700–12300	16 100 15 2000 600	15 bits	(Strobl et al. 1996; Ben-Dor et al. 2002) German Aerospace Center (www.op.dlr.de/dais/)

Hyperspectral Sensors and Applications

Sensor	Year	No.	Spectral range (nm)	Bandwidth	Digitization	Reference
EPS-H[1]	1995	76 32 32 12	430– 1050 1500– 1800 2000– 2500 8000–12500	8 50 16 600	16 bits	GER Corporation (www.ger.com)
EPS-A	1998	28 1 1 1	400– 1000 1550– 1750 2100– 2350 8000–12000	Customized to the requirements of the user	16 bits	GER Corporation (www.ger.com) (Vos et al. 2003)
HYDICE Hyperspectral Digital Imagery Collection Experiment	1995	210	400– 2500	7.6–14.9		(Resmini et al. 1997)
HYMAP	1996	126	450– 2500	15–20	12–16 bits	(Cocks et al. 1998) Intspec (www.intspec.com) Hyvista (www.hyvista.com)
MAIS Modular Airborne Imaging Spectrometer	1991	32 32 7	450– 1100 1400– 2500 8200–12200	20 30 400–800	12 bits	NASA (ltpwww.gsfc.nasa.gov/ISSSR-95/modulara.htm) (van der Meer et al. 1997; VanderMeer et al. 1997; Yang et al. 2000)
MAS MODIS Airborne Simulator	1993	9 16 16 9	529– 969 1395– 2405 2925– 5325 8342–14521	31–55 47–57 142–151 352–517	12 bits (pre-1995: 8 bits)	(King et al. 1996) MODIS Airborne Simulator (mas.arc.nasa.gov)

[1] Number of spectrometers, detector types and band placements are tailored to user requirements. GER GER EPS-H Airborne Imaging Spectrometer System Technical Description, 2003.

Sensor	Availability	Number of Bands	Spectral range (nm)	Band width at FWHM (nm)	Notes	Technical Reference
MIVIS Multispectral Infrared and Visible Spectrometer	1993	20 8 64 10	433– 833 1150– 1550 2000– 2500 8200–12700	20 50 8 400–500	12 bits	Sensytech (www.sensystech.com)
OMIS Operative Modular Airborne Imaging Spectrometer	1999	64 16 32 8 8	460– 1100 1060– 1700 2000– 2500 3000– 5000 8000–12500	10 40 15 250 500	12 bits	(Huadong et al. 2001)
PROBE-1		100–200	440– 2543	11– 18		(McGwire et al. 2000)
ROSIS Reflective Optics System Imaging Spectrometer	1993	128 selectable spectral bands	440– 850	5	12 bits	(Kunkel et al. 1991; Su et al. 1997) DLR Institute of Optoelectronics www.op.dlr.de/ne-oe/fo/rosis/home.html
SASI Shortwave Infrared Airborne Spectrographic Sensor	2002	160	850– 2450	10	14 bits	www.itres.com telsat.belspo.be/documents/casi2003.html
SFSI Short Wavelength Infrared Full Spectrum Imager	1994	22–120	1230– 2380	10.3		(Nadeau et al. 2002) Canada Center for Remote Sensing www.ccrs.nrcan.gc.ca
VIFIS Variable Interference Filter Imaging Spectrometer	1994	64	420– 870	10.12	8 bits	(Sun and Anderson 1993; Gu et al. 1999) www.aeroconcepts.com/Tern/Instrument.html

A significant advancement in hyperspectral remote sensing occurred when the Airborne Visible/Infrared Imaging Spectrometer (AVIRIS) was proposed to NASA in 1983 (Vane et al. 1993). This sensor was the first to acquire image data in continuous narrow bands simultaneously in the visible to SWIR wavelength regions. Following approval in 1984, the first engineering flights occurred in early 1987 with data provided to scientific investigators shortly after. The AVIRIS was modified subsequently and flown throughout North America and Eurasia in the following 4 years and, since this period, this sensor has continued to be improved and is still providing some of the highest quality data to the scientific community today.

Parallel to NASA's development of the AVIRIS sensor have been a number of commercial ventures where hyperspectral sensors have been designed and deployed. Since 1978, the Canadian company ITRES has paved the way for airborne hyperspectral remote sensing through the development of the Compact Airborne Spectrographic Imager (CASI), which was based on the earlier FLI (Hollinger et al. 1987). This sensor has been in operation since 1989 and was used not only for commercial operations in the private sector but also for research projects. The CASI-2 and CASI-3 instruments were replacements, with the latter released in 2002. The particular advantage of the CASI-3 is that the swath width (i.e., width of an individual scene) was increased by a factor of 2.9. However, all CASI instruments observe only in the VNIR and, as such, have been found to be of reduced value in many terrestrial applications. In response to this limitation, ITRES also developed the SASI (Shortwave Infrared Airborne Spectrographic Imager) which, when coupled with the CASI, extended the spectral range to the SWIR.

A number of sensors were also developed to provide bidirectional viewing capability. For example, the Advanced Solid-State Array Spectroradiometer (ASAS) has been maintained and operated by the NASA Goddard Space Flight Centre (GSFC) since 1984, although evolved from technology developed in the early 1970s. This sensor provides directional viewing capability (from 70° forward to 55° aft along the direction of flight) in the VNIR regions (Irons et al. 1991; Dabney et al. 1994; Ranson et al. 1994).

In 1994, the first HyMap series of hyperspectral sensors was commissioned by De Beers and the sensor has been in operation since 1996. The HyMap series of airborne hyperspectral sensors have been deployed internationally, undertaking remote sensing surveys to support an extremely disparate range of applications, including mineral exploration, defense research and forest inventory. In many ways, the Australian-based HyMap systems have set the benchmark for commercially available hyperspectral sensors in terms of their SNR, image quality, stability, adaptability and ease of use.

1.3.1.1
Spectral Regions, Resolutions and Bidirectional Capability

Although hyperspectral remote sensing implies observation across the full spectral region (i.e. 400–2500 nm), many sensors operate only within the VNIR regions. As examples, the ASAS and CASI record reflectance in the

404–1020 nm and 400–1050 nm regions respectively. The SASI operates only in the SWIR (from 820–2500 nm). Sensors operating across the full spectral range include the AVIRIS (224 wavebands) and HyMap (126 wavebands). Each of these sensors provides contiguous spectral coverage, including those regions where absorption by atmospheric water occurs, facilitating atmospheric absorption effects to be calculated. More recently, sensors designed to observe across the full reflected spectral range *plus* the TIR wavelength range have become operational, including the DAIS-7915 (Ben-Dor et al. 2002).

The spectral resolution and sampling interval are critical in determining the accuracy to which features in the radiation spectra can be measured. The spectral resolution of the sensor is defined as the narrowest bandwidth over which radiation is recorded, and is measured based on the FWHM of the instrument response. The spectral sampling interval is the distance (in units of wavelength) between spectral measurement points. Sensors with spectral resolutions of 10 nm or smaller were designed largely for aquatic and atmospheric applications and include the CASI-series of instruments, whilst those with bandwidths greater than 10 nm were designed primarily for terrestrial applications (Curran 1994) and include the AVIRIS (FWHM of 10 nm with bands spaced $\sim$ 10 nm apart) and HyMap (FWHM of 15–20 nm). Spectral sampling intervals in the order of 2 to 3 nm magnitude are commonly provided by hyperspectral sensors. For many sensors, the band positions and widths are fixed although some (e.g., the CASI-series and AISA) offer programmable bands depending upon the application. The ASAS was one of the few sensors that recorded reflectance at varying viewing geometries, thereby allowing measurement of the bi-directional reflectance characteristics of surfaces. Airborne sensors operating in the short wave infrared (SWIR) and TIR regions are listed in Table 1.2.

1.3.1.2
Spatial Resolution and Coverage

The spatial resolution of an observing sensor refers to a distance between the nearest objects that can be resolved, is given in units of length (e.g., meter) and depends on the instantaneous field of view (IFOV). For many airborne sensors, the spatial resolution is dictated largely by the flying height of the aircraft as well as the configuration of the sensor and, in some cases, the aircraft platform needs to be changed to achieve the required resolution (Vane et al. 1993). The flying height of the aircraft also influences the width of the scan and hence the extent of coverage. The lens optics and the integration time will similarly limit the potential spatial resolution of the image, although data can be acquired at spatial resolutions finer than 1 m.

The level of detail able to be resolved at different spatial resolutions is indicated in Fig. 1.1a–d, which compares data acquired by several airborne and also spaceborne sensors over an area of subtropical woodland in central Queensland, Australia. Using aerial photography, tree crowns can be resolved easily through differentiation between photosynthetic vegetation (PV), non-photosynthetic vegetation (NPV; e.g., branches) and soil background, and different tree species can be distinguished. Using 1 m spatial resolution CASI

Hyperspectral Sensors and Applications

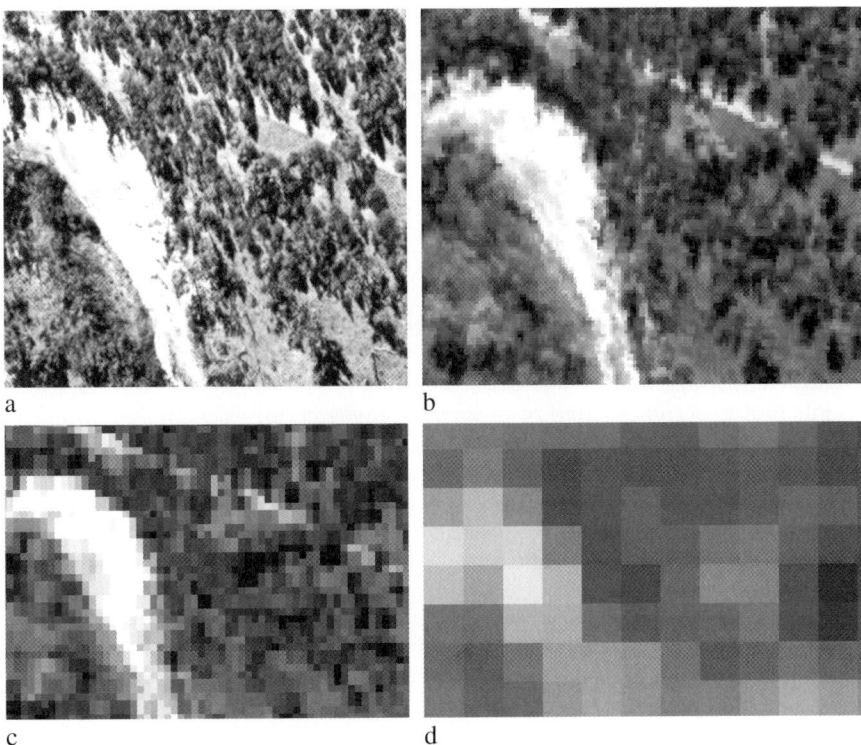

Fig. 1.1a–d. Observations of the same area of mixed species subtropical woodlands near Injune, central Queensland, Australia, observed using **a** stereo colour aerial photography, **b** CASI, **c** HyMap, and **d** Hyperion data at spatial resolutions of < 1 m, 1 m, 2.8 m and 30 m respectively. For a colored version of this figure, see the end of the book

data, tree crowns are still resolved and species can be differentiated although within-canopy components cannot be isolated and pixels containing a mix of ground and vegetation spectra are common. Even so, relatively pure spectra relating to ground (e.g., bare soil) and vegetation (e.g., leaf components) can be extracted. At $\sim$ 2.5–5 m spatial resolution, which is typical to the HyMap sensor, only larger tree crowns are discernible and, due to the openness of the woodland canopy, most pixels contain a mix of ground and vegetation spectra. At coarser ($\sim$ 5–> 20 m) spatial resolutions, tree crowns cannot be differentiated and the averaging of the signal is such that pure spectral signatures for specific surfaces cannot be extracted from the imagery, and only broad vegetation and surface categories can be distinguished. Sensors observing at resolutions from 20 to 30 m include the airborne AVIRIS (Vane et al. 1993), DAIS-7915 (Ben-Dor et al. 2002) and MODIS-ASTER simulator (MASTER) as well as the spaceborne Hyperion. These observations illustrate the difficulties associated with obtaining *"pure"* spectral reflectance data from specific surfaces and materials, particularly in complex environments.

1.3.1.3
Temporal Resolution

The temporal resolution refers to the frequency of observation by sensors. Until the advent of hyperspectral data from spaceborne sensors, observations were obtained on an 'as needs' basis. Due primarily to limitations of cost, few multitemporal datasets have been produced. As a result, the complex time-series of hyperspectral dataset acquisitions targeting geology, soil, and plant applications constructed primarily of AVIRIS, HyMap, and CHRIS data (1992–current) over Jasper Ridge in California (Merton 1999) perhaps represents one of the few comprehensive datasets available.

The lack of temporal observations of hyperspectral sensors has proved particularly limiting for understanding, mapping or monitoring dynamic environments that experience rapid or marked changes in, for example, seasonal leaf cover and chemistry (e.g., temperate woodlands), water status (e.g., wetlands) or snow cover (e.g., high mountains). In many environments, such as arid or semi-arid zones, temporal changes are less significant which accounts partly for the success of hyperspectral data in, for example, geological exploration and mapping.

1.3.1.4
Radiometric Resolution

Radiometric resolution, or quantization, is defined as the sensitivity of a sensor to differences in strength of the electromagnetic radiation (EMR) signal and determines the smallest difference in intensity of the signal that can be distinguished. In other words, it is the amount of energy required to increase a pixel value by one count. In contrast to many MSS, which typically recorded data up to 8-bit in size, hyperspectral sensors have been optimised to record data to use at least 10 to 12 bits. As an example, AVIRIS recorded data in 10 bits prior to the 1993 flight season and 12 bits thereafter (Vane et al. 1993). The consequence of increasing the quantization has been to increase the sensitivity of the sensor to variations in the reflected signal, thereby allowing more subtle reflectance differences from surfaces to be detected and recorded. If the quantization level is too small then these differences can be lost. The acquisition of data by sensors that support a larger number of quantization levels is therefore important, especially when dealing with some of the applications discussed later in this chapter (e.g., determination of vegetation stress and health and discrimination of minerals).

A fundamental requirement of hyperspectral remote sensing has been the need to maximise the signal as opposed to the noise; in other words, the SNR. The SNR is a measure of how the signal from surfaces compares to the background values (i.e., noise) and is determined typically by estimating the signal from pixel values averaged over a homogenous target and dividing this by the noise estimated from the standard deviation of the pixel values. SNR vary considerably between sensors and spectral regions. Manufacturers of CASI, for example, state that SNR values are greater than 480:1. Manufacturers of HyMap, state that SNR values are above 500:1 at 2200 nm, and 1000:1 in the VNIR region. For the GER sensor, values of 5000:1 for the visible and NIR regions and 500:1 for the SWIR region, have been reported (Mackin and Munday 1988). The SNR also varies depending upon the nature of the surface. In a study of soils in Israel (Ben-Dor and Levin 2000), the SNR for sensors varied from 20:1 to 120:1 for light targets and 1:1 to 5:1 for dark targets.

The development and use of airborne hyperspectral sensors has continued as they provide flexibility in the acquisition of data, particularly in terms of temporal frequency and spatial coverage. Specifically, data can be acquired when conditions (e.g., local weather) are optimal, and parameters (e.g., the area of coverage, waveband configurations and spatial resolutions) can be determined prior to acquisition. Even so, the expense of acquiring data from airborne sensors still limits the construction of multi-temporal datasets and generally reduces coverage to relatively restricted areas. Many studies using such data have therefore been developed for specific and often narrow applications. A further disadvantage of using airborne sensors has been that aircraft motion and sun-surface-sensor geometries impact negatively on the quality of the data acquired. For these reasons, there has been a drive to advance spaceborne missions with hyperspectral capability.

1.3.2
Spaceborne sensors

Although airborne hyperspectral sensors have undoubtedly made a valuable contribution to Earth observation research and monitoring and satisfied many of the demands of the user community, the arguments for spaceborne observations have been compelling. Firstly, such sensors allow regular, repeated observations and coverage over wider areas. Secondly, variations in both atmospheric and bi-directional effects across the scans, and distortions resulting from sensor motion, are also reduced.

The drive towards spaceborne hyperspectral data was hampered initially as a planned Space Shuttle imaging spectrometer was cancelled following the Challenger accident. Delays also occurred in implementing the Earth Observing System (EOS) High Resolution Imaging Spectrometer (HRIS) (Goetz and Herring 1989) due to EOS budget constraints and the failure of the LEWIS satellite to achieve orbit. The first satellite-based imaging spectrometer available for general usage was therefore the Moderate Resolution Imaging Spectrometer (MODIS) on both the EOS TERRA-1 and AQUA satellites, which were launched in 1999 and 2002 respectively. The TERRA-1 Advanced Spaceborne Thermal Emission and Reflection (ASTER) sensor and the ENVISAT Medium Resolution Imaging Spectrometer (MERIS) sensor (launched 2002) also provide hyperspectral capability, albeit in selected wavelength regions only. Both the MODIS and MERIS sensors provide near global coverage at coarse ($>$ 250 m) spatial resolution.

Hyperion was the first hyperspectral sensor providing data across the full reflectance spectral region and was launched in November, 2000, onboard the NASA Earth Observing-1 (EO-1) platform. In October, 2001, the Compact

High Resolution Imaging Spectrometer (CHRIS) was also deployed onboard the Project for On-Board Autonomy (PROBA) small satellite mission. CHRIS was designed for a broad base of applications providing hyperspectral data within the 410–1050 nm wavelength regions and over a wide range of spectral and spatial configurations, thereby providing hyperspectral Bidirectional Reflectance Distribution Function (BRDF) observations of the Earth's surface at −55°, −36°, 0° (nadir), +36°, +55° along track look angles. Orbview-4 was to provide 1 m panchromatic and 4 m multispectral data as well as a 200 channel hyperspectral sensor, although the sensor was lost in a launch failure. A number of satellite systems also remain in the development phase (Table 1.3), including ESA's Visible and Infrared Thermal Imaging Spectrometer (VIRTIS-H) and Land Surface Processes and Interactions Mission (LSPIM) (Buckingham et al. 2002), and SPECTRA, whilst research is underway into the development of ultraspectral sensors which will provide data in thousands of very narrow spectral bands giving sufficient spectral sensitivity for identifying specific constituents (e.g. aerosols and gas plumes). The Canadian Space Agency (CSA) along with the Canadian Centre for Remote Sensing (CCRS) and industry is also developing a targeted hyperspectral imager for the International Space Station (ISS) for launch on a small satellite (Buckingham et al. 2002).

The benefits of using data from spaceborne hyperspectral sensors are only just being realised with the recent launch of satellite-based sensors. The Hyperion mission has provided the first opportunity to explore the potential of multi-temporal hyperspectral data for Earth surface observation and to test synergy with other satellite systems (e.g., LANDSAT-7 ETM+). Even so, this sensor remains more an experimental research tool rather than one that is ready for general applications development, due largely to the limited swath width and the generally poorly understood radiometry. The CHRIS sensor, although not observing in SWIR regions, has given a greater opportunity to explore the bidirectional capabilities and information content of hyperspectral data with applications in land surface biochemical and biophysical property retrieval and water characterisation.

At more synoptic spatial scales, the coarser spatial resolution sensors (e.g., MODIS) with hyperspectral capability provide a greater opportunity for multi-temporal observations globally. A particular advantage of these sensors is that they provide consistent observations over an extended time period, relative consistency and/or knowledge of viewing angles, and uniformity in illumination and image geometry (van der Meer 2001). The following section provides an overview of the characteristics of these sensors that renders them unique in the context of Earth observation.

1.3.2.1
Spectral Regions, Resolutions and Bidirectional Capability

Although not providing a continuous spectral profile, the MODIS and ASTER image the spectral region from the visible through to the SWIR and TIR regions and at variable sampling intervals. The MERIS, although observing in only the 390–1040 nm region and at band widths of 2.5–12.5 nm, is fully programmable

Hyperspectral Sensors and Applications

(in terms of width and location) such that observations can be made in up to 15 moveable bands with these selected through telecommands. This increases the utility of the instrument as selected portions of the spectrum with contiguous channels can be viewed. The ASTER sensor also has channels centred on selected absorption features. Specifically, band 6 (SWIR) is centred on a clay absorption feature (associated with hydrothermal alteration), whilst bands 8 (SWIR) and 14 (TIR) are centred on a carbonate feature, thereby allowing discrimination of limestones and dolomites. Bands 10 to 12 (TIR) are designed to detect sulphate and silica spectral features, whilst band 10 (with 6) allows discrimination of common minerals, including alunite and anhydrite (Ellis 1999).

Only the Hyperion and CHRIS systems provide a continuous spectral profile for each pixel value (Table 1.2). The Hyperion observes in 220 bands extending from 400 to 2500 nm with a FWHM of 10 nm. The CHRIS observes in the 410–1050 nm range in 63 bands with bandwidths ranging variably from 1–3 nm at the blue end of the spectrum to about 12 nm at the NIR end. The red-edge region ($\sim$ 690–740 nm) is sampled at a bandwidth of approximately 7 nm. Although a continuous spectral profile is useful, it can be argued that for many applications, the information required can be easily extracted from a few specific spectral bands, particularly since many are highly correlated and often redundant. Feature extraction and selection therefore becomes an important image processing issue under such circumstances. Aspects of over sampling are discussed in more detail in Chapters 8 and 9.

A particular benefit of CHRIS is that multi-angle viewing can be achieved in the five along track view angles. For example, using PROBA's agile steering capabilities in both along and across track directions, observations of targets outside the nominal field of view of 1.3° can be obtained. From CHRIS, surface biophysical parameters are being estimated using a number of techniques ranging from traditional vegetation indices to more advanced techniques such as BRDF model inversion. MERIS also provides directional viewing capability.

1.3.2.2
Spatial Resolution and Coverage

The spatial resolution of spaceborne hyperspectral sensors is coarser compared to most airborne sensors, due largely to their greater altitude. MODIS and MERIS sensors provide data at spatial resolutions ranging from 250–300 m to 1000 m. Hyperion observes at 30 m spatial resolution whilst CHRIS acquires data at spatial resolutions of 36 m and 19 m with 63 and 18 spectral bands respectively. MODIS provides data of the same location at variable spatial resolutions of 250 m–1000 m (Table 1.3). As suggested earlier, coarse spatial resolution can limit the ability to extract pure spectral signatures associated with different surfaces. Orbview 4 (launch failure) was to include a finer (8 m) spatial resolution 200 band hyperspectral sensor.

The swath width, and hence the area of coverage, also varies between satellite sensors. MODIS and MERIS have swath widths of 2330 km and 1150 km respectively, allowing wide area coverage within single scenes. Such data are

Table 1.3. Basic characteristics of selected spaceborne hyperspectral and multispectral sensors

Sensor	Launch	Platform	Number of Bands	Spectral range (nm)	Band width at FWHM	Spatial Resolution (m, unless stated)	Technical reference
ASTER	Dec 1999	Terra (EOS)	14	520–1165	40–100	VNIR 15 SWIR 30 TIR 90	terra.nasa.gov/
CHRIS	Oct 2001	ESA PROBA	up to 62	410–1050	5–12	18–36	www.chris-proba.org
GLI	Dec 2002	NASDA ADEOS-II	23 VNIR; 6 SWIR; 7 MTIR	380–1200	10–220	250–1km	www.nasda.go.jp
HYPERION	Nov 2002	EO-1	220	400–2500	10	30	eo1.gsfc.nasa.gov/
MERIS	Feb 2002	ENVISAT	15	390–1040	programmable width and position	300	www.esa.int
MODIS	Dec 1999	Terra (EOS)	36	405–14385	VNIR-SWIR: 10-50	Bands 1+2 VNIR 250; Bands 3–7 VNIR+SWIR 500; TIR 1 km	terra.nasa.gov
HRVIR	May 2002	SPOT 5	4	500–1750	90–170	10	www.spotimage.fr

therefore more suitable for use in regional to global studies. Hyperion has a nominal altitude of 705 km giving a narrow (7.5 km) swath width, whilst CHRIS has a nominal 600 km orbit and images the Earth with a 14 km swath (with a spatial resolution of 18 m). This coverage generally limits use to local or landscape-scale studies.

1.3.2.3
Temporal Resolution

Repeat observations with spaceborne sensors, can be achieved to yield a significant advantage for many applications, particularly vegetation mapping and monitoring. The Hyperion sensor acquires data over the same area (albeit approximately 1 minute after) as the LANDSAT – 7 ETM+ and the Argentinean SAC-C as part of the AM Constellation and therefore has a similar repeat period of 16 days. CHRIS also has a repeat period of 16 days. MODIS and MERIS with repeat periods of 1–2 days and 3 days respectively are more suited for various land applications with a regional or global context. Although long time-series of hyperspectral data have not been adequately constructed, the potential benefits of these archives for land applications is enormous, particularly in the context of vegetation species/community discrimination and understanding the seasonal variability in biophysical and biochemical attributes.

1.4
Ground Spectroscopy

Data acquired by airborne or spaceborne sensors cannot be considered in isolation since effective data interpretation requires a detailed understanding of the processes and interactions occurring at the Earth's surface. In this respect, a fundamental component of understanding hyperspectral sensors is the laboratory and field measurement of the spectral reflectance of different surfaces. A number of portable field and laboratory spectroradiometers have been developed for this purpose, ranging from the Milton spectroradiometer to the more advanced spectroradiometers that include the Analytical Spectral Devices (ASD) Fieldspec Pro FR and the IRIS spectroradiometer developed by GER (Table 1.4).

Technological advances in a number of areas have led to improvements in the field of spectroscopy. Advances in spectroradiometer technology, specifically with respect to the electro-optical systems, have resulted in an increase in sensor sensitivity and a decrease in scan times, permitting greater data collection in a shorter period of time. This has enabled researchers to acquire high quality reflectance data rapidly, both in the field and under laboratory conditions. A second advancement has been the increase in the processing sophistication of computer technology and the reduction in the cost of data storage. These technological improvements have also reduced the overall costs so that a greater number of users are able to acquire portable spectrometers, particularly for use in the field.

Table 1.4. Ground spectrometers

Instrument	Spectral range (nm)	Band width over specified ranges (nm)	No. of bands	Technical reference
ASD Fieldspec FR	350–2500	1.4 (350–1050) 2.0 (1000–2500)	512	www.asdi.com
ASD Fieldspec JR	350–2500	1.4 (350–1050) 2.0 (1000–2500)	512	www.asdi.com
ASD Fieldspec VNIR	350–1050	1.4	512	www.asdi.com
GER 1500	350–1050	1.5	512	www.ger.com
GER 2600	350–2500	1.5 (350–1050) 11.5 (1050–2500)	640	www.ger.com
GER 3700	350–2500	1.5 (350–1050) 6.2 (1050–1900) 9.5 (1900–2500)	704	www.ger.com
GER Hi-RES	350–2500	1.5 (350–1000) 3.0 (1000–2500)	970	www.ger.com
LICOR 1800 Portable	300–1100	1, 2, 5 Variable with sampling interval	N/A	www.glenspectra.co.uk
OKSI	400–1600	5 (visible) 50 (NIR)	N/A	www.oksi.com
PIMA SP	1300–2500	2 or 7	N/A	www.intspec.com
PP Systems Unispec	300–1100	3.7	N/A	www.ppsystems.com/

Field spectroscopic studies are important in hyperspectral remote sensing for a number of reasons. The first is calibration of airborne and satellite-based image products. All remotely sensed datasets have been imaged through a variety of atmospheric conditions, which usually incorporate varying amounts of water vapour, CO_2, O_2, and particulate matter. In addition, imaging is carried out at various times of the year and, in some cases, during different times of the day and under sometimes highly variable atmospheric conditions. To account for this variability, image spectra can generally be compared to spectra that have not been affected by these conditions. One method is to capture spectra from a disparate range of pseudo-invariant targets (reference targets) at the time of remote sensing data acquisition. As these ground-based spectra are not collected through a large atmospheric path, they should be relatively free of the adverse "noise" influences that are common in spectra recorded by airborne and spaceborne sensors. By comparing spectra to those retrieved from the imaging spectrometers, these noise influences can be largely removed and the data can be converted from units of radiance to that of scale surface reflectance. The latter represents a simple standardization that allows comparison of spatially and/or temporally diverse data.

The second reason is that field and laboratory spectra can be collated into spectral libraries which can be useful for a number of purposes. For example, field spectra of certain mineral assemblages can be used to train the classification of geologically diverse areas and have been used frequently to derive geological maps (Kruse and Hauff 1991; Kruse et al. 1993).

Spectral libraries can also play a key role in selecting endmembers for subsequently spectral unmixing of hyperspectral data. This process decomposes individual pixels into their reflectance components based on a knowledge of and comparison with the spectral characteristics of known target materials. These known materials, also referred to as *endmembers*, are then used as references to decompose the pixels. For a more extensive description of this process see (Adams and Smith 1986; Elvidge 1990; Gaddis et al. 1993; Aber et al. 1994; Kupiek and Curran 1995; Levesque et al. 2000). Endmembers for spectral decomposition can be extracted by locating "pure" targets in the image through manual or automated techniques. Such an approach assumes that such pure targets exist in the image, which may be the case for very fine ($< 1-2$ m) spatial resolution images but becomes less likely as the spatial resolution coarsens (*esp.* mixed pixels). An alternative approach therefore is to use spectral libraries generated from field-based spectroscopy or even from the finer spatial resolution data themselves. Much of the theory related to spectral endmembers has originated from minerals exploration applications with pure targets that do not vary greatly. However, it should be noted that the spectral signatures of vegetation are dynamic in spectral-, spatial-, and temporal-space, and should cautiously be constructed into spectral libraries.

1.5
Software for Hyperspectral Processing

The datasets originating from hyperspectral sensors are complex and cannot be adequately analyzed using the more traditional image processing packages. Significant research has therefore been carried out to develop algorithms that can be incorporated into commercially available software. At the time of writing two of the more commonly used packages for hyperspectral image processing have been produced by Research Systems Inc. (RSI), U.S.A. and PCI Geomatics, Canada. Other examples of software that state hyperspectral capability are listed in Table 1.5. It should be noted that many of the algorithms used for hyperspectral imaging applications presented in this book are of an advanced nature and are still under research and development.

Table 1.5. Commercially available software for hyperspectral applications

Software	Developer/Distributor
Environment for Visualizing Images (ENVI)	Research Systems Inc., USA
EASI-PACE	PCI Geomatics, Canada
Imagine	ERDAS, USA
Hyperspectral product Generation System (HPGS)	Analytical Imaging and Geophysics, USA
Hyperspectral Image Processing and Analysis System (HIPAS)	Chinese Academy of Sciences, China
Spectral Image Processing System (SIPS)	University of Colorado, USA

1.6
Applications

Although imaging spectrometry has been used in military applications (e.g., distinguishing between camouflage and actual vegetation) for many years, the classified nature of the information has resulted in few published papers regarding their origins (van der Meer and de Jong 2001). Therefore, as hyperspectral remote sensing data became available to the civilian community, the early phases of analysis focused largely on understanding the information content of the data over a disparate range of environments. Subsequently, the development of algorithms specifically designed to more effectively manipulate the enormous quantities of data generated became a priority. As familiarity with the data increased, the potential benefits of using hyperspectral imaging became apparent. Today, hyperspectral data are increasingly used for applications ranging from atmospheric characterisation and climate research, snow and ice hydrology, monitoring of coastal environments, understanding the structure and functioning of ecosystems, mineral exploration and land use, land cover and vegetation mapping. The following provides a brief overview of these applications.

1.6.1
Atmosphere and Hydrosphere

Within the atmospheric sciences, hyperspectral remote sensing has been used primarily to investigate the retrieval of atmospheric properties, thereby allowing the development and implementation of correction techniques for airborne and spaceborne remote sensing data (Curran 1994; Green et al. 1998b; Roberts et al. 1998). In particular, certain regions of the electromagnetic spectrum are sensitive to different atmospheric constituents. For example, absorption of water vapour occurs most strongly at 820 nm, 940 nm, 1130 nm, 1380 nm and 1880 nm (Gao and Goetz 1991) whilst CO_2 absorption is prominent at

1600 nm but particularly at ~ 2080 nm. O_2 absorption occurs at 760 nm and 1270 nm. Using this information, estimates of these constituents can be derived using hyperspectral data. As an example, water vapour can be estimated using the Continuum Interpolated Band Ratio (CIBR) and Narrow/Wide and Atmospheric Pre-Corrected Differential Absorption (APDA), with these measures utilising differential absorption techniques and applied across specific atmospheric water vapour absorption features (e.g. 940 and 1130 nm) (Rodger and Lynch 2001). The EO-1 also supported an Atmospheric Corrector for facilitating correction for atmospheric water vapour, thereby optimising retrieval of surface features. Coarser spatial resolution datasets have also been used to characterise the atmosphere. For example, MODIS data have been used in the measurement of cloud optical thickness, cloud top pressures, total precipitable water and both coarse and fine aerosol contents in the atmosphere (Baum et al. 2000; Seemann et al. 2003).

For studies of the cryosphere, hyperspectral techniques have been used to retrieve information relating to the physical and chemical properties of snow including grain size, fractional cover, impurities, and snow mass liquid water content (Dozier and Painter 2004). Few airborne studies have been undertaken in high latitude regions due to the difficulty in acquiring data. Snow products are also routinely and automatically derived from orbital platforms such as MODIS, and include snow cover maps from individual scenes to spatial and temporal composites (Hall et al. 1995; Klein and Barnett 2003).

Within marine and freshwater environments, hyperspectral data have been used to characterize and map coral reefs and submerged aquatic vegetation (Kutser et al. 2003; Malthus and Mumby 2003), including sea grasses (Fyfe 2003), and for quantifying sediment loads and water quality (Fraser 1998). At the land-sea interface, applications have included characterization of aquatic vegetation (Zacharias et al. 1992), salt marshes (Schmidt and Skidmore 2003; Silvestri et al. 2003) and mangroves (Held et al. 2003). As an example, Zacharias et al. (1992) were able to detect submerged kelp beds using airborne hyperspectral data at several spatial resolutions (Fig. 1.2). The benefits of using hyperspectral CASI data for identifying and mapping different mangrove species and communities have been highlighted in studies of Kakadu National Park, northern Australia (Fig. 1.3). When the derived maps were compared in a time-series with those generated from historical black and white and true colour stereo aerial photography, changes in both the extent of mangroves and their contained species and communities were evident (Mitchell 2004). These changes suggested a long-term problem of saltwater intrusion as a result of coastal environmental change. The study also emphasized the benefits of using hyperspectral data at fine (< 1) spatial resolution for characterizing and mapping mangrove communities.

The application of remotely sensed data to the study of coral reefs was proposed in the mid 1980s (Kuchler 1986) although difficulties associated with wavelength-specific penetration of light in water, mixed pixels and atmospheric attenuation caused initial disillusionment amongst some users (Green et al. 1996; Holden and LeDrew 1998). Even so, the science developed dramatically in the 1990s, due partly to concerns arising from the impacts on coral reefs

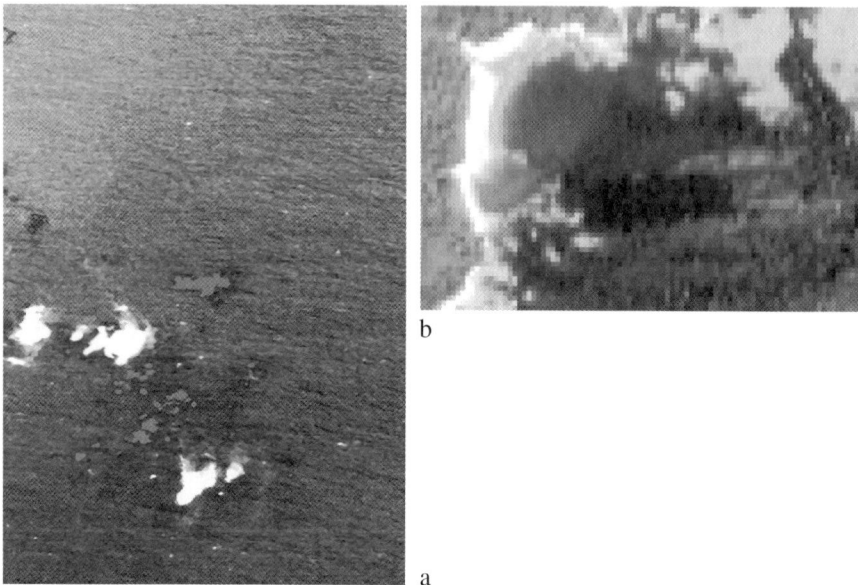

Fig. 1.2. Example of detection of aquatic vegetation using airborne imaging spectrometers: partially submerged kelp beds (*bright red*) as observed using **a** 2 m CASI and **b** 20 m AVIRIS data. For a colored version of this figure, see the end of the book

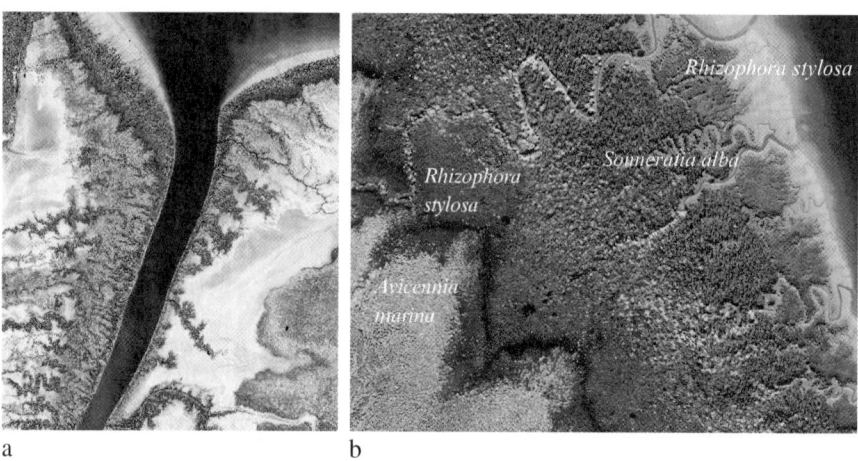

Fig. 1.3. a Colour composite (837 nm, 713 nm and 446 nm in RGB) of fourteen 1 km × ~ 15 km strips of CASI data acquired over the West Alligator River mangroves, Kakadu National Park, Australia by BallAIMS (Adelaide). The data were acquired at 1 m spatial resolution in the visible and NIR wavelength (446 nm–838 nm) region. **b** Full resolution image of the west bank near the river mouth with main species/communities indicated (Mitchell 2004). For a coloured version of this figure, see the end of the book

of anthropogenic activity (Yentsch et al. 2002) and changes in regional and global climate and particularly the El Niño Southern Oscillation (ENSO). Similarly, recognition that the sensitivity of corals to temperature (Riegl 2002), light (Sakami 2000), salinity (Ferrier-Pages et al. 1999), turbidity (Perry 2003) and oxygen availability actually rendered them as suitable "environmental barometers".

Hyperspectral sensors have enabled new avenues of coral reef research, particularly in relation to the critical differentiation between healthy and environmentally stressed coral (Hochberg et al. 2003). Furthermore, some of the difficulties associated with the use of conventional satellite sensors and even IKONOS data (e.g., coarse spatial or spectral resolution) for discriminating substrate features and principal geomorphological zones associated with coral reefs (Maeder et al. 2002; Mumby and Edwards 2002) have also been partly overcome using hyperspectral data. As with many applications, monitoring of coral reefs using remote sensing data has provided a cost-effective and time-efficient tool for reef surveys, change detection, and management (Maeder et al. 2002).

1.6.2
Vegetation

Hyperspectral sensors are well suited for vegetation studies as reflectance/absorption spectral signatures from individual species as well as more complex mixed-pixel community scale spectra can be utilised as a powerful diagnostic tool for vegetation sciences. The spectral geometry of vegetation signatures vary according to, for example, biochemical content and the physical structure of plant tissues, and is further influenced by phenologic factors and natural and anthropogenically induced factors creating a complex matrix of influencing variables. To "unscramble" the weighted influence of these variables upon the resultant vegetation spectral signature remains the perennial challenge to the monitoring of vegetation with hyperspectral systems.

1.6.2.1
Reflectance Characteristics of Vegetation

Knowledge of the causes of variation in the spectral reflectance of vegetation (Fig. 1.4) has been fundamental to understanding the information content of spectra derived from ground-based or airborne/spaceborne spectrometers. In the visible (red, green, and blue) wavelength regions, plant reflectance is dictated by the amount and concentration of photosynthetic pigments, namely chlorophyll a, chlorophyll b, xanthophylls, anthocyanins and carotenoids (Guyot et al. 1989; Cochrane 2000; Chisholm 2001). In the NIR wavelength region, the internal structure of leaves and, in particular, the size, shape and distribution of air spaces and also the abundance of air-water interfaces (and hence refractive index discontinuities) within the mesophyll layers exert the greatest influence on reflectance (Knox 1997). Much of the radiation

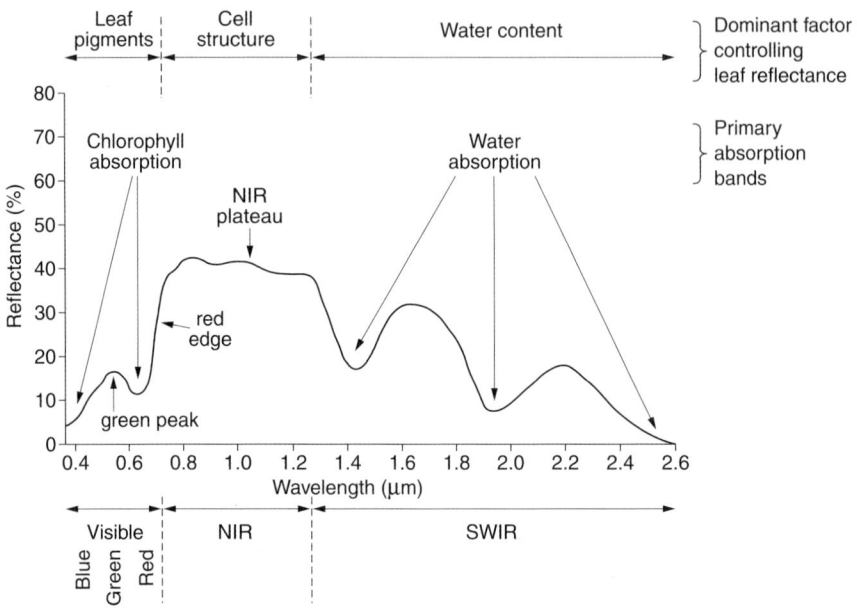

Fig. 1.4. Typical spectral reflectance profile of vegetation

that is scattered within the leaf is reflected back through the leaf surface, with a proportion transmitted through the leaf (Kumar 1998). SWIR reflectance is determined largely by moisture content (Kaufman and Remer 1994), with water absorption features centred primarily at 2660 nm and 2730 nm (Kumar et al. 2001). SWIR reflectance is also influenced by leaf biochemicals such as lignin, cellulose, starch, proteins and nitrogen (Guyot et al. 1989; Kumar et al. 2001) but absorption is relatively weak and often masked by the more dominant water absorption features (Kumar 1998).

Key descriptive elements of the vegetation spectral reflectance signatures include the green peak, the chlorophyll well, the red-edge, the NIR plateau, and the water absorption features (Kumar 1998). Of particular importance is the red-edge, which is defined as the rise of reflectance at the boundary between the chlorophyll absorption feature in red wavelengths and leaf scattering in NIR wavelengths (Treitz and Howarth 1999). The red-edge, which displays the greatest change in reflectance per change in wavelength of any green leaf spectral feature in the VNIR regions (Elvidge, 1985), is identified typically using the red-edge inflection point (REIP) or point of maximum slope and is located between 680 and 750 nm regardless of species (Kumar et al. 2001).

Using hyperspectral data, key factors that determine the likelihood of retrieving a 'pure' reflectance spectra include the spatial resolution of the sensor, the density of the canopy, the area and angle distribution of leaves and branches, and the proportion of shade in the canopy (Adams et al. 1995). In most cases, the spectral signature becomes mixed with that of other surfaces, although the influence of multiple reflective elements within a pixel can be removed or mitigated using techniques such as spectral mixture analysis (SMA). Furthermore, information on both the biochemical and biophysical properties of canopies can still be retrieved, despite the degradation of the 'pure' spectral reflectance feature. Differences in these attributes can also be used to discriminate vegetation species and/or communities using either single date or multi-temporal imagery. Such applications are outlined in more detail in the following sections.

1.6.2.2
Foliar Biochemistry and Biophysical Properties

A major focus of hyperspectral remote sensing has been the retrieval of foliage chemicals, particularly those that play critical roles in ecosystem processes, such as chlorophyll, nitrogen and carbon (Daughtry et al. 2000; Coops et al. 2001). Most retrievals have been based on simple, conventional stepwise or multiple regression between chemical quantities (measured using wet or dry chemistry based techniques) and reflectance (R) or absorbance (calculated as $\log 10(1/R)$) data or band ratios of spectral reflectance, which themselves are usually correlated to measures of green vegetation amount, cover or functioning (Gitelson and Merzlyak 1997). Modified partial least squares (MPLS), neural network, statistical methods (Niemann and Goodenough 2003) and model inversion techniques (Demarez and Gastellu-Etchegorry 2000; Jacquemoud et al. 2000) have also been used to assist retrieval. MPLS has proved particularly useful as the information content of hundreds of bands can be concentrated within a few variables, although the optimal predictive bands still need to be identified using correlograms of spectra and biochemical concentration or regression equations from the MPLS analysis.

Although a wide range of foliar chemicals exist in the leaf, hyperspectral remote sensing has shown greatest promise for the retrieval of chlorophyll a and b (Clevers 1994; Gitelson and Merzlyak 1997), nitrogen (Curran 1989; Matson et al. 1994; Gastellu-Etchegorry et al. 1995; Johnson and Billow 1996), carbon (Ustin et al. 2001), cellulose (Zagolski et al. 1996), lignin (Gastellu-Etchegorry et al. 1995), anthocyanin, starch and water (Curran et al. 1992; Serrano et al. 2000), and sideroxylonal-A (Blackburn 1999; Ebbers et al. 2002). Other biochemicals which exhibit clearly identifiable absorption features and are becoming increasingly important to the understanding of photosynthetic and other leaf biochemical processes include: yellow carotenes and pale yellow xanthophyll pigments (strong absorptions in the blue wavelengths), carotene absorption (strong absorption at ~ 450 nm), phycocyanin pigment (absorbs primarily in the green and red regions at ~ 620 nm) and phycoerythrin (strong absorption at ~ 550 nm). However, chlorophyll pigments are dominant and normally mask these pigments. During senescence or severe stress, the chlorophyll dominance may be lost, causing the other pigments to become dominant (e.g. at leaf fall). Anthocyanin may also be produced in autumn causing leaves to appear bright red when observed in visible wavelengths.

A number of studies have successfully retrieved foliar chemicals from airborne and spaceborne hyperspectral data. As examples, foliar nitrogen concentration has been retrieved using AVIRIS, HyMap and Hyperion data (Johnson 1994; Matson et al. 1994; LaCapra et al. 1996; Martin and Aber 1997) and lignin, among other biochemicals, has been quantified using AVIRIS data (Johnson 1994). Indices such as the Water Band Index (WBI) (Penuelas et al. 1997) and the Normalised Difference Water Band Index (NDWBI) have also been used as indicators of leaf and canopy moisture content (Ustin et al. 2001).

Despite successes with the retrieval of foliar chemicals, the process is complicated by the distortion effects of the atmosphere, low SNR, complicated tree canopy characteristics and the influence of the underlying soils and topography. Many understorey species, which exhibit wide ranging foliar chemical diversity, also cannot be observed. Even so, continued efforts at retrieving algorithms for quantifying foliar biochemicals are advocated given the importance of many (e.g., N and C) in global cycles.

Key biophysical attributes that can be retrieved using hyperspectral data include measures of foliage and canopy cover (e.g., FPC or LAI) (Spanner et al. 1990a; Spanner et al. 1990b; Gong et al. 1995), the fraction of absorbed photosynthetically active radiation (fAPAR) and also measures of canopy architecture, including leaf angle distributions. Although woody attributes (e.g., branch and trunk biomass) cannot be sensed directly, these can often be inferred (Asner et al. 1999). Approaches to retrieving biophysical properties from spectral reflectance measurements include the integration of canopy radiation models with leaf optical models (Otterman et al. 1987; Franklin and Strahler 1988; Goel and Grier 1988; Ustin et al. 2001) or the use of empirical relationships and vegetation indices (Treitz and Howarth 1999). Other specific techniques include derivative spectra, continuum removal, hierarchical foreground/background analysis and SMA (Gamon and Qiu 1999), with most associated with either physically-based canopy radiation models (Treitz and Howarth 1999) or empirical spectral relationships (indices).

Key biophysical attributes that can be retrieved using hyperspectral data include measures of foliage and canopy cover (e.g., Foliage Projected Cover (FPC) or Leaf Area Index (LAI)) (Spanner et al. 1990a; Spanner et al. 1990b; Gong et al. 1995), the fraction of absorbed photosynthetically active radiation (fAPAR) and also measures of canopy architecture, including leaf angle distributions. Although woody attributes (e.g., branch and trunk biomass) cannot be sensed directly, these can often be inferred (Asner et al. 1999). Approaches to retrieving biophysical properties from spectral reflectance measurements include the integration of canopy radiation models with leaf optical models (Otterman et al. 1987; Franklin and Strahler 1988; Goel and Grier 1988; Ustin et al. 2001) or the use of empirical relationships and vegetation indices (Treitz and Howarth 1999). Other specific techniques include derivative spectra, continuum removal, hierarchical foreground/background analysis and SMA (Gamon and Qiu 1999), with most associated with either physically-based canopy radiation models (Treitz and Howarth 1999) or empirical spectral relationships (indices).

1.6.2.3
Vegetation Mapping

Vegetation discrimination has been highlighted as one of the benefits of using hyperspectral data and the establishment of spectral libraries for different species has been suggested (Roberts et al. 1998). However, discrimination of species from single date imagery is only achievable where a combination of leaf chemistry, structure and moisture content culminate to form a unique spectral signature. Discrimination from imagery then relies on the extraction of the pure spectral signature for each species which is dictated by the spatial resolution of the observing sensor and also the timing of observation (Asner and Heidebrecht 2002) and requires consideration of the spectral differences and variations that occur within and between species. Identification of species and communities can be further enhanced through observation over the same area at different times within the year. As an example, studies using AVIRIS data acquired on several dates over Jasper Ridge, California (Merton 1999), confirmed the existence of distinct hysteresis loops for different vegetation classes that could be used to refine discrimination. Moreover, these multi-temporal hysteresis loops can identify the timing of critical phenological or environmental-induced events and assist researchers in targeting subsequent image acquisitions to specific "indicator times" of the year (Merton 1999; Merton and Silver 2000).

Fundamental to the discrimination of tree/shrub species from fine (< 1 m) spatial resolution imagery is the extraction of signatures from the crown. For this purpose, a number of procedures have been developed for the delineation of individual tree crowns/clusters within fine spatial resolution hyperspectral data (Ticehurst 2001; Culvenor 2002; Held et al. 2003). From these delineated crowns, representative reflectance data or indices can be derived subsequently to facilitate discrimination and mapping.

For regional to global mapping, the enhanced spectral characteristics and radiometric quality of the MODIS sensor has allowed routine mapping of land cover types. Using 1 km data acquired between November 2000 and October 2001, 17 different land cover types have been classified, following the International Geosphere-Biosphere Program (IGBP) scheme and using decision tree and artificial neural network techniques (see Chapter 2). These land cover types include eleven classes of natural vegetation (e.g., savannas, wetlands and evergreen/deciduous woodlands), several classes of developed and mosaic land, and areas with snow and ice (Friedl et al. 2002). Validation of the algorithm has been based on a network of sites observed using finer spatial resolution remote sensing data. These maps provide a better understanding of global ecosystems and land cover, particularly as they are updated routinely, and contribute to, for example, global carbon science initiatives.

1.6.3
Soils and Geology

Soils are complex and typically heterogeneous, so their properties therefore cannot be assessed easily and directly as a function of their composite spectral reflectance profiles, even under controlled laboratory conditions (Ben-Dor and Banin 1994). Furthermore, the entire vertical soil profile cannot be imaged using multispectral or hyperspectral sensors and inferences have to be made about the soil based on the surface veneer. In most cases, vegetation (dry or green) obscures the soil surface either partly or completely. Despite these limitations, certain variables associated within soils have been quantified using remote sensing data. As examples, hyperspectral remote sensing (e. g., AVIRIS) has shown promise in retrieving soil moisture content, temperature, texture/surface roughness, iron content and soil organic carbon (SOC) (Palacios-Orueta and Ustin 1998; Ahn et al. 1999). Using reference reflectance spectra from soil samples and established procedures, the relationships of SOC to moisture, exchangeable calcium and magnesium cation exchange capacity and also soil colour were able to be quantified (Schreier et al. 1988). Many of these variables actively influence the response across the reflectance region and, for this reason, only a small selection of spectral channels is required for their quantification (Ben-Dor et al. 2002).

Colour is also a key attribute as it allows characterisation, differentiation and ultimately classification of soil types. In general, colour is determined by the amount and state of iron and/or organic matter (OM) contained, with darker soils tending to be more indicative of higher OM contents, although manganese can also produce dark colouration. The colour of soils with low OM is often dominated by pigmenting agents such as secondary iron oxides. Comparisons with airborne MIVIS visible data acquired in farmed areas of Italy suggested that soil colour could be retrieved, allowing estimation of both chroma and hue, although mapping across the image was considered difficult. As many mineral components of rocks and soils have distinct spectral signatures, hyperspectral data have also played a key role in identifying and mapping expansive soils (Chabrillat et al. 2002).

Geological applications have been the principal driving force behind much of the early development phases of imaging spectrometry. The development can be attributed partly to the recognition that a wide range of rock forming minerals have distinct spectral signatures (Curran 1994; Escadafel 1994; Goetz 1995). Following the work of Hunt (1989), numerous publications have provided detailed laboratory spectra of rocks and minerals and their mixtures as well as accurate analyses of absorption features, many of which have been obtained from laboratory measurements. These serve as references for spectra obtained using ground radiometers or airborne/spaceborne sensors, with spectroscopic criteria applied to identify and map minerals and alteration zones (Hunt 1989; Christensen et al. 2000; Longhi et al. 2001).

Despite the availability of suitable reference sources, their use in rock-type matching has been complicated by several interacting variables including original parent mineral associations, diagenesis, transformation/alteration through time and micro- to macro-scale morphology. A further limitation is that the dominant mineral components of a rock, which generally determine their petrographic classification, do not necessarily produce the most prominent spectral features. Even so, dominant spectral patterns produced by minor mineral components with strong absorption features can significantly assist in characterising rocks, at least within well constrained geological settings (Longhi et al. 2001).

A range of multispectral sensors have been used for mapping a variety of different minerals including the LANDSAT (Carranza and Hale 2002) and SPOT sensors (Cole 1991; Chica-Olmo et al. 2002). However, given the spatial resolutions of such orbital sensors, techniques such as SMA are often required. As imaging spectrometers such as AVIRIS and HyMap provide higher spectral resolution data, these sensors have frequently been preferred for geological mapping and mineral exploration although the extent of coverage is often reduced. As examples, AVIRIS has been used to map mineral assemblages in areas of the United States (Mackin et al. 1991; Kokaly et al. 1998), including areas of ammonium in hydrothermal rocks (Baugh et al. 1998), and as a tool in gold exploration (van der Meer and Bakker 1998; Rowan 2000). Similarly, HyMap has been utilised to map gold mineralization in Western Australia (Bierwirth et al. 2002) and for detecting hydrocarbons (Horig et al. 2001). Simulations have also suggested that different types of metamorphic rocks could be further discriminated using AVIRIS or MIVIS data. Other applications have included the measurement of exposed lava and volcanic hotspots (Oppenheimer et al. 1993; Green et al. 1998b), the study of petroleum geology (van der Meer and de Jong 2001) and the application of AVIRIS to study mineral-induced stress on vegetation (Merton 1999; Merton and Silver 2000). Recent studies (Kruse et al. 2003) have also investigated the use of Hyperion for geological mapping and monitoring the processes that control the occurrence of non-renewable mineral resources, especially mineral deposits.

The difficulties of applying remote sensing to the mapping of minerals are illustrated in heavily forested regions, where reflectance based data are less useful because of the lack of direct observations of the ground surface. Steep terrain, heavy rainfall and cloud cover as well as high population densities (and hence changes in land use) in many regions (particularly in the tropics) also limit exploration using remote sensing data (Cervelle, 1991).

1.6.4
Environmental Hazards and Anthropogenic Activity

Hyperspectral remote sensing offers many opportunities for monitoring natural hazards such as bushfires, volcanic activity and observing anthropogenic-induced activities and impacts such as acidification, land clearing and degradation (de Jong and Epema, 2001), biomass burning (Green et al. 1998a), water pollution (Bianchi et al. 1995a; Bianchi et al. 1995b), atmospheric fallout of dust and particulate matter emitted by industry (Ong et al. 2001) and soil salinity (Metternicht and Zinck 2003). As an example, Chisholm (2001) indicated that spectral indices could be used to detect vegetation water content at the canopy level which could be used subsequently to assess fire fuel loading and hence assist in the prediction of wildfires. Asner et al. (1999) also indicated that observed variations in AVIRIS hyperspectral signatures were indicative of the 3-dimensional variation in LAI and dry carbon and that simultaneous observations of both LAI and dry carbon area index (NPVAI) allowed the production of maps of both structural and functional vegetation types as well as fire fuel load.

1.7
Summary

This chapter has provided a brief history of hyperspectral sensors and an overview of the main airborne and spaceborne instruments. The particular spatial, spectral and temporal benefits of both airborne and spaceborne hyperspectral sensors, which make them well suited to observation of Earth surfaces, have also been highlighted. Applications of hyperspectral remote sensing have been reviewed in brief and readers are encouraged to consult the cited literature for more detailed information.

The review comes at a time when the remote sensing community is at the start of a new era in which spaceborne observations of hyperspectral data, often acquired at different view angles, will be acquired routinely. The information content within these new datasets is enormous and will present many challenges to scientists and land managers charged with analysing and transforming these to derive practical output products. New research will not only need to focus on the nature of information that can be extracted across a wide and disparate range of disciplines, but also attention will be drawn towards the way in which the vast data-streams are handled, particularly with the advent of multi-angle and ultraspectral sensors. The development of algorithms and models to optimise processing and analysis of these data will also be a significant growth area of research, and methods for efficient and appropriate extraction of data, especially from time-series datasets, will need to be developed as historical archives of hyperspectral data increase. The following chapters in this book discuss a selection of data processing and extraction tools which can be applied to both existing and forthcoming hyperspectral data.

References

Aber JD, Bolster KL, Newman SD, Soulia M, Martin ME (1994) Analysis of forest foliage II: Measurement of carbon fraction and nitrogen content by end-member analysis. Journal of Near Infrared Spectroscopy 2: 15–23

Adams JB, Sabol DE, Kapos V, Filho RA, Roberts DA, Smith MO (1995) Classification of multispectral images based on fractions of endmembers: application to land-cover change in the Brazilian Amazon. Remote Sensing of Environment 52: 137–154

Adams JB, Smith MO (1986) Spectral mixture modeling: a new analysis of rock and soil types at the Viking Lander 1 site. Journal of Geophysical Research 91B8: 8098–8112

References

Ahn CW, Baumgardner MF, Biehl LL (1999) Delineation of soil variability using geostatistics and fuzzy clustering analyses of hyperspectral data. Soil Science Society of America Journal 63(1): 142–150

Asner GP, Heidebrecht KB (2002) Spectral unmixing of vegetation, soil and dry carbon cover in arid regions: comparing multispectral and hyperspectral observations. International Journal of Remote Sensing 23(19): 3939–3958

Asner GP, Townsend AR, Bustamante MMC (1999) Spectrometry of pasture condition and biogeochemistry in the Central Amazon. Geophysical Research Letters 26(17): 2769–2772

Aspinall RJ, Marcus WA, Boardman JW (2002) Considerations in collecting, processing, and analysing high spatial resolution hyperspectral data for environmental investigations. Journal of Geographical Systems 4: 15–29

Baugh WM, Kruse FA, Atkinson WW (1998) Quantitative geochemical mapping of ammonium minerals in the southern Cedar Mountains, Nevada, using the Airborne Visible Infrared Imaging Spectrometer (AVIRIS). Remote Sensing of Environment 65(3): 292–308

Baum BA, Kratz DP, Yang P, Ou SC, Hu YX, Soulen PF, Tsay SC (2000) Remote sensing of cloud properties using MODIS airborne simulator imagery during SUCCESS 1. Data and models. Journal of Geophysical Research-Atmospheres 105(D9): 11767–11780

Ben-Dor E, Banin A (1994) Visible and near-infrared (0.4–1.1 μm analysis of arid and semi-arid soils. Remote Sensing of Environment 48(3): 261–274

Ben-Dor E, Levin N (2000). Determination of surface reflectance from raw hyperspectral data without simultaneous ground data measurements: a case study of the GER 63-channel sensor data acquired over Naan, Israel. International Journal of Remote Sensing 21(10): 2053–2074

Ben-Dor E, Patkin K, Banin A, Karnieli A (2002) Mapping of several soil properties using DAIS-7915 hyperspectral scanner data – a case study over clayey soils in Israel. International Journal of Remote Sensing 23(6): 1043–1062

Bianchi R, Castagnoli A, Cavalli RM, Marino CM, Pignatti S, Poscolieri M (1995a) Use of airborne hyperspectral images to assess the spatial distribution of oil spilled during the Trecate blow-out (Northern Italy). Remote Sensing for Agriculture, Forestry and Natural Resources. E. T. Engman, Guyot, G. and Marino, C. M., SPIE Proceedings 2585, pp 352–362

Bianchi R, Castagnoli A, Cavalli RM, Marino CM, Pignatti S, Zilioli E (1995b) Preliminary analysis of aerial hyperspectral data on shallow lacustrine waters. Remote Sensing for Agriculture, Forestry and Natural Resources. E. T. Engman, Guyot, G. and Marino, C. M., SPIE Proceedings 2585, pp 341–351

Bierwirth P, Huston D, Blewett R (2002) Hyperspectral mapping of mineral assemblages associated with gold mineralization in the Central Pilbara, Western Australia. Economic Geology and the Bulletin of the Society of Economic Geologists 97(4): 819–826

Blackburn GA (1999) Relationships between spectral reflectance and pigment concentrations in stacks of deciduous broadleaves. Remote Sensing of Environment 70(2): 224–237

Buckingham B, Staenz K, Hollinger A (2002) Review of Canadian airborne and space activities in hyperspectral remote sensing. Canadian Aeronautics and Space Journal 48(1): 115–121

Carranza EJM, Hale M (2002) Mineral imaging with Landsat Thematic Mapper data for hydrothermal alteration mapping in heavily vegetated terrain. International Journal of Remote Sensing 23(22): 4827–4852

Cervelle B. (1991) Application of mineralogical constraints to remote-sensing. European Journal of Mineralogy 3(4): 677–688

Chabrillat S, Goetz AFH, Krosley L, Olsen HW (2002) Use of hyperspectral images in the identification and mapping of expansive clay soils and the role of spatial resolution. Remote Sensing of Environment 82(2–3): 431–445

Chica-Olmo M, Abarca F, Rigol JP (2002) Development of a Decision Support System based on remote sensing and GIS techniques for gold-rich area identification in SE Spain. International Journal of Remote Sensing 23(22): 4801–4814

Chisholm LA (2001) Characterisation and evaluation of moisture stress in E. camalduensis using hyperspectral remote sensing. Sydney, University of New South Wales

Christensen PR, Bandfield JL, Hamilton VE, Howard DA, Lane MD, Piatek JL, Ruff SW, Stefanov WL (2000) A thermal emission spectral library of rock-forming minerals. Journal of Geophysical Research-Planets 105(E4): 9735–9739

Clevers JPGW (1994) Imaging spectrometry in agriculture – plant vitality and yield indicators. Imaging Spectrometry – a Tool for Environmental Observations. Hill JAM (ed), Kluwer Academic Publishers, Dordrecht The Netherlands, pp 193–219.

Cochrane MA (2000) Using vegetation reflectance variability for species level classification of hyperspectral data. International Journal of Remote Sensing 21(10): 2075–2087

Cocks T, Jenssen R, Stewart A, Wilson I, Shields T (1998) The HYMAP^{TM} airborne hyperspectral sensor:the system, calibration and performance. 1st EARSEL Workshop on Imaging Spectroscopy, Zurich.

Cole MM (1991) Remote-sensing, geobotany and biogeochemistry in detection of Thalanga zinc lead copper-deposit near Charters-Towers, Queensland, Australia. Transactions of the Institution of Mining and Metallurgy Section B-Applied Earth Science 100: B1-B8

Coops NC, Smith M-L, Martin ME, Ollinger SV, Held A, Dury SJ (2001) Assessing the performance of HYPERION in relation to eucalypt biochemistry: preliminary project design and specifications. Proceedings of International Geoscience and Remote Sensing Symposium (IGARSS 2001), CD

Culvenor DS (2002) TIDA: an algorithm for the delineation of tree crowns in high spatial resolution remotely sensed imagery. Computers and Geosciences 28(1): 33–44

Curran PJ (1989) Remote-sensing of foliar chemistry. Remote Sensing of Environment 30(3): 271–278

Curran PJ (1994) Imaging spectrometry. Progress in Physical Geography 18(2): 247–266

Curran PJ, Dungan JL (1989) Estimation of signal-to-noise – a new procedure applied to AVIRIS data. IEEE Transactions on Geoscience and Remote Sensing 27(5): 620–628

Curran PJ, Dungan JL (1990) An image recorded by the Airborne Visible Infrared Imaging Spectrometer (AVIRIS). International Journal of Remote Sensing 11(6): 929–931

Curran PJ, Dungan JL, Macler BA, Plummer SE, Peterson DL (1992) Reflectance spectroscopy of fresh whole leaves for the estimation of chemical concentration. Remote Sensing of Environment 39(2): 153–166

Dabney PW, Irons JR, Travis JW, Kasten MS, Bhardwaj S (1994) Impact of recent enhancements and upgrades of the Advanced Solid-state Array Spectroradiometer (ASAS). Proceedings of International Geoscience and Remote Sensing Symposium (IGARSS), Pasadena, US, Digest, 3, pp 649–1651.

Daughtry CST, Walthall CL, Kim MS, de Colstoun EB, McMurtrey JE (2000) Estimating corn leaf chlorophyll concentration from leaf and canopy reflectance. Remote Sensing of Environment 74(2): 229–239

de Jong SM, Epema GF (2001) Imaging spectrometry for surveying and modeling land degradation. In: van der Meer FD, de Jong SM, Imaging spectrometry: basic principles and prospective applications, Kluwer Academic Publishers, Dordrecht, The Netherlands, pp 65–86

Demarez V, Gastellu-Etchegorry JP (2000) A modeling approach for studying forest chlorophyll content. Remote Sensing of Environment 71: 226–238

Dozier J (2004) Multispectral and hyperspectral remote sensing of alpine snow properties. Annual Review of Earth and Planetary Sciences 32

Ebbers MJH, Wallis IR, Dury S, Floyd R, Foley WJ (2002) Spectrometric prediction of secondary metabolites and nitrogen in fresh Eucalyptus foliage: towards remote sensing of the nutritional quality of foliage for leaf-eating marsupials. Australian Journal of Botany 50(6): 761–768

Ellis J (1999) NASA satellite syems for the non-renewable resources sector. Earth Observation Magazine, 47, EOM, Littleton, Colarado, USA

Elvidge CD (1990) Detection of trace quantities of green vegetation in 1989 AVIRIS data. 2nd Airborne Visible/Infrared Imaging Spectrometer (AVIRIS) Workshop, JPL, Pasadena, CA

Escadafel R (1994) Soil spectral properties and their relationships with environmental parameters – examples from arid regions. In: Hill JAM, (ed), Imaging spectrometry – a tool for environmental observations, Kluwer Academic Publishers, Dordrecht, The Netherlands, pp 71–87

Ferrier-Pages C, Gattuso JP, Jaubert J (1999) Effect of small variations in salinity on the rates of photosynthesis and respiration of the zooxanthellate coral Stylophora pistillata. Marine Ecology-Progress Series 181: 309–314

Franklin J, Strahler AH (1988) Invertible canopy reflectance modeling of vegetation structure in semiarid woodland. IEEE Transactions on Geoscience and Remote Sensing 26(6): 809–825

Fraser RN (1998) Hyperspectral remote sensing of turbidity and chlorophyll a among Nebraska sand hills lakes. International Journal of Remote Sensing 19(8): 1579–1589

Friedl, MA, McIver DK, Hodges JCF, Zhang XY, Muchoney D, Strahler AH, Woodcock CE, Gopal S, Schneider A, Cooper A, Baccini A, Gao F, Schaaf C (2002). Global land cover mapping from MODIS: algorithms and early results. *Remote Sensing of Environment* 83(1–2): 287–302.

Fyfe SK (2003) Spatial and temporal variation in spectral reflectance: Are seagrass species spectrally distinct? Limnology and Oceanography 48(1): 464–479

Gaddis L, Soderblom L, Kieffer H, Becker K, Torson J, Mullins K (1993) Spectral decomposition of AVIRIS data. Proceedings of 4th Airborne Visible/Infrared Imaging Spectrometer (AVIRIS) Workshop, JPL, Pasadena, CA, pp 57–60

Gamon JA, Qiu H (1999) Ecological applications of remote sensing at multiple scales. In: Pugnaire FI, Valladares F (eds), Handbook of Functional Plant Ecology, Marcel Dekker, Inc, New York, pp 805–846

Gao BC, Goetz AFH (1991) Cloud area determination from AVIRIS data using water-vapor channels near 1 µm. Journal of Geophysical Research-Atmospheres 96(D2): 2857–2864

Gastellu-Etchegorry JP, Zagolski F, Mougin E, Marty G, Giordano G (1995) An assessment of canopy chemistry with AVIRIS – a case-study in the Landes Forest, South-West France. International Journal of Remote Sensing 16(3): 487–501

GER GER EPS-H Series Airborne Imaging Spectrometer System Technical Description. (2003).

Gitelson AA, Merzlyak MN (1997) Remote estimation of chlorophyll content in higher plant leaves. International Journal of Remote Sensing 18(12): 2691–2697

Goel NS, Grier T (1988). Estimation of canopy parameters for inhomogeneous vegetation canopies from reflectance data. 3. Trim – a model for radiative-transfer in heterogeneous 3-dimensional canopies. Remote Sensing of Environment 25(3): 255–293

Goetz AFH (ed) (1995) Imaging spectrometry for remote sensing: vision to reality in 15 years. Proceedings of SPIE Society for Optical Engineers, Orlando, Florida

Goetz AFH, Herring M (1989) The High-Resolution Imaging Spectrometer (HIRIS) for EOS. IEEE Transactions on Geoscience and Remote Sensing 27(2): 136–144

Gong P, Pu RL, Miller JR (1995) Coniferous forest leaf-area index estimation along the oregon transect using Compact Airborne Spectrographic Imager data. Photogrammetric Engineering and Remote Sensing 61(9): 1107–1117

Green EP, Clark CD, Mumby PJ, Edwards AJ, Ellis AC (1998a) Remote sensing techniques for mangrove mapping. International Journal of Remote Sensing 19(5): 935–956

Green EP, Mumby PJ, Edwards AJ, Clark CD (1996) A review of remote sensing for the assessment and management of tropical coastal resources. Coastal Management 24(1): 1–40

Green RO, Eastwood ML, Sarture CM, Chrien TG, Aronsson M, Chippendale BJ, Faust JA, Pavri BE, Chovit CJ, Solis MS, Olah MR, Williams O (1998b) Imaging spectroscopy and the Airborne Visible Infrared Imaging Spectrometer (AVIRIS). Remote Sensing of Environment 65(3): 227–248

Gu Y, Anderson JM, Monk JGC (1999) An approach to the spectral and radiometric calibration of the VIFIS system. International Journal of Remote Sensing 20(3): 535–548

Guyot G, Guyon D, Riom J (1989) Factors affecting the spectral response of forest canopies: a review. Geocarto International 4(3): 3–18

Hall DK, Riggs GA, Salomonson VV (1995) Development of methods for mapping global snow cover using moderate resolution imaging spectroradiometer data. Remote Sensing of Environment 54: 127– 140

Held A, Ticehurst C, Lymburner L, Williams N (2003) High resolution mapping of tropical mangrove ecosystems using hyperspectral and radar remote sensing. International Journal of Remote Sensing 24(13): 2739–2759

Hochberg EJ, Atkinson MJ (2003) Capabilities of remote sensors to classify coral, algae, and sand as pure and mixed spectra. Remote Sensing of Environment 85(2): 174–189

Hochberg EJ, Atkinson MJ, Andrefouet S (2003) Spectral reflectance of coral reef bottom-types worldwide and implications for coral reef remote sensing. Remote Sensing of Environment 85(2): 159–173

Holden H, LeDrew E (1998) The scientific issues surrounding remote detection of submerged coral ecosystems. Progress in Physical Geography 22(2): 190–221

Hollinger AB, Gray LH, Gowere JFR, Edel H (1987) The fluorescence line imager: an imaging spectrometer for land and ocean remote sensing. Proceedings of the SPIE, 834

Horig B, Kuhn F, Oschutz F, Lehmann F (2001) HyMap hyperspectral remote sensing to detect hydrocarbons. International Journal of Remote Sensing 22(8): 1413–1422

Huadong G, Jianmin X, Guoqiang N, Jialing M (2001) A new airborne earth observing system and its applications. International Geoscience and Remote Sensing Symposium (IGARSS), Sydney, Australia, CD

Hunt GR (1989) Spectroscopic properties of rock and minerals. In: Carmichael RC (ed), Practical handbook of physical properties of rocks and minerals, C.R.C. Press Inc., Boca Raton, Florida, pp 599–669

Irons JR, Ranson KJ, Williams DL, Irish RR, Huegel FG (1991) An off-nadir-pointing imaging spectroradiometer for terrestrial ecosystem studies. IEEE Transactions on Geoscience and Remote Sensing 29(1): 66–74

Jacquemoud S, Bacour C, Poilve H, Frangi JP (2000) Comparison of four radiative transfer models to simulate plant canopies reflectance-direct and inverse mode. Remote Sensing of Environment 74: 741–781

Johnson LF, Billow CR (1996) Spectrometric estimation of total nitrogen concentration in Douglas-fir foliage. International Journal of Remote Sensing 17(3): 489–500

Johnson LF, Hlavka CA, Peterson DL (1994) Multivariate analysis of AVIRIS data for canopy biochemical estimation along the Oregon transect. Remote Sensing of Environment 47: 216–230

Kallio K, Kutser T, Hannonen T, Koponen S, Pulliainen J, Vepsalainen J, Pyhalahti T (2001) Retrieval of water quality from airborne imaging spectrometry of various lake types in different seasons. Science of the Total Environment 268(1–3): 59–77

Kaufman YJ, Remer LA (1994) Detection of forests using mid-IR reflectance: An application for aerosol studies. IEEE Transactions on Geoscience and Remote Sensing 32: 672–683

King MD, Menzel WP, Grant PS, Myers JS, Arnold GT, Platnick SE, Gumley LE, Tsay SC, Moeller CC, Fitzgerald M, Brown KS, Osterwisch FG (1996) Airborne scanning spectrometer for remote sensing of cloud, aerosol, water vapor, and surface properties. Journal of Atmospheric and Oceanic Technology 13(4): 777–794

Klein AG, Barnett AC (2003) Validation of daily MODIS snow cover maps of the Upper Rio Grande River Basin for the 2000–2001 snow year. Remote Sensing of Environment 86(2): 162–176

Knox B, Ladiges P, Evans B (1997) Biology. McGraw-Hill, Sydney.

Kokaly RF, Clark RN, Livo KE (1998). Mapping the biology and mineralogy of Yellowstone National Park using imaging spectroscopy. Proceedings of the Seventh JPL Airborne Earth Science Workshop, Pasadena, CA, pp 245–254

Kruse FA, Boardman JW, Huntington JF (2003) Comparison of airborne hyperspectral data and EO-1 Hyperion for mineral mapping. IEEE Transactions on Geoscience and Remote Sensing 41(6): 1388–1400

Kruse FA, Hauff PL (1991) Identification of illite polytype zoning in disseminated gold deposits using reflectance spectroscopy and X-ray-diffraction – potential for mapping with imaging spectrometers. IEEE Transactions on Geoscience and Remote Sensing 29(1): 101–104

Kruse FA, Lefkoff AB, Dietz JB (1993) Expert system-based mineral mapping in Northern Death-Valley, California Nevada, using the Airborne Visible Infrared Imaging Spectrometer (AVIRIS). Remote Sensing of Environment 44(2–3): 309–336

Kuchler DA (1986) Remote-sensing – what can it offer coral-reef studies. Oceanus 29(2): 90–93

Kumar L (1998) Modeling forest resources using geographical information systems and hyperspectral remote sensing, Unpublished PhD Thesis, University of New South Wales, Sydney

Kumar L, Schmidt K, Dury S, Skidmore A (2001) Imaging spectrometry and vegetation science. In: van der Meer FD, de Jong SM, Imaging spectrometry: basic principles and prospective applications, Kluwer Academic Publishers, Dordrecht, The Netherlands, pp 111–155

Kunkel B, Blechinger F, Viehmann D, Vanderpiepen H, Doerffer R (1991) ROSIS imaging spectrometer and its potential for ocean parameter measurements (airborne and spaceborne). International Journal of Remote Sensing 12(4): 753–761

Kupiek JA, Curran PJ (1995) Decoupling effects of the canopy and foliar biochemicals in AVIRIS spectra. International Journal of Remote Sensing 16(9): 1713–1739

Kutser T, Dekker AG, Skirving W (2003) Modeling spectral discrimination of Great Barrier Reef benthic communities by remote sensing instruments. Limnology and Oceanography 48(1): 497–510

LaCapra VC, Melack JM, Gastil M, Valeriano D (1996) Remote sensing of foliar chemistry of inundated rice with imaging spectrometry. Remote Sensing of Environment 55(1): 50–58

Levesque J, Staenz K, Szeredi T (2000) The impact of spectral band characteristics on unmixong of hyperspectral data for monitoring mine tailings site rehabilitation. Canadian Journal of Remote Sensing 26(3): 231–240

Longhi I, Sgavetti M, Chiari R, Mazzoli C (2001) Spectral analysis and classification of metamorphic rocks from laboratory reflectance spectra in the 0.4–2.5 mu m interval: a tool for hyperspectral data interpretation. International Journal of Remote Sensing 22(18): 3763–3782

Mackin S, Drake N, Settle J, Briggs S (1991) Curve shape matching, end-member selection and mixture modeling of AVIRIS and GER data for mapping surface mineralogy and vegetation communities. Proceedings of the 2nd JPL Airborne Earth Science Workshop, JPL Publication, Pasadena, CA, pp 158–162

Mackin S, Munday TJ (1988) Imaging spectrometry in environmental science research and applications – preliminary results from teh anaglysis of GER II imaging spectrometer data – Australi and the USA, Departmnet of Geological Sciences, University of Durham

Maeder J, Narumalani S, Rundquist DC, Perk RL, Schalles J, Hutchins K, Keck J (2002) Classifying and mapping general coral-reef structure using Ikonos data. Photogrammetric Engineering and Remote Sensing 68(12): 1297–1305

Malthus TJ, Mumby PJ (2003) Remote sensing of the coastal zone: an overview and priorities for future research. International Journal of Remote Sensing 24(13): 2805–2815

Martin ME, Aber JD (1997) High spectral resolution remote sensing of forest canopy lignin, nitrogen, and ecosystem processes. Ecological Applications 7(2): 431–443

Matson P, Johnson L, Billow C, Miller J, Pu RL (1994) Seasonal patterns and remote spectral estimation of canopy chemistry across the Oregon transect. Ecological Applications 4(2): 280–298

Mauser W (2003) The airborne visible/infrared imaging spectrometer AVIS-2 – multiangular and hyperspectral data for environmental analysis. International Geoscience and Remote Sensing Symposium (IGARSS), Toulouse, France, CD

McGwire K, Minor T, Fenstermaker L (2000) Hyperspectral mixture modeling for quantifying sparse vegetation cover in arid environments. Remote Sensing of Environment 72(3): 360–374

Merton RN (1999) Multi-temporal analysis of community scale vegetation stress with imaging spectroscopy. Unpublished PhD Thesis, Department of Geography, The University of Auckland.

Merton RN, Silver E (2000) Tracking vegetation spectral trajectories with multi-temporal hysteresis models. Proceedings of the Ninth Annual JPL Airborne Earth Science Workshop, Jet Propulsion Laboratory, Pasadena, CA.

Metternicht GI, Zinck JA (2003) Remote sensing of soil salinity: potentials and constraints. Remote Sensing of Environment 85(1): 1–20

Mills F, Kannari Y, Watanabe H, Sano M, Chang SH (1993) Thermal Airborne Multispectral Aster Simulator and its preliminary-results. Remote Sensing of Earths Surface and Atmosphere. 14: 49–58

Mitchell, A. (2004). Remote sensing techniques for the assessment of mangrove forest structure, species composition and biomass, and response to environmental change. School of Biological Earth Environmental Sciences. Sydney, University of New South Wales.

References

Mumby, PJ, Edwards AJ (2002) Mapping marine environments with IKONOS imagery: enhanced spatial resolution can deliver greater thematic accuracy. Remote Sensing of Environment 82(2–3): 248–257

Nadeau C, Neville RA, Staenz K, O'Neill NT, Royer A (2002) Atmospheric effects on the classification of surface minerals in an arid region using Short-Wave Infrared (SWIR) hyperspectral imagery and a spectral unmixing technique. Canadian Journal of Remote Sensing 28(6): 738–749

Niemann KO (1995) Remote-sensing of forest stand age using airborne spectrometer data. Photogrammetric Engineering and Remote Sensing 61(9): 1119–1127

Niemann KO, Goodenough DG (2003) Estimation of foliar chemistry in Western Hemlock using hyperspectral data. In: Wulder M, Franklin S (eds), Remote sensing of forest environments: concepts and case Studies, Kluwer Academic Publishers, Norwell, MA, pp 447–467

Ong C, Cudahy T, Caccetta M, Hick P, Piggott M (2001) Quantifying dust loading on mangroves using hyperspectral techniques. International Geoscience and Remote Sensing Symposium (IGARSS), Sydney, Australia, CD

Oppenheimer C, Rothery DA, Pieri DC, Abrams MJ, Carrere V (1993) Analysis of Airborne Visible Infrared Imaging Spectrometer (AVIRIS) data of volcanic hot-spots. International Journal of Remote Sensing 14(16): 2919–2934

Otterman J, Strebel DE, Ranson KJ (1987) Inferring Spectral Reflectances of Plant-Elements by Simple Inversion of Bidirectional Reflectance Measurements. Remote Sensing of Environment 21(2): 215–228

Palacios-Orueta A, Ustin SL (1998) Remote sensing of soil properties in the Santa Monica Mountains I. Spectral analysis. Remote Sensing of Environment 65(2): 170–183

Penuelas J, Pinol J, Ogaya R, Filella I (1997) Estimation of plant water concentration by the reflectance water index WI (R900/R970). International Journal of Remote Sensing 18(13): 2869–2875

Perry CT (2003) Reef development at Inhaca Island, Mozambique: Coral communities and impacts of the 1999/2000 Southern African floods. Ambio 32(2): 134–139

Ranson, KJ, Irons JR, Williams DL (1994) Multispectral Bidirectional Reflectance of Northern forest canopies with the advanced solid-state array spectroradiometer (Asas). Remote Sensing of Environment 47(2): 276–289

Resmini RG, Kappus ME, Aldrich WS, Harsanyi JC, Anderson M (1997) Mineral mapping with HYperspectral Digital Imagery Collection Experiment (HYDICE) sensor-data at Cuprite, Nevada, USA. International Journal of Remote Sensing 18(7): 1553–1570

Riegl B (2002) Effects of the 1996 and 1998 positive sea-surface temperature anomalies on corals, coral diseases and fish in the Arabian Gulf (Dubai, UAE). Marine Biology 140(1): 29–40

Roberts DA, Batista G, Pereira J, Waller E, Nelson B (1998) Change identification using multitemporal spectral mixture analysis: applications in eastern Amazonia. In: Elvidge C, Ann Arbor LR, Remote sensing change detection: environmental monitoring applications and methods, Ann Arbor Press, Michigan, pp 137–161

Rodger A, Lynch MJ (2001) Determining atmospheric column water vapour in the 0.4–2.5 μm spectral region. Proceedings of the 10th JPL Airborne Earth Science Workshop, Pasadena, CA, pp 321–330

Rowan LC, Crowley JK, Schmidt RG, Ager CM, Mars JC (2000) Mapping hydrothermally altered rocks by analyzing hyperspectral image (AVIRIS) data of forested areas in the Southeastern United States. Journal of Geochemical Exploration 68(3): 135–166

Sakami T (2000) Effects of temperature, irradiance, salinity and inorganic nitrogen concentration on coral zooxanthellae in culture. Fisheries Science 66(6): 1006–1013

Schaepman ME, Itten KI, Schläpfer D, Kaiser JW, Brazile J, Debruyn W, Neukom A, Feusi H, Adolph P, Moser R, Schilliger T, De Vos L, Brandt G, Kohler P, Meng M, Piesbergen J, Strobl P, Gavira J, Ulbrich G, Meynart R (2003) Status of the airborne dispersive pushbroom imaging spectrometer APEX (Airborne Prism Experiment). International Geoscience and Remote Sensing Symposium (IGARSS), Toulouse, France, CD

Schmidt KS, Skidmore AK (2003) Spectral discrimination of vegetation types in a coastal wetland. Remote Sensing of Environment 85(1): 92–108

Schreier H, Wiart R, Smith S (1988) Quantifying organic-matter degradation in agricultural fields using PC-based image-analysis. Journal of Soil and Water Conservation 43(5): 421–424

Seemann SW, Li J, Menzel WP, Gumley LE (2003) Operational retrieval of atmospheric temperature, moisture, and ozone from MODIS infrared radiances. Journal of Applied Meteorology 42(8): 1072–1091

Serrano L, Ustin SL, Roberts DA, Gamon JA, Penuelas J (2000) Deriving water content of chaparral vegetation from AVIRIS data. Remote Sensing of Environment 74(3): 570–581

Silvestri S, Marani M, Marani A (2003) Hyperspectral remote sensing of salt marsh vegetation, morphology and soil topography. Physics and Chemistry of the Earth 28(1–3): 15–25

Spanner MA, Pierce LL, Peterson DL, Running SW (1990a) Remote-sensing of temperate coniferous forest Leaf-Area Index – the influence of canopy closure, understory vegetation and background reflectance. International Journal of Remote Sensing 11(1): 95–111

Spanner MA, Pierce LL, Running SW, Peterson DL (1990b) The seasonality of AVHRR data of temperate coniferous forests – relationship with Leaf-Area Index. Remote Sensing of Environment 33(2): 97–112

Strobl P, Nielsen A, Lehmann F, Richter R, Mueller A (1996) DAIS system performance, first results from the 1995 evaluation campaigns. 2nd International Airborne Remote Sensing Conference and Exhibition, San Francisco, USA.

Su Z, Troch PA, DeTroch FP (1997) Remote sensing of bare surface soil moisture using EMAC/ESAR data. International Journal of Remote Sensing 18(10): 2105–2124

Sun XH, Anderson JM (1993) A spatially-variable light-frequency-selective component-based, airborne pushbroom imaging spectrometer for the water environment. Photogrammetric Engineering and Remote Sensing 59(3): 399–406

Thenkabail PS, Smith RB, de Pauw E (2000) Hyperspectral vegetation indices and their relationships with agricultural crop characteristics. Remote Sensing of Environment 71: 158–182

Ticehurst C, Lymburner L, Held A, Palylyk C, Martindale D, Sarosa W, Phinn S, Stanforf M (2001) Mapping tree crowns using hyperspectral and high spatial resolution imagery. Proceedings od 3rd International Conference on Geospatial Information in Agriculture and Forestry, Denver, Colorado, CD

Treitz PM, Howarth PJ (1999) Hyperspectral remote sensing for estimating biophysical parameters of forest ecosystems. Progress in Physical Geography 23(3): 359–390

Ustin, SL, Zarco-Tejada PJ, Asner GP (2001) The role of hyperspectral data in understanding the global carbon cycle. Proceedings of the 10th JPL Airborne Earth Science Workshop, Pasadena, CA, pp 397–410

References

van der Meer F, Bakker W (1998) Validated surface mineralogy from high-spectral resolution remote sensing: a review and a novel approach applied to gold exploration using AVIRIS data. Terra Nova 10(2): 112–119

van der Meer F, Fan LH, Bodechtel J (1997) MAIS imaging spectrometer data analysis for Ni-Cu prospecting in ultramafic rocks of the Jinchuan group, China. International Journal of Remote Sensing 18(13): 2743–2761

van der Meer FD, de Jong SM (2001). Introduction. In: van der Meer FD, de Jong SM (eds), Imaging spectrometry: basic principles and prospective applications, Kluwer Academic Publishers, Dordrecht, The Netherlands, pp xxi-xxiii

van der Meer, FD, de Jong SM, Bakker W (Eds) (2001) Imaging spectrometry: basic analytical techniques. Imaging spectrometry: basic principles and prospective applications, Kluwer Academic Publishers, Dordrecht, The Netherlands

Vane G, Goetz AFH, Wellman JB (1984) Airborne imaging spectrometer – a new tool for remote-sensing. IEEE Transactions on Geoscience and Remote Sensing 22(6): 546–549

Vane G, Green RO, Chrien TG, Enmark HT, Hansen EG, Porter WM (1993) The Airborne Visible Infrared Imaging Spectrometer (AVIRIS). Remote Sensing of Environment 44(2–3): 127–143

Vos RJ, Hakvoort JHM, Jordans RWJ, Ibelings BW (2003) Multiplatform optical monitoring of eutrophication in temporally and spatially variable lakes. Science of the Total Environment 312(1–3): 221–243

Yang H, Zhang J, van der Meer F, Kroonenberg SB (2000) Imaging spectrometry data correlated to hydrocarbon microseepage. International Journal of Remote Sensing 21(1): 197–202

Yentsch CS, Yentsch CM, Cullen JJ, Lapointe B, Phinney DA, Yentsch SW (2002) Sunlight and water transparency: cornerstones in coral research. Journal of Experimental Marine Biology and Ecology 268(2): 171–183

Zacharias, M, Niemann KO, Borstad G (1992) An assessment and classification of a multispectral bandset for the remote sensing if intertidal seaweeds. Canadian Journal of Remote Sensing 18(4): 263–273

Zagolski F, Pinel V, Romier J, Alcayde D, Fontanari J, Gastellu-Etchegorry JP, Giordano G, Marty G, Mougin E, Joffre R (1996) Forest canopy chemistry with high spectral resolution remote sensing. International Journal of Remote Sensing 17(6): 1107–1128

Zhou Y, Yan G, Zhou Q, Tang S (2003) *New airborne multi-angle high resolution sensor AMTIS LAI inversion based on neural network.* International Geoscience and Remote Sensing Symposium (IGARSS), Toulouse, France, CD

CHAPTER 2
Overview of Image Processing

Raghuveer M. Rao, Manoj K. Arora

2.1 Introduction

The current mode of image capture in remote sensing of the earth by aircraft or satellite based sensors is in digital form. The pixels correspond to localized spatial information while the quantization levels in each spectral band correspond to the quantized radiometric measurements. It is most logical to regard each image as a vector array, that is, the pixels are arranged on a rectangular grid but the value of the image at each pixel is a vector whose elements correspond to radiometric levels (also known as intensity values or digital numbers) of the different bands. The image formed for each band is a monochrome image and for this reason, we refer to the quantized values of such an image as gray levels. The images may be acquired in a few bands (i.e. multi-spectral image) or in hundreds of bands (i.e. hyperspectral image). Image processing operations for both multispectral and hyperspectral images can therefore be either scalar image oriented, that is, each band is processed separately as an independent image or vector image oriented, where the operations take into account the vector nature of each pixel. Image processing can take place at several different levels. At its most basic level, the processing enhances an image or highlights specific objects for the analyst to view. At higher levels, processing can take on the form of automatically detecting objects in the image and classifying them.

The aim of this chapter is to provide an overview of image processing of multispectral and hyperspectral data from the basic to the advanced techniques. While a few excellent books are available on the subject (Jensen 1996; Mather 1999; Richards and Jia 1999; Campbell 2002), the purpose here is to provide a ready reference to go with the other chapters of the book. It will be seen that the bulk of processing techniques for image enhancement are those that have been developed for consumer and document image processing and are thus generic. They do not differ substantially in their applicability as a function of the imaging platform be it hyperpsectral or otherwise and many of the illustrations in this chapter use images in the visible band.

2.2
Image File Formats

Digital images are organized into records. Appropriate header data indicate the organization. The earliest storage medium for remotely sensed images was a magnetic tape. Consequently, several image format schemes are tailored towards the sequential access mode of this medium. With the increased use of optical storage media, some of the schemes have fallen out of favor. Because hyperspectral data are gathered in terms of spatial location, in the form of pixel coordinates, and spectral bands (i.e. gray values of the pixel in different bands), image format schemes typically differ in how they handle data with respect to location and bands.

The band interleaved by pixel format (BIP) is pixel oriented in the sense that all data corresponding to a pixel are first stored before moving on to the next pixel. Thus, the first record corresponds to the first pixel of the first band of the first scan line. The second record corresponds to the first pixel of the first line but data in band 2 and so on. After data from all the bands are recorded sequentially, we move on to pixel 2 of the first line.

In the band interleaved by line format (BIL) the data are organized in line-oriented fashion. The first record is identical to that of the BIP image. However, the second record corresponds to the second pixel of band 1 of the first line. Once all pixel values of band 1 of line 1 have been recorded, we begin recording data of pixel 1 of band 2 followed by pixel 3 of band 2 and so on. We begin recording of line 2 data after completion of recording of all line 1 data.

The band sequential format (BSQ) requires storing of the data of each band as a separate file. This format is most useful when one might process just one or a few bands of data. With the availability of random access capability in optical storage media, such a separation into band-based image files does not provide a substantial increase in search speed over the other formats when one is interested in a small number of bands.

A versatile data format called the hierarchical data format (HDF) has been developed for the National Center for Supercomputing Applications with the aim of facilitating storage and exchange of scientific data. This format provides for different data models and supports single file and multifile modes. The following different data structures are supported in HDF4: raster images, color palettes, texts, scientific data sets (which are multidimensional arrays), Vdata (2D arrays of different data types) and Vgroup (a structure for associating sets of data objects). In the newer HDF5, only two fundamental structures are supported: groups and dataspaces. HDF-EOS an extension of HDF5 developed by NASA for the Earth Observing Systems, additional structures are used. These are point, swath and grid. Some advantages of HDF are: portability to platform independence, self-documentation, efficient storage and access, accommodation of multiple data types within the same file and ease of extending the format to suit specific needs (Folk et al. 1999; Ullman 1999).

2.3
Image Distortion and Rectification

Obviously, the ideal image is an accurate reproduction of ground reflectance in the various spectral bands. However, there are various factors that make this impossible. Thus, most recorded images suffer from various degrees of distortion. There are two main types of distortion, radiometric and geometric. The former is the inaccurate translation of ground reflectance to gray values, typically arising due to atmospheric interference and/or sensor defects. The latter refers to shape and scale distortions owing usually to errors due to perspective projection and instrumentation. We provide an overview of these distortions along with schemes used to rectify the errors.

2.3.1
Radiometric Distortion

This refers to inaccurate representation of relative brightness levels (or intensity values) either in a given band across pixels or across bands for a given pixel. The ideal measured brightness by a sensor at a pixel in a given band is proportional to the product of the surface reflectance and the sun's spectral irradiance assuming uniform surface reflectance over the area contributing to the pixel. Thus the ideal image captures the relative brightness as measured at the ground. However, the intervening atmosphere between the ground and a space based sensor causes inaccuracies in the recording of the relative levels.

Atmosphere induced radiometric distortion is due primarily to scattering of electromagnetic radiation by either the air molecules or suspended particles. Part of the down-going radiation from the sun is scattered back and away from the earth by the atmosphere itself. Also, apart from the radiation reflected by the area covered by the pixel of interest, radiation reflected from neighboring regions is scattered by the atmosphere into the pixel of interest. The main effect of both types of scattering is loss of image detail. It also causes inaccurate representation of radiometric levels across the bands due to the wavelength dependent nature of scattering.

Correction of atmospheric distortion can be done in either a detailed or gross fashion. Detailed correction requires information regarding the incident angle of the sunlight and atmospheric constituents at the time of imaging. The scattering effects are then modeled into atmospheric attenuation as a function of path length and wavelength. There are a number of atmospheric models that allow the conversion of at-sensor image data to transform into ground reflectance. Some of the methods are flat field correction, internal average relative reflectance and empirical line method (Rast et al. 1991; Smith and Milton 1999). Because detailed atmospheric information is hard to obtain, gross or bulk corrections are more common. Bulk brightness correction attempts mainly to compensate for at-sensor radiance or path radiance. Path radiance effect for a given pixel refers to interfering scatter from the atmosphere as well as from other pixels. A typical approach, referred to as dark object subtraction, is to subtract the lowest gray value of each band from the values of other pixels in that band. The lowest gray value in each band provides an estimate of the contribution to each pixel's gray value in that band from neighboring pixels. In the visible region, atmospheric scattering injects extraneous radiation at higher frequencies. Consequently the bulk correction just described tends to reduce the blue component.

The radiometric distortion may also occur due to the image capture system itself. If the response of the detector is a nonlinear function of the input radiance, it results in nonlinear distortion. Another source of instrumentation error in multi-detector cameras is mismatch of detector characteristics, that is, each detector may have a slightly different transfer characteristic (even if linear) and offset. Offset refers to the fact that each detector outputs a different dark current, which is the current from a radiation-sensing device even when there is no radiation impinging on it. Yet another common source of contribution of instruments to radiometric distortion is sensor noise, which shows up as random deviations from the true intensity values.

Radiometric distortion due to detector mismatch in multidetector systems typically shows up as a striping artifact. The correction usually performed is that of adjusting the output gray values of the sensors to match, in mean and variance to, that of one chosen as a reference. For example, let μ_i; $i = 1,\ldots,N$, be the respective means of the outputs of N sensors and σ_i; $i = 1,\ldots,N$, be the standard deviations. Suppose we choose sensor 1 as the reference. Then, if I_i denotes the image from the ith sensor, the corrections are implemented as

$$I_i = \frac{\sigma_1}{\sigma_i}(I_i - \mu_i) + \mu_1 \qquad (2.1)$$

for $i = 1,\ldots,N$. Mitigation of noise effects is done through filtering and is discussed in a later section.

2.3.2
Geometric Distortion and Rectification

The rotation of the Earth beneath the satellite as the satellite traverses causes images to be rhombus-shaped when mapped back to the imaged region. In other words, even if the acquired image is treated as square from the point of view of storage and display, the actual region of the Earth that it corresponds to is a rhombus. This is when the area covered in the image is small enough to ignore the Earth's curvature. There is a further distortion introduced when this curvature cannot be ignored. This happens with large swath widths on high altitude satellites. The ground pixel size is larger at a swath edge than in the middle and happens because of the geometry as depicted in Fig. 2.1. Given that the angular field of view (FOV) of space imaging systems is constant, there is another type of geometric distortion called panoramic distortion, which refers to pixel sizes being larger at the edge of a scan than at the nadir. This occurs because the footprint of the FOV directly underneath the air or space platform is smaller than that several hundred meters away owing to the fact that the footprint size is proportional to the tangent of the scan angle rather than to the scan angle itself.

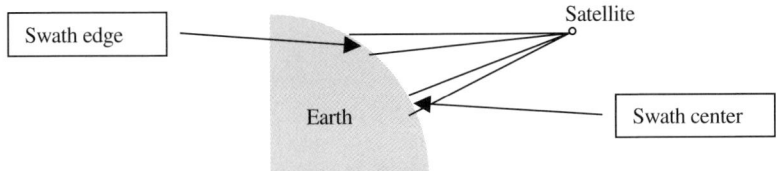

Fig. 2.1. Effect of earth curvature on pixel size

The above distortions are systematic in the sense that the errors are predictable. For example, once a satellite's glide path is known, it is easy to predict the square to rhombus transformation due to the Earth's rotation. Often, the agency-supplied data are systematic corrected. However, there are also non-systematic distortions. These arise mainly due to unpredictable occurrences such as random variations in platform velocity, altitude and attitude. Obviously, such changes result in changing the FOV thus leading to distortion.

A common approach to correcting shape distortions is to geometrically transform the coordinates of the image to coincide with ground control points (GCP). A geo-referenced data set such as a map of the imaged area or an actual ground survey, conducted these days using global positioning systems (GPS), is required to collect the GCP. A GCP is a point on the reference data set that is also identifiable in the recorded image. For example, this could be a mountain peak or a cape or human-made features such as road intersections. Several GCPs are identified and their geographical coordinates are noted. A coordinate transformation is determined to relate the image coordinates to the true coordinates,

$$R : (x_I, y_I) \rightarrow (x_T, y_T) . \tag{2.2}$$

Here R is the rectification mapping that maps the set of image coordinates (subscripted I) to the true coordinates (subscripted T). The simplest coordinate transformation that one can choose is linear or affine. While this works for simple distortions such as the square to rhombus alluded to above, it is insufficient for other distortions. Polynomial transformations are more general. For example, a second order transformation has the form

$$\begin{bmatrix} x_I \\ y_I \end{bmatrix} = \begin{bmatrix} a_1 \\ a_2 \end{bmatrix} + \begin{bmatrix} b_{11} & b_{12} \\ b_{21} & b_{22} \end{bmatrix} \begin{bmatrix} x_T \\ y_T \end{bmatrix} + \begin{bmatrix} c_{11} & c_{12} & c_{13} \\ c_{21} & c_{22} & c_{23} \end{bmatrix} \begin{bmatrix} x_T^2 \\ x_T y_T \\ y_T^2 \end{bmatrix} . \tag{2.3}$$

If we choose exactly 12 GCPs we can solve for the constants in (2.3) through 12 simultaneous linear equations. If we choose more GCPs we will have an overdetermined set of linear equations for which we can get a least squares solution.

Once the parameters of the mapping have been determined, we need to adjust the intensity values in the rectified image. This is because pixel centers in the original image do not necessarily map to pixel centers in the rectified image. The widely used nearest neighbor resampling technique simply chooses the gray value of that pixel of the original image whose center is closest to the new pixel center. Other graphical interpolation techniques such as bilinear or cubic interpolation may also be used (Richards and Jia 1999).

The above approach may also be used to register images acquired from two different sensors or from two different sources or at two different times. The process of image registration has widespread application in image fusion, change detection and GIS modeling, and is described next.

2.4
Image Registration

The registration of two images is the process by which they are aligned so that overlapping pixels correspond to the same entity that has been imaged. Either feature based or intensity based image registration may be performed. The feature based registration technique is based on manual selection of GCP as described in the previous section. The feature-based technique is, however, laborious, time intensive and a complex task. Some automatic algorithms have been developed to automate the selection of GCP to improve the efficiency. However, the extraction of GCP may still suffer from the fact that sometimes too few a points will be selected, and further the extracted points may be inaccurate and unevenly distributed over the image. Hence, intensity based registration techniques may be more appropriate than the feature based techniques.

We provide a basic overview of the principles of intensity based image registration in this chapter. An advanced intensity based technique using mutual information as the similarity measure has been dealt in Chaps. 3 and 7.

The problem of intensity based registration of two images containing translation errors can be posed for two scalar-valued images $I_1(m, n)$ and $I_2(m, n)$ as finding integers k and l such that $I_1(m-k, n-l)$ is as close as possible to $I_2(m, n)$. If the images are of the same size and the relative camera movement between the two image-capture positions is mostly translational, then registration can be attempted by shifting until a close match is obtained.

If the mean squared error is used as a criterion of fit, it is then equivalent to maximizing the cross-correlation between the two images, that is, k and l are chosen to maximize the correlation between the images I_1 and I_2. Finding the cross-correlation over all possible offsets is computationally expensive. Therefore, knowing approximately the range of offsets within which best registration is achieved helps to keep the computational load small. In many instances, a control point is chosen in the scene and a window of pixels around the control point is extracted from the image to be registered and is then correlated with the other image in the vicinity of the estimated location of the control point in it. Instead of computing the correlation, one can also compute the sum of the least absolute differences between corresponding pixels of the two images over the window. The window position for which the summed difference is the smallest corresponds to the best translation.

An advantage of working with control points and windows centered on them lies not only in the reduced computation but also in enabling registration of relevant portions of the images. This is important when the images are not of the same size or the two images have only a partial overlap in the scenes captured. The correlation approach is obviously most applicable when one image or a portion of it can be regarded as a shifted version of another image or a portion thereof. However, there are several situations where this is not true. For example, even if a geographical area is captured with the same imaging system with perfect camera registration but at widely separated times, say in summer and winter, the intensity values will vary substantially due to seasonal variations. Trying to register images through the maximization of correlation would then be meaningless. An alternative approach in such cases is shape matching which correlates features such as coastlines that do not show seasonal variations.

A more general model is to account for differences in images that arise not only due to translation but also due to scaling, rotation or shear. Such effects complicate the registration process substantially. Feature based techniques that apply polynomial coordinate transformation identical to that described in the context of rectifying geometrical distortion, may be useful here. Multiresolution techniques based on the wavelet transform have also been developed to perform feature-based registration of images. For a review on registration methods the reader is referred to (Brown 1992; Zitova and Flusser 2003).

2.5
Image Enhancement

The aim of image enhancement is to improve the quality of the image so that it is more easily interpreted (e.g. certain objects become more distinct). The enhancement may, however, occur at the expense of the quality of other objects, which may become subdued in the enhancement operation. Image enhancement can be performed using point operations and geometric operations, which are described next. It may, however, be mentioned that before applying these operations, the images are assumed to have been corrected for radiometric distortions. Otherwise, these distortions will also be enhanced.

2.5.1
Point Operations

We begin by focusing on scalar image oriented processing, in particular point operations where modification of a pixel value is on the basis of its current value independent of other pixel values. A simple example is thresholding where pixels of a specified value or greater are forced to one gray level and the rest to another as shown in Fig. 2.2. Point operations are typically used to enhance an image or to isolate pixel values on the basis of the range of values

Fig. 2.2. *Left*: Input image. Conesus Lake Upstate New York. Modular imaging spectrometer Gray version. http://www.cis.rit.edu/research/dirs/pages/images.html. *Right*: Result after thresholding pixel values below 60 to 0 and those above 60 to 255

they fall into. In the latter case, it is usually done because objects of interest often contribute to a specific range of radiometric values in the image. For example, the black pixels resulting after thresholding in Fig. 2.2. belong almost entirely to the water body in the original image.

2.5.1.1
Image Histograms

Typically a point operation is influenced by the image histogram, which is a bar plot of the number of pixels at a given gray value as a function of gray value. Thus, the input image of Fig. 2.2 has the histogram shown in Fig. 2.3. It shows bimodal characteristics, that is, there are two peaks. Choosing a value of 60 from the valley between the two peaks yields the image on the right in Fig. 2.2, which shows that the large collection of pixels in the neighborhood of the first peak in the histogram is due to the water body in the input image.

Manipulation of image histograms to enhance radiometric images is a fairly common operation. In most instances, the intended result of histogram ma-

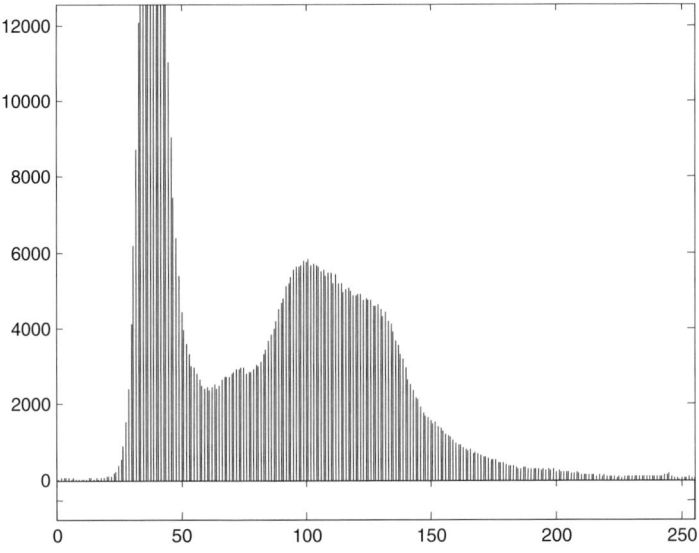

Fig. 2.3. Histogram of image shown in Fig. 2.2

nipulation is contrast enhancement. A histogram may be viewed as an approximation to the probability density function (PDF) of a random variable. Each gray value in the image would then be an instance of a realization of a random variable. The image, therefore, is a collection of realized values of the random variable. A histogram thus does not reveal spatial information. Whereas there is only one histogram corresponding to an image, there may be multiple images corresponding to a given histogram. Suppose we have an image of size $M \times N$ pixels where $MN = 256$ and there are n_k pixels of intensity value k for $k = 0, 1, \ldots, 255$. Obviously $\sum n_k = MN$. Then there are

$$\binom{MN}{n_0} \binom{MN - n_0}{n_1} \cdots \binom{MN - \sum_{k=0}^{253} n_k}{n_{254}} \qquad (2.4)$$

distinct images with the same histogram. In spite of this limitation, image histograms turn out to be very useful (Gonzalez and Woods 1992; Richards and Jia 1999).

Histogram equalization is a common procedure for contrast enhancement. The process creates a new image whose histogram is nearly flat across all gray levels. An example is shown in Fig. 2.4 where the top panel shows the original image along with its histogram and the bottom panel shows resulting image and histogram after equalization.

There are several objects, for example, the river that is visible in the histogram-equalized image is not in the original image. The equalization process is based on transforming the PDF of a continuous random variable to the uniform distribution. Suppose a random variable x has a PDF $f_x(x)$ and

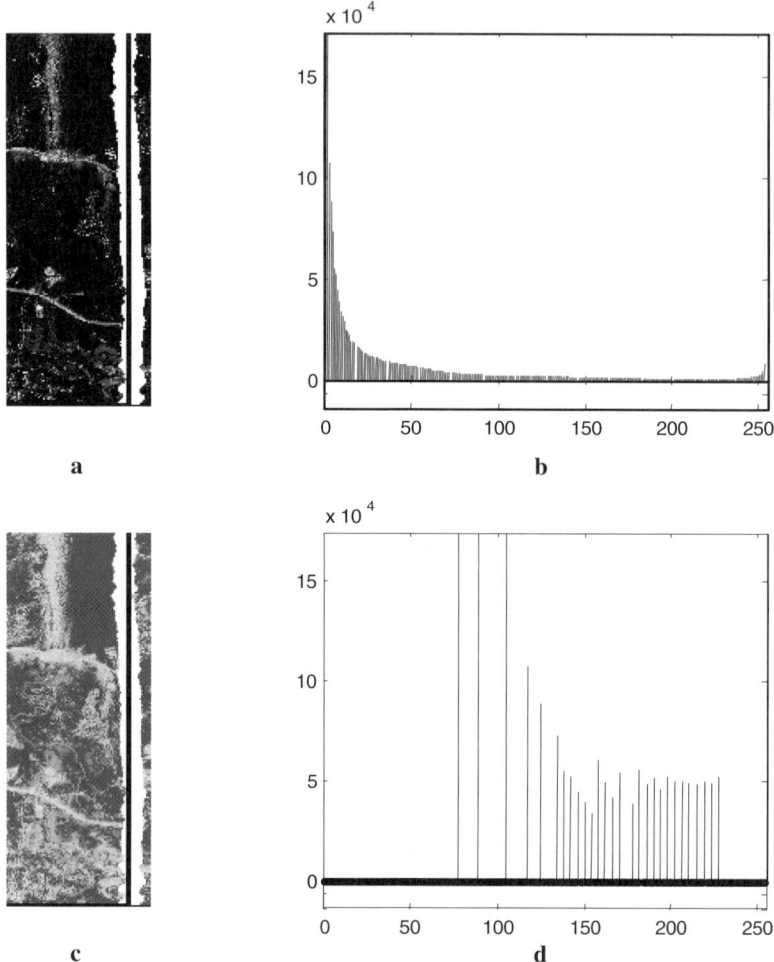

Fig. 2.4. a Original image. **b** Histogram. **c** Histogram-equalized image. **d** Equalized histogram

a cumulative distribution function $F_x(x)$. The random variable can be transformed to a new random variable y that has a uniform PDF over 0 to 1 through the transformation

$$y = F_x(x) \,. \tag{2.5}$$

Thus, histogram equalization is a linearization in some sense of the cumulative distribution function. Because digital images are not continuous-valued, only an approximation to this process is implemented. Consequently, the resulting histogram is rarely perfectly uniform. Histogram equalization is most effective when the histogram of the input image is not dominated by one prominent gray level to the exclusion of others.

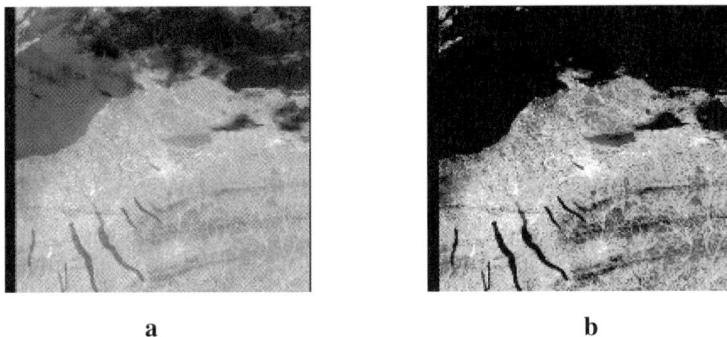

Fig. 2.5a,b. Landsat ETM+ image of the Finger Lakes region, upstate New York. **a** Original image. **b** Piecewise contrast enhanced image

Contrast enhancement can be done through other procedures as well. A particularly common process is to truncate extreme low and high values to 0 and 255 respectively and then stretch other values in a linear or piecewise linear fashion. For example, Fig. 2.5 shows a Landsat ETM+ image on which such contrast enhancement has been applied. The histogram of the original image is generated as shown in Fig. 2.6a. The dip at an intensity value of 125 is used as a threshold to force the pixels over the lake area to become zero. That is, all pixels with intensity value < 125 are set to zero. At the other end, pixels with

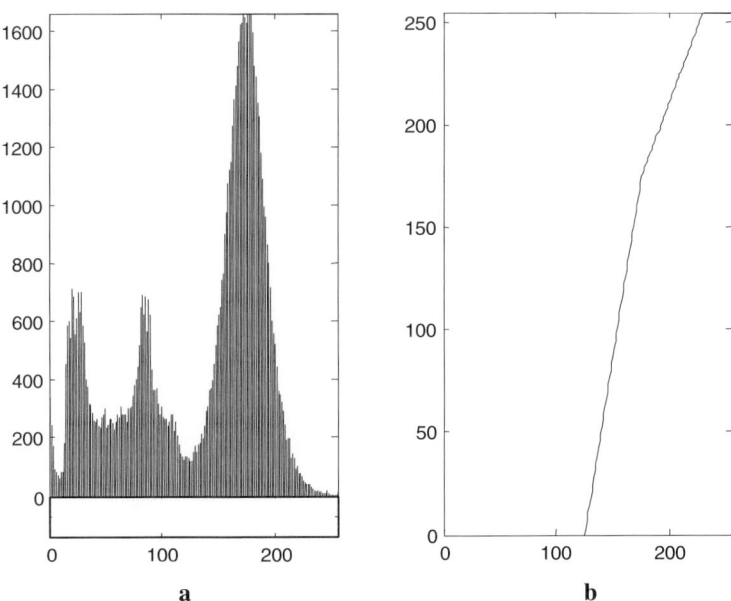

Fig. 2.6. a Histogram. **b** Piecewise linear contrast enhancement curve

intensity values above 230 are set to 255. Intensity value transformations are then applied as

$$y = 3.5(x - 125) \quad \text{for} \quad 125 \leq x \leq 175 \tag{2.6}$$

and

$$y = \frac{16}{11}(x - 175) + 175 \quad \text{for} \quad 175 \leq x \leq 230. \tag{2.7}$$

The effect of these transformations is to map values between 125 and 175 to 0 and 175, and to map values between 175 and 230 to 175 and 255. The value of 175 itself is chosen because of it being the third mode as shown in the histogram. The piecewise linear mapping corresponding to these operations is also shown in Fig 2.6b. The operation can be seen as a form of stretching in the sense that values between 125 and 230 are stretched to a new set of values.

To the extent that a pointwise operator depends only on the local pixel value, it can be implemented as a lookup table. Any such lookup table is by definition a pointwise operator and for gray values in the range 0 to 255, there are 256 factorial such tables. However, only a few will lead to meaningful results. Apart from the histogram equalization or the linear stretching for contrast enhancement, which have been illustrated above and are dependent on the

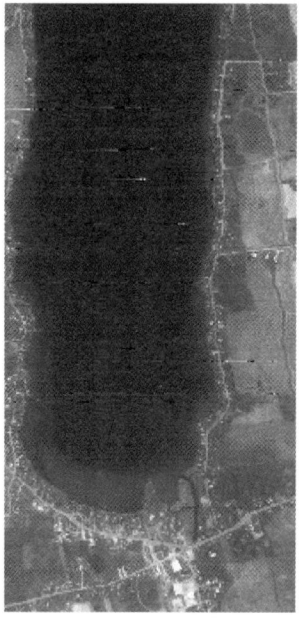
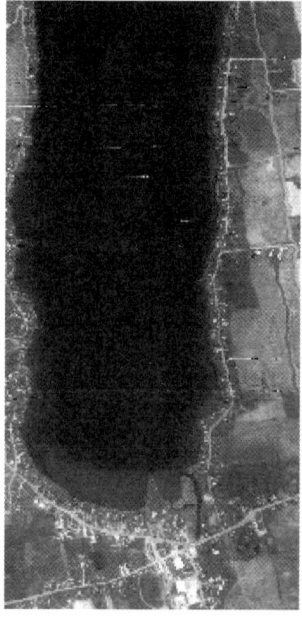

a b

Fig. 2.7. a Original image. b Image after nonlinear contrast enhancement

characteristics of the actual image and its histogram, one can apply generic operators for contrast enhancement. For example, the mapping

$$y = \frac{255}{2}\left[1 - \cos\left(\frac{\pi x}{255}\right)\right] \qquad (2.8)$$

results in the stretching of the mid-level gray values and compression of the extremes. The result is illustrated in Fig. 2.7. Notice that the contrast modification curve given by the above expression is nonlinear.

2.5.2
Geometric Operations

Geometric operations involve modification of pixel gray values based not only on the individual pixel value but also on the pixel values in its neighborhood. In the extreme case, each new pixel value is a function of all pixel values in the original image.

2.5.2.1
Linear Filters

The simplest and most common operation is linear finite impulse response (FIR) filtering. Given an image $I(m, n)$, a linear FIR filtered version $J(m, n)$ is obtained through the operation

$$J(m, n) = \sum\sum_{k,\ell \in A} h(k, \ell) I(m - k, n - \ell), \qquad (2.9)$$

where A is a finite set of paired indices. It is the finite support of the weighting sequence $h(k, \ell)$ that gives the name finite impulse response filtering, that is, $h(k, \ell) = 0$ for $(k, \ell) \notin A$. Most often A is symmetric and square. In other words, it is of the form

$$A = \{(m, n) : -P \leq m \leq P, -P \leq n \leq P\}. \qquad (2.10)$$

However, other geometries are also possible for A. Thus, as seen from (2.9), at any pixel location (m, n), the linear filtered output is a weighted sum of the input pixels at (m, n) and a local neighborhood as determined by A. The filter form given in (2.9) is not only linear but is also space-invariant, that is, the filter coefficients $h(k, \ell)$ are independent of the pixel location (m, n). It is possible to have coefficients that are dependent on pixel location in which case the filter becomes space varying (Gonzalez and Woods 1992).

A particularly simple example of linear filtering is smoothing. Suppose we average the values of an image over a 3×3 window, that is we have

$$h(k, \ell) = \begin{cases} 1/9 & -1 \leq k, \ell \leq 1 \\ 0 & \text{otherwise} \end{cases}. \qquad (2.11)$$

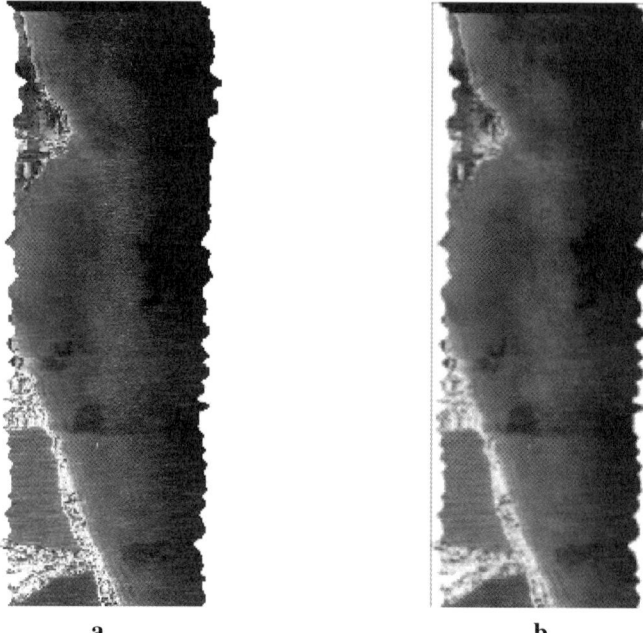

Fig. 2.8. a Input image. **b** Result after smoothing

Then, the effect is one of smoothing. This operation is most commonly employed for suppressing the effects of noise as demonstrated in Fig. 2.8. In this instance, the filter (kernel) in (2.11) was employed. However, one may choose other regions of support and a different set of coefficients. Smoothing is invariably accompanied by a loss of high-frequency information in the image. Here frequency refers to spatial, not optical, frequency (or gray levels).

The operation in (2.9) is a linear convolution in two dimensions regardless of the extent of support of A. If A is not only finite but also small, the sequence $h(k, \ell)$, also known as the convolution kernel, is represented as a matrix, with k indexing the rows and ℓ indexing the columns. For example, the uniform 3×3 smoothing kernel becomes

$$h(k, \ell) = \begin{bmatrix} 1/9 & 1/9 & 1/9 \\ 1/9 & 1/9 & 1/9 \\ 1/9 & 1/9 & 1/9 \end{bmatrix}.$$

2.5.2.2
Edge Operations

The human visual system is sensitive to edges (Gonzalez and Woods 1992). As such edge detection, extraction and enhancement are employed for various purposes including assisting manual interpretation of images, automatic detection of region boundaries, shape and object classification and so on. The edge related operations might prepare an image for further higher-level operations such as pattern recognition.

An edge is defined as a significant transition from an image region at one gray level to another. Clearly, this is a subjective definition. To arrive at an objective definition one has to quantify the degree of transition. If images are acquired in continuous space then the sharpness of transition is measured by the magnitude of the gradient operator. Given an image $I(x,y)$ in continuous space, the gradient operation is given by

$$\nabla I(x,y) = \left(\frac{\partial}{\partial x}\mathbf{i} + \frac{\partial}{\partial y}\mathbf{j}\right) I, \qquad (2.12)$$

where $\mathbf{i}$ and $\mathbf{j}$ are the unit vectors along the x and y directions respectively. For digital images, several approximations to the gradient have been proposed.

The Sobel edge operator is a common and simple approximation to the gradient. This approximates the horizontal and vertical derivatives using filter kernels

$$h_1(k,\ell) = \begin{bmatrix} -1 & 0 & 1 \\ -2 & 0 & 2 \\ -1 & 0 & 1 \end{bmatrix} \text{ and } h_2(k,\ell) = \begin{bmatrix} 1 & 2 & 1 \\ 0 & 0 & 0 \\ -1 & -2 & -1 \end{bmatrix}, \qquad (2.13)$$

respectively. Suppose $I_1(m,n)$, and $I_2(m,n)$ are the respective outputs of applying these filters to an image $I(m,n)$. The edge intensity is calculated as $\sqrt{I_1^2(m,n) + I_2^2(m,n)}$.

An alternative to gradient approximations is provided by the Laplacian of Gaussian (LoG) approach. For a continuous image $I(x,y)$, the Laplacian is provided by the operation

$$L\{I\} = \left(\frac{\partial^2}{\partial x^2} + \frac{\partial^2}{\partial y^2}\right) I(x,y). \qquad (2.14)$$

Unlike the gradient, the Laplacian is a scalar operator and is isotropic. A discrete space approximation is provided by the filter kernel

$$h(k,\ell) = \begin{bmatrix} 0 & 1 & 0 \\ 1 & -4 & 1 \\ 0 & 1 & 0 \end{bmatrix}. \qquad (2.15)$$

By itself, the Laplacian is sensitive to noise. Therefore, a smoothing filter is first applied and then the Laplacian is performed. Because of its isotropic nature and smooth falloff, the Gaussian function is employed as a smoothing kernel. In continuous-space, the filter kernel for the combined Gaussian filtering and Laplacian, otherwise known as the LoG operation is given by

$$h_{\text{LoG}}(x,y) = \frac{1}{\pi\sigma^4}\left[\frac{x^2+y^2}{2\sigma^2} - 14\right] e^{-\frac{x^2+y^2}{2\sigma^2}}. \qquad (2.16)$$

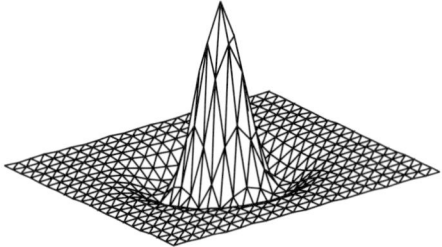

Fig. 2.9. Laplacian of Gaussian kernel

A surface plot of the function is shown in Fig. 2.9. The parameter σ controls the spread of the kernel. Discrete approximations can be generated for different values of σ.

2.5.2.3
Rank-Ordered Filtering

Filtering operations need not be linear. Nonlinear approaches are useful as well. In particular, rank-ordered filtering of which median filtering is a well-known example, is widely used. Rank-ordered filtering replaces each pixel value in an image by that pixel value in its specified neighborhood that occupies a specific position in a list formed by arranging the pixel gray values in the neighborhood in ascending or descending order. For example, one may choose to pick the maximum in the neighborhood. Median filtering results when the value picked is in the exact middle of the list.

Formally, given an image $I(m, n)$, the median filter amounts to

$$I_{\text{median}}(m, n) = \text{median}\,\{I(m - k, n - \ell)\}; \quad (k, \ell) \in A\,, \tag{2.17}$$

where A is the neighborhood over which the median is taken. Median filters are most useful in mitigating the effects of "salt and pepper" noise that arises typically due to isolated pixels incorrectly switching to extremes of opposite intensity. This can happen during image acquisition or transmission where black pixels erroneously become white and vice-versa due to overflows and saturation.

2.6
Image Classification

Image classification procedures help delineate regions in the image on the basis of attributes of interest. For example, one might be interested in identifying regions on the basis of vegetation or inhabitation. The classification problem can be stated in formal terms as follows. Suppose we want to classify each pixel in an image into one of N classes, say $C_1, C_2, \ldots, C_N$. Then, decision rules must be established to enable assignment of any given pixel to these classes. The procedures that have been developed are grounded to a significant degree in statistical decision theory, and are regarded as parametric classifiers. Two classification approaches are followed – supervised and unsupervised.

2.6.1
Supervised Classification

We illustrate the principles involved by first setting $N = 2$, that is, we will initially look at assigning each pixel to one of two classes. This is referred to as the binary hypothesis-testing problem or binary classification. Suppose x represents the intensity value of a pixel. Clearly, x is a vector for a multispectral and hyperspectral image. We represent the PDF of x for Class 1 by $f_{x|C_1}(y)$. Thus, the probability that pixel values from an object of Class 1 fall into a region $x \in A$ is given by

$$P(x \in A | C_1) = \int_{y \in A} f_{x|C_1}(y) \, dy, \tag{2.18}$$

where the integral is multidimensional. The PDF for measurements from Class 2 are defined accordingly by subscripting with the numeral 2.

The widely used maximum-likelihood (ML) decision rule assigns a pixel with value y to Class 1 if and only if

$$f_{x|C_1}(y) > f_{x|C_2}(y). \tag{2.19}$$

Otherwise, the rule assigns the pixel to Class 2. The principle for the rule is as follows. If the two classes contribute equally in the image, then there is a greater probability of Class 1 pixels being in the neighborhood of y than Class 2 pixels for any y satisfying (2.19). However, the two classes may not contribute pixels in equal proportion in which case one has to take their individual probabilities of occurrence into account.

The maximum *a posteriori* (MAP) classification rule picks Class 1 if for a measured pixel value of y,

$$P(C_1|y) > P(C_2|y). \tag{2.20}$$

This means that given the measurement, the probability that it came from a particular class is evaluated for purpose of classification. This is a reverse or *a posteriori* probability as opposed to the forward probability in (2.18), which gives the probability of obtaining a value for the measurement given the pixel class. By Bayes theorem, which relates forward and reverse probabilities,

$$P(C_1|y) = \frac{P(C_1) f_{x|C_1}(y)}{f_x(y)}, \tag{2.21}$$

where $f_x(y)$ is the PDF of the pixel measurement regardless of the class to which it belongs and $P(C_1)$ is the *a priori* probability that a randomly chosen pixel belongs to Class 1. A similar relationship obviously holds for $P(C_2|y)$. Thus, (2.20) yields, in conjunction with Bayes theorem, the MAP classification rule:

Assign pixel measurement y to Class 1 if

$$P(C_1) f_{x|C_1}(y) > P(C_2) f_{x|C_2}(y) \qquad (2.22)$$

and to Class 2 otherwise.

The MAP rule reduces to the maximum-likelihood rule if the *a priori* probabilities of belonging to either of the two classes are the same since the relationship in (2.22) then reduces to the one in (2.19).

Since a priori probabilities and conditional probability density functions are required for application of the MAP classification procedure, one has to form estimates of these probabilities from actual data. This requires a training process where the particular class a pixel belongs to is known beforehand (from the reference data such as field surveys, aerial photographs, existing maps etc.) during the training period and the various probabilities are estimated. Thus, this approach falls under the category of a supervised classification approach. Prior probabilities are more difficult to estimate reliably. If one is trying to distinguish between vegetation and sand, for example, the percentage of image area covered by vegetation varies not only by geographic region but also from image to image in the same geographic region. A conditional PDF such as $f_{x|C_1}(y)$ tends to be more reliable and is estimated through conditional histograms. Thus, the ML classification rule is easier to apply than the MAP rule.

Generally, the higher the number of spectral bands, the more difficult the procedure becomes. For this reason, the process is typically restricted to using data from a small number of bands (i.e. multispectral data only). A simplification that is often done in lieu of estimating the actual PDF is to regard the PDF as multivariate Gaussian. This requires estimating the mean vector and covariance matrix of multispectral data from each class. If the classes differ only in the mean of their measurements and have symmetric covariance matrices[1], then the ML rule reduces to a pixel being classified as belonging to a particular class if the pixel measurement is closest in Euclidean distance to the mean of that class when compared to the distance between the measurement and the means of other classes. When this rule is adopted for pixel classification regardless of whether the class distributions are jointly normal with identical, symmetric distributions but for their means, the approach is called nearest neighbor (or minimum distance to mean) classification. This rule is very easy to implement but when the distribution is asymmetric it does not conform to the maximum likelihood principle.

---

[1] A symmetric distribution about a mean of $\bar{x}$ implies that the PDF $f_x(y)$ is a function only of the Euclidean distance between y and $\bar{x}$.

2.6.2
Unsupervised Classification

Supervised classification may be impractical in various situations such as due to imprecise knowledge of number and type of classes, for example. Unsupervised approaches permit the formation of classes from the input data. While several methods have been developed, we focus here on the issue of clustering which forms the basis of many of these approaches. In any clustering approach, the final output is a set of clusters with each input pixel assigned uniquely to a cluster. Most clustering approaches are iterative. Obviously, there has to be a logical basis for the clustering and this is provided by the principle that a pixel belongs to that cluster whose representative value is closest to the pixel with respect to a distance measure. Thus, at the end of the clustering process, we should be presented with a representative value for each cluster formed.

A classic approach for cluster formation is the K-means algorithm, which is described next. Suppose we set a target for the number of clusters, say K. Then, K pixels are randomly chosen. Let their corresponding values be $x_1^1, \ldots, x_K^1$ where the superscript 1 indicates that these are the initial values. Each of the remaining pixels is assigned uniquely to the closest of these initial pixels to form the initial clusters. The most popular distance measure in practice is the absolute distance or L_1-distance which, for a pair of N–dimensional vectors $\boldsymbol{x} = [x_1 \ldots x_N]$ and $\boldsymbol{y} = [y_1 \ldots y_N]$, is defined as

$$|\boldsymbol{x} - \boldsymbol{y}| = |x_1 - y_1| + \cdots + |x_N - y_N| . \tag{2.23}$$

This distance measure is easy to calculate and is robust with respect to outliers. Thus, cluster C_k is formed initially as

$$C_k = \left\{ \boldsymbol{x} : |\boldsymbol{x} - \boldsymbol{x}_k^1| < |\boldsymbol{x} - \boldsymbol{x}_j^1| \text{ for all } j \neq k \right\}; \quad 1 \leq j, k \leq K . \tag{2.24}$$

The next step is the location of the centroid within each cluster. The centroid is defined as that value for which the sum of its distances from each member of the cluster is minimized. For the L_1-distance this turns out to be the vector each of whose components is the median of that component taken over the members of the cluster. For example, suppose we are working with just two bands and there is a cluster with three members, say, $\boldsymbol{a}_1 = [2, 34]$, $\boldsymbol{a}_2 = [10, 22]$ and $\boldsymbol{a}_3 = [17, 36]$. Then the centroid is the vector $[10, 34]$, which is obtained as the median of the two components of these three vectors. Notice that the centroid of a cluster may not be an actual input value. Following the calculation of the centroids, the clustering process is repeated as in (2.24) with the centroids replacing the initial values. A new set of centroids is then calculated following the clustering. The two steps of clustering and centroid calculations are performed until either the clusters are stable or there is no significant change in clusters.

If the Euclidean or L_2-distance is used instead of the absolute distance, then the centroid simply becomes the arithmetic mean of all vectors. This measure is appropriate if the distribution is Gaussian. It is close to the ML rule in that sense. However, it is susceptible to impulsive disturbance if the distribution is non-Gaussian with significant outlier probability.

Once the clustering is done, it may be observed that the clustering is not entirely satisfactory in the sense that there may be too fine or too coarse a clustering. Under such circumstances close clusters may be merged if the clustering is too fine or the clustering procedure may be repeated requesting a higher number of clusters if it is too coarse. These decisions are driven by

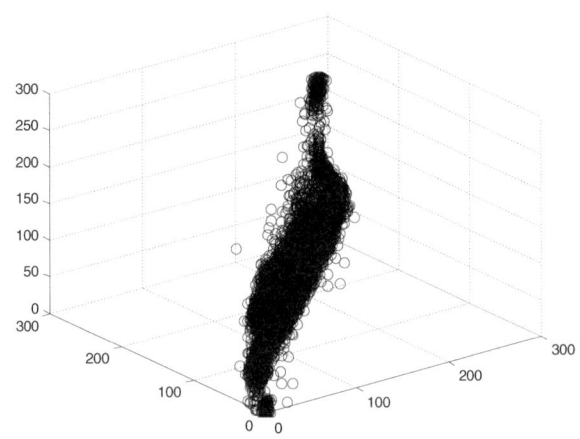
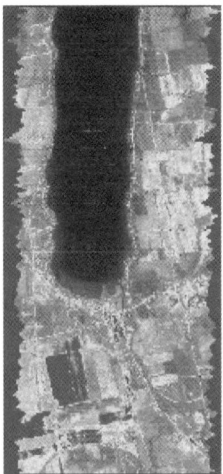

Fig. 2.10. a Input color image. **b** 3D scatter plot of R, G and B components. **c** Pseudocolor map of unsupervised classification. For a colored version of this figure, see the end of the book

heuristics. We have explained just one clustering approach. There are other approaches to be found in the literature (Gersho and Gray 1992).

An illustration of forming classes through clustering is shown in Fig. 2.10. A color image has been divided into five classes through unsupervised clustering. A pseudocolor image is then created where each class is assigned a color. The resulting classes reflect to a large degree actual ground objects such as water, buildings and fields.

2.6.3
Crisp Classification Algorithms

Clearly, the objective of classification is to allocate each pixel of an image to one of the classes. This is known as per-pixel, crisp or hard classification. Maximum likelihood decision rule is most commonly used for per pixel based supervised classification when the data satisfy Gaussian distribution assumption. However, often the spectral properties of the classes are far from the assumed distribution. For example, images acquired in complex and heterogeneous environments actually record the reflectance from a number of different classes. Many pixels are, therefore, mixed because the boundaries of mutually exclusive classes meet within a pixel (boundary pixels) or a small proportion of the classes exist within the major classes (sub-pixel phenomenon). The occurrence of mixed pixels is a major problem and has been discussed under fuzzy classification algorithms.

Another issue associated with the statistical algorithm is the selection of training data of appropriate size for accurate estimation of parameters governing that algorithm. In the absence of sufficient number of pure pixels to define training data (generally 10 to 30 times the number of bands (Swain and Davis 1978)), the performance of parametric or statistical classification algorithms may deteriorate. Further, these classification algorithms have limitations in integrating remote sensing data with ancillary data. Digital classification performed solely on the basis of the statistical analysis of spectral reflectance values may create problems in classifying certain areas. For example, remote sensing data are an ideal source for land cover mapping in hilly areas as it minimizes accessibility problems. Hilly areas, on the other hand, are covered with shadows due to high peaks and ridges thereby resulting in the change of the reflectance characteristics of the objects underneath. To reduce the effect of shadows in remote sensing image, the inclusion of information from the digital elevation model (DEM) can be advantageous (Arora and Mathur 2001). Therefore, the ancillary (non-spectral) information from other sources such as DEM (Bruzzone et al. 1997; Saha et al. 2004), and/or geological and soil maps (Gong et al. 1993) may be more powerful in characterizing the classes of interest. Moreover, due to the widespread availability of GIS, digital spatial data have become more accessible than before. Therefore, greater attention is now being paid to the use of ancillary data in classification using remote sensing imagery, where parametric classifiers may result into inappropriate classification.

Non-parametric classification may be useful when the data distribution assumptions are not satisfied. A number of nonparametric classification al-

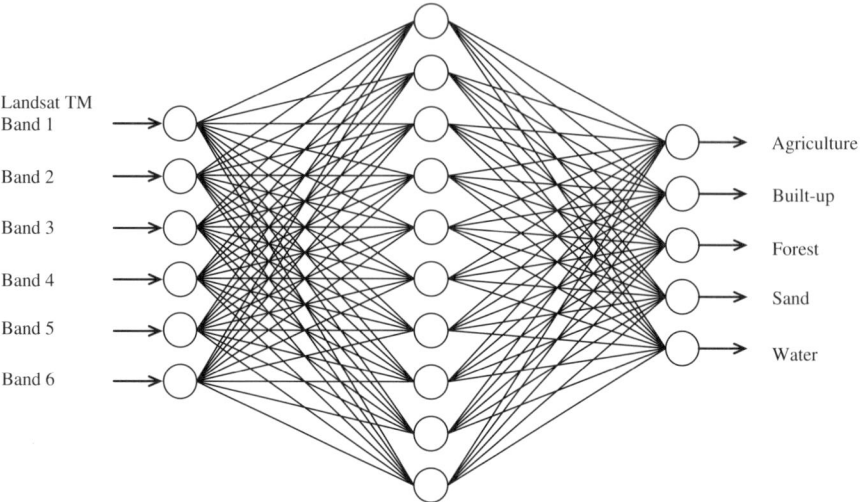

Fig. 2.11. A typical neural network architecture

gorithms such as neural network (Heermann and Khazenie 1992; Cipollini et al. 2001), knowledge based (Kimes 1991), decision tree (Brodley et al. 1996), genetic algorithms (Roberts 2000) and support vector machines (Melgani and Bruzzone 2002) are in vogue. In fact, the neural network approach has been viewed as a replacement of the most widely used MLC in the remote sensing community. A brief description of a neural network classifier is provided here. Details on other classifiers can be found in the references cited. Since support vector machine is a recent development in image classification, focus has been placed on this classifier. As a result two chapters (Chap. 5 and Chap. 10) have been included in this book to discuss this classifier in greater detail.

Typically, a neural network consists of an input layer, a hidden layer and an output layer (Fig. 2.11). The input layer is passive and merely receives the data (e.g. the multispectral remote sensing data). Consequently, the units in the input layer equal the number of bands used in classification. Unlike the input layer, both, hidden and output layers actively process the data. The output layer, as the name suggests, produces the neural network results. The number of units in the output layer is generally kept equal to the number of classes to be mapped. Hence, the number of units in the input and the output layers are typically fixed by the application designed. Introducing hidden layer between input and output layer increases the network's ability to model complex functions (Mueller and Hammerstorm 1992). Selection of appropriate number of hidden layers and their units is critical for the successful operation of the neural network. With too few hidden units, the network may not be powerful enough to process the data. On the other hand, with a large number of hidden units, computation is expensive and the network may also memorize the training samples instead of learning from the training samples. This may also result in the phenomenon called overtraining (Chi 1995). The optimal number of hidden units is often determined experimentally using a trial and error method although some basic geometrical arguments may be used to give an approximate indication (Bernard et al. 1995).

Thus, the units in the neural network are representation of the biological concept of a neuron and weighted paths connecting them (Schalkoff 1997). The data supplied to a unit's input are multiplied by the path's weight and are summed to derive the net input to that unit. The net input (NET) is then transformed by an activation function (f) to produce an output for the unit. The most common form of the activation function is a Sigmoid function, defined as

$$f(NET) = \frac{1}{1 + \exp(-NET)} \tag{2.25}$$

and accordingly,

$$\text{Output} = f(NET), \tag{2.26}$$

where NET is the sum of the weighted inputs to the processing unit and may be expressed as

$$NET = \sum_{i=1}^{n} x_i w_i, \tag{2.27}$$

where x_i is the magnitude of the ith input and w_i is the weight of the interconnected path (Foody 1998).

The determination of the appropriate weights is referred to as learning or training. Learning algorithms may be supervised and unsupervised, as discussed earlier. Generally a backpropagation algorithm is applied which is a supervised algorithm and has been widely used in applications related to neural network classification in remote sensing. The algorithm iteratively minimizes an error function over the network outputs and target outputs for a sample of training pixels c. The process continues until the error value converges to a minimum. Conventionally, the error function is given as

$$E = 0.5 \sum_{i=1}^{c} (T_i - O_i)^2, \tag{2.28}$$

where T_i is the target output and O_i is the network output, also known as the activation levels.

Thus, the magnitudes of the weights are determined by an iterative training procedure in which network repeatedly tries to learn the correct output for each training sample. The procedure involves modifying the weights of the layers connecting units until the network is able to characterize the training data (Foody 1995b). Once the network is trained, the adjusted weights are used to classify the unknown dataset.

The successful implementation of the neural network classifier, however, depends on a range of parameters related to the design of the neural network architecture (Arora and Foody 1997). A recent review on setting of different parameters in neutral network classification can be found in Kavzoglu and Mather (2003). Moreover, neural networks are very slow in the learning phase of the classification, which is a serious drawback when dealing with images of very large sizes such as hyperspectral data. Further, apart from their complexities, neural network and other non-parametric approaches are restricted to the analysis of data that is inherently in numerical form. Nevertheless, a GIS database may consist of many spatial datasets that are non-numeric, which may enhance the classification performance if they can be readily incorporated (Richards and Jia 1999). Thus, a method involving qualitative reasoning would be of particular value. The knowledge based approach offers great promise in this regard. A number of decision tree (Hansen et al. 2000) and expert system classifiers (Huang and Jensen 1997) have been developed to implement the knowledge based approach. The applicability of these classifiers, however, depends upon the formation of suitable production rules, which are meaningful and easy to understand. The generation of crisp, clear and objective rules is, thus, the crux of the problem in any knowledge based approach.

2.6.4
Fuzzy Classification Algorithms

For a large country, mapping from remote sensing is often carried out at the regional level. This necessitates the use of coarse spatial resolution images from a number of satellite based sensors such as MODIS, MERIS and AVHRR that provide data ranging from 250 m to 1.1 km spatial resolution. The images obtained from these sensors are frequently contaminated by mixed pixels. These are the pixels that do not represent a single homogeneous class but instead two or more classes are present in a single pixel area. The result is large variations in the spectral reflectance value of the pixel. Thus, occurrence of mixed pixels in an image is a major problem.

The crisp classification algorithms force the mixed pixels to be allocated to one and only one class thereby resulting in erroneous classification. The problem may be resolved either by ignoring or removing the mixed pixels from the classification process. This may, however, be undesirable since it may result in loss of pertinent information hidden in these pixels. Moreover, since the mixed pixel displays a composite spectral response, which may be dissimilar to each of its component classes, the pixel may not be allocated to any of its component classes. Therefore, error is likely to occur in the classification of images that contain a large proportion of mixed pixels (Foody and Arora 1996). Alternative approaches for classifying images dominated by mixed pixels are, therefore, highly desirable.

The output of the maximum likelihood classification may be fuzzified to represent multiple class memberships for each pixel. Thus, for example, the *a posteriori* probabilities from MLC may reflect, to some extent, the class composition of a mixed pixel (Foody 1992). Another widely used method to determine the class composition of a mixed pixel is linear mixture modeling (LMM), also referred to as spectral mixture analysis (Settle and Drake 1993). The objective of mixture modeling is to unmix the classes that are identifiable at some finer spatial scale. It is based on the assumption that the spectral response of an individual pixel is the linear sum of the spectral responses of the component classes of the pixels weighted by their relative proportions. The details on LMM algorithm can be found in Settle and Drake (1993).

However, to adopt fuzzy classification outputs from either MLC or LMM, the data need to satisfy statistical distribution assumptions, which as discussed before are often untenable. Alternative methods such as fuzzy set and neural network based classification, which are non-parametric in nature, may be more appropriate. Besides providing crisp classification outputs, these approaches may also be used to extract information at sub-pixel level. For instance, class membership values from the most widely used fuzzy c-means (FCM) clustering (described briefly here) and the activation levels of the output unit of a neural network may be used to represent the class composition of a mixed pixel (Foody and Arora 1996).

FCM clustering is an iterative clustering algorithm where class membership values are obtained by minimizing the generalized least-square error function given by (Bezdek et al. 1984)

$$J_m(U, V) = \sum_{i=1}^{n} \sum_{j=1}^{c} (\mu_{ij})^m \left\| X_i - \mu_j^* \right\|_A^2, \qquad (2.29)$$

where μ_i^* is the vector of cluster centers (i.e. class means), μ_{ij} are class membership values of a pixel, c and n are number of classes and pixels respectively, $\|X_i - \mu_j^*\|_A^2$ is the squared distance (d_{ij}) between spectral response of a pixel X_j and the class mean μ_i^* and m is a weighting exponent, which controls the degree of fuzziness. The value of m varies from 1 (no fuzziness) to ∞ (complete fuzziness). Earlier studies have shown that there is no optimal value of m but a value in the range 1.5 to 3 can generally be adopted. The class membership μ_{ij} is computed from

$$\mu_{ij} = \frac{1}{\sum_{k=1}^{c} \left(d_{ij}^2 / d_{ik}^2 \right)^{1/(m-1)}}. \qquad (2.30)$$

These class membership values of a pixel denote the class proportions and are treated as sub-pixel classifications. The sub-pixel classifications are represented in the form of fraction or proportion images, equal to the number of classes being mapped. In Chap. 11, a Markov random field (MRF) model based approach for sub-pixel classification of multi and hyperspectral data is presented.

2.6.5
Classification Accuracy Assessment

Accuracy assessment is an essential component of an image classification. A typical strategy to assess the accuracy of a crisp classification begins with the selection of a sample of pixels (known as testing samples) in the classified image based on a sampling design procedure (Stehman 1999), and verifying their class allocation from the reference data. The pixels of agreement and disagreement are summarized in the form of a contingency table (known as error or confusion matrix), which can be used to estimate various accuracy measures (Congalton 1991). Table 2.1 shows a typical $c \times c$ error matrix (c is the number of classes) with columns representing the reference data and rows the classified image albeit both are interchangeable. For an ideal classification, it is expected that all the testing samples would lie along the diagonal of the matrix indicating the perfect agreement. The off-diagonal elements indicate the disagreements referred to as the errors of omission and commission (Story and Congalton 1986).

Preferably, a single accuracy measure should express classification accuracy. However, error matrix has been used to derive a plethora of measures (Arora and Ghosh 1998). Overall accuracy (OA) or percent correct allocation is used to determine the accuracy of whole classification and is computed from the ratio of sum of diagonal entries to the total number of testing samples. Producer's accuracy (PA) and User's accuracy (UA) are used to determine the accuracy of individual classes. PA is so aptly called, since the producer of the classified image is interested in knowing how well the samples from the reference data can be mapped using remotely sensed data. It is the ratio of correctly classified samples of a class to the total number of testing samples of that class in the reference data. In contrast, UA indicates the probability or reliability that a sample from the classified image represents an actual class on the ground (Story and Congalton 1986). It is computed from the ratio of correctly classified samples of a class to the total number of testing samples of that class in the classified image.

Table 2.1. A typical error matrix (n_{ij} are the pixels of agreement and disagreement, N is the total number of testing pixels)

		Reference data				Row Total
		Class 1	Class 2	...	Class c	
	Class 1	n_{11}	n_{12}	...	n_{1c}	N_1
Classified	Class 2	n_{21}	n_{22}	...	n_{2c}	N_2
image	...	...	...	...	...	...
	Class c	n_{c1}	n_{c2}	...	n_{cc}	N_c
Column total		M_1	M_2	...	M_c	$N = \sum_{i=1}^{c} N_i$

OA, PA and UA do not take into account the agreement between the data sets (i.e. classified image and reference data) that arises due to chance alone. Thus, these measures tend to overestimate the classification accuracy (Ma and Redmond 1995). The kappa coefficient of agreement (κ) has the ability to account for chance agreement (Rosenfeld and Fitzpatrick-Lins 1986). The proportion of agreement by chance is the result of the misclassifications represented by the off-diagonal elements of the error matrix. Therefore, κ uses all the elements of the error matrix, and not just the diagonal elements (as is the case with OA). When some classes have more confusion than others, weighted kappa may be implemented since it does not treat all the misclassifications (disagreements) equally and tends to give more weight to the confusions that are more serious than others (Cohen 1968; Naesset 1996). To determine the accuracy of individual classes, a conditional kappa may be computed (Rosenfeld and Fitzpatrick-Lins 1986).

It may thus be seen that there are a number of measures that may be computed from an error matrix. Each measure may however be based on different assumptions about the data and thus may evaluate different components of accuracy (Lark 1995; Stehman 1999). Therefore, in general, it may be expedient to provide an error matrix with the classified image and report more than one measure of accuracy to fully describe the quality of that classification (Stehman 1997).

The error matrix based measures inherently assume that each pixel is associated with only one class in the classified image and only one class in the reference data. Use of these measures to assess the accuracy of fuzzy classification may therefore under or over estimate the accuracy since a fuzzy classification has to be degraded to adhere to this assumption. Moreover, in addition to the classification output, often ambiguities exist in the reference data, which may therefore be treated as fuzzy and thus other alternative measures need to be adopted (Foody 1995a; Foody 1995b).

One of the simplest measures to assess the accuracy of fuzzy classification is the entropy, which shows how the strength of class membership (i.e. fuzzy memberships) in the classified image is partitioned between the classes for each pixel. The entropy for a pixel is maximum when the pixel has equal class memberships for all the classes. Conversely, its value is minimum, when the pixel is entirely allocated to one class. It thus tells us the degree to which a classification output is fuzzy or crisp. The utility of entropy may, however, be appropriate for situations in which the output of the classification is fuzzy (Binaghi et al. 1999). Often, there exist ambiguities in the reference data, which may therefore be treated as fuzzy or uncertain (Bastin et al. 2002). Other measures may have to be applied when the reference data and classified image are fuzzy. Under these circumstances, measures such as Euclidian distance, L_1 distance and cross-entropy or directed divergence may be adopted. These measures estimate the separation of two data sets based on the relative extent or proportion of each class in the pixel (Foody and Arora 1996). Lower the values of these measures, higher is the accuracy of classification. To indicate the accuracy of an individual class in a fuzzy classification, correlation coefficient (R) may be used (Maselli et al. 1996).

The higher the correlation coefficient better is the classification accuracy of a class.

Thus, it can be seen that a number of measures may be adopted to evaluate the accuracy of a fuzzy classification. However, as pointed out by Binaghi et al. (1999), though each measure has something to offer, none is universally applicable. In their opinion, a common constraint with them is that they do not preserve the location information and therefore propose a fuzzy error matrix based approach.

The layout of a fuzzy error matrix is similar to an error matrix (see Table 2.1), to evaluate crisp classifications, with the exception that elements of a fuzzy error matrix can be any non-negative real numbers instead of non-negative integer numbers. The elements of the fuzzy error matrix represent class proportions corresponding to reference data (i.e. fuzzy reference data) and classified outputs (i.e. fuzzy classified image) respectively.

Let R_n and C_m be the sets of reference and classification data assigned to class n and m, respectively where the values of n and m are bounded by one and the number of classes L. Note here that R_n and C_m are fuzzy sets and $\{R_n\}$ and $\{C_m\}$ form two fuzzy partitions of the sample data set X where x denotes a sample element in X. The membership functions of R_n and C_m are given by

$$\mu_{R_n} : X \to [0, 1] \tag{2.31}$$

and

$$\mu_{C_m} : X \to [0, 1], \tag{2.32}$$

where [0, 1] denotes the interval of real numbers from 0 to 1 inclusive. Here, $\mu_{R_n}(x)$ and $\mu_{C_m}(x)$ is the gradual membership of the testing sample x in R_n and C_m, respectively. Since, in the context of fuzzy classification, these membership functions also represent the proportion of a class in the testing sample, the *orthogonality* or *sum-normalization* is often required, i.e.

$$\sum_{l=1}^{c} \mu_{R_l}(x) = 1. \tag{2.33}$$

The procedure used to construct the fuzzy error matrix M employs fuzzy *min* operator to determine the element $M(m,n)$ in which the degree of membership in the fuzzy intersection $C_m \cap R_n$ is computed as

$$M(m, n) = |C_m \cap R_n| = \sum_{x \in X} \min\left(\mu_{C_m}, \mu_{R_n}\right). \tag{2.34}$$

Once the fuzzy error matrix is generated, conventional error matrix based measures such as OA, PA and UA may be computed in the similar fashion to indicate the accuracy of a fuzzy classification and of individual class. Thus, the use of fuzzy error matrix based measures to evaluate fuzzy classification have conformity with the measures based on conventional error matrix based measures for crisp classification, and therefore may be more appropriate than the distance and entropy based measures. However, further research needs to be conducted to operationalize these measures in remote sensing community.

2.7
Image Change Detection

The objects on the earth surface change with time due to the economic, social and environmental pressures. These changes need to be recorded periodically for planning, management and monitoring programs. Some programs require estimation of changes over long time intervals whereas others need the recording of short-term changes. The change in time is typically referred to as temporal resolution.

To perform a systematic change detection study, two maps prepared at different times are required. Aerial photographs and remote sensing images that provide synoptic view of the terrain are regarded as attractive sources for studying the changes on the earth's surface. This is due the availability of these data at many different temporal resolutions. While the temporal resolution of data from aircrafts may be altered as per the needs of the project, satellite based data products have fixed temporal resolution. Once the data at two different times are available, these can be analyzed to detect the changes using three approaches – manual compilation, photographic compilation and digital processing. Manual and photographic compilations are laborious and time consuming. Therefore, digital methods of change detection are gaining a lot of importance.

Before detecting changes using digital images, these need to be accurately registered with each other (see Sect. 2.4). For example, registration accuracy of less than one-fifth of a pixel is required to achieve a change detection error of less than 10% (Dai and Khorram 1999). Either feature based or intensity based registration may be performed.

The fundamental assumption in applying any digital change detection algorithm is that there exists a difference in spectral response of a pixel (i.e. intensity value) on images of two dates if there is a change in two objects from one date to the other. There are many digital change detection algorithms that can be used to derive change maps; three widely used ones are,

1. Image differencing
2. Image ratioing
3. Post classification comparisons

In image differencing, one digital image is simply subtracted from another digital image of the same area acquired at different times. The difference of intensity values is stored as a third image in which features with no change will have near zero values. The third image is thresholded in such a way that pixels showing change have a value one and the pixels with no change have a value zero thereby creating a binary change image.

The image ratioing approach involves the division of intensity values of pixels in one image to the other image. The ratios are used to generate a third image. In this image, if the intensity values are close to 1, there is no change in objects. This method has an important advantage in that by taking ratios the variations in the illumination can be minimized.

In post classification comparison, the two images are first classified using any of the techniques mentioned in Sect. 2.6. These two classified images are then compared pixel by pixel to produce a change image representing pixels placed in different classes. For this method to be successful, the classification of images should be performed very accurately.

Change detection has also been performed using many other algorithms such as change vector analysis (Lambin and Strahler 1994), principal component analysis (Mas 1999) and neural networks (Liu and Lathrop 2002). A recently developed MRF model based change detection algorithm is presented in Chap. 12.

2.8
Image Fusion

During the last several years, enormous amount of data from a number of remote sensing sensors and geo-spatial databases has become available to the user community. Often, the information provided by each individual sensor may be incomplete, inconsistent and imprecise for a given application (Simone et al. 2002). The additional sources of data may provide complementary information to remote sensing data analysis. Therefore, fusion of different information may result in a better understanding of the environment. This is primarily due to the fact that the merits of each source may be tapped so as to produce a better quality data product. For example, a multispectral sensor image (at fine spectral resolution) may be fused with a panchromatic sensor image (at high spatial resolution) to generate a product with enhanced quality, as spectral characteristics of one image are merged with spatial characteristics of another image to increase the accuracy of detecting certain objects, which otherwise may not be possible from individual sensor data. Thus, image fusion may be defined as the process of merging data from multiple sources to achieve refined information. The fusion of images may be performed to (Gupta 2003):

1. sharpen the images

2. improve geometric quality of images

3. improve classification accuracy

4. enhance certain features

5. detect changes

6. substitute missing information in one image with information from another image

The complementary information about the same scene may be available in different forms and therefore different kinds of fusion may be adopted. For example, data from different sensors (Multi-sensor fusion), at different times (multi-temporal fusion), different spectral bands (multi-frequency fusion), different polarizations (multi-polarization fusion), different sources (multi-source fusion) can be fused to obtain remote sensing products of enhanced quality.

Fusion of data may be performed at different levels – pixel level, image level, feature level and decision level (Varshney et al. 1999). The basis of fusion at any level is again a good quality image registration without which we can not fully exploit the benefits of fusion. Pixel level fusion requires images to be registered at sub-pixel accuracy. In this type of fusion, original pixel values take part in the fusion process thereby retaining the measured physical parameters about the scene. An example of pixel level fusion is simple image averaging where the two pixel intensities are averaged. When larger regions of the two images are used for fusion by applying region based fusion rules, we refer to it as image level fusion. In the extreme case of image fusion, entire images are used for fusion. Fusion techniques based on pyramidal decomposition employ image level fusion (Zhang and Blum 1999). In feature level fusion, important features are extracted from the two images and then fused so that features are more clearly visible for visual inspection or for further processing. In decision level fusion, individual images are first classified individually using any image classification method, and then a decision rule is applied to generate a fused classified image that resolves ambiguities in class allocation to increase classification accuracy.

There exist a number of image fusion methods that may be categorized into several groups such as transformation based fusion (e.g. intensity-hue-saturation transformation (IHS), principal component analysis (PCA) transformation), addition and multiplication fusion (e.g. intensity modulation), filter fusion, wavelet decomposition fusion, statistical methods, evidential reasoning methods, neural network and knowledge based methods (Bretschneider and Kao 1999; Simone et al. 2002; Gupta 2003).

In a typical IHS transformation fusion, first the spatial (I) and spectral (H and S) components are generated from standard RGB image. The I component is then replaced with the high spatial resolution panchromatic image to generate new RGB image, which is referred to as fused or sharpened image (Gupta 2003). A novel method based on MRF models to perform image fusion using hyperspectral data has been described in Chap. 12 of this book.

2.9
Automatic Target Recognition

Automatic target recognition (ATR) consists of detection and identification of objects of interest in the image by processing algorithms without human intervention. In general, ATR is very difficult. A related and supposedly less difficult task is automatic target cuing where the algorithm indicates areas or objects of interest to a human operator. The targets themselves might be actual targets such as missile batteries in a military setting or areas such as vegetation plots.

The key contribution of hyperspectral imaging to ATR stems from the fact that target-background separation is seen in the spectral components. It is almost impossible for a target and the background to offer identical reflectance in each spectral band. Whereas natural backgrounds typically offer highly correlated images in narrowly spaced spectral bands (Yu et al. 1997), human-made objects are less correlated.

ATR may generally be seen as a two-stage operation: i) Detection of anomalies and ii) Recognition. In the detection stage, pixels composing different targets or a single target are identified. This requires classifying each pixel as target or non-target using classification approaches described earlier. Thus the detection stage requires operating on the entire image. On the other hand, the recognition stage is confined to processing only those regions that have been classified as target pixels. Recognition also involves prior knowledge of targets. In many instances, the latter can also be cast as a classification problem. While there is this broad categorization of ATR into two stages, actual application might involve a hierarchy of classification stages. For example, one might go through classifying objects first as human-made or natural (the detection of potential targets), then classify the human-made objects into buildings, roads and vehicles, and finally classify vehicles as tanks, trucks and so on.

The early approaches to the ATR problem in images consisted roughly of thresholding single sensor imagery, doing a shape analysis on the objects of interest and then comparing the extracted shape information to stored templates. It is easy to appreciate the complexity of the problem, given the many views, a single 3D object can provide as a function of its position relative to the sensor with factors such as scaling and rotation thrown in. Furthermore, one has to operate in poor signal to noise ratio (SNR) and contrast in highly cluttered backgrounds making traditional approaches such as matched filtering virtually useless. Thus, we may conclude that approaches begun in the early 1980s have reached their limits.

On the other hand, multiband imagery offers many advantages, the chief one, as already mentioned being the spectral separation that is seen between human-made and natural objects. An important theme of current investigations is to collect spectral signatures of targets in different bands. Classifier schemes are then trained with target and background data for discriminating target pixels across bands against background. The performance of hyperspectral or multispectral ATR is yet to reach acceptable levels in field conditions. An Independent Component Analysis (ICA) based feature extraction approach, as discussed in Chap. 8, may be used for the ATR problem.

2.10
Summary

This chapter has provided an overview of image processing and analysis techniques used with multiband image data. The intent of this chapter is to provide a ready reference within the book for the basics of image analysis.

References

Arora MK, Foody GM (1997) Log-linear modeling for the evaluation of the variables affecting the accuracy of probabilistic, fuzzy and neural network classifications. International Journal of Remote Sensing 18(4): 785–798

Arora MK, Ghosh SK (1998) A comparative evaluation of accuracy measures for the classification of remotely sensed data. Asian Pacific Remote Sensing and GIS Journal 10(2): 1–9

Arora MK, Mathur S (2001) Multi-source image classification using neural network in a rugged terrain. Geo Carto International 16(3): 37–44

Bastin LPF, Fisher P, Wood J (2002) Visualizing uncertainty in multi-spectral remotely sensed imagery. Computers and Geosciences 28: 337–350

Bernard AC, Kanellopoulos I, Wilkinson GG (1995) Neural network classification of mixtures. Proceedings of the International Workshop on Soft Computing in Remote Sensing Data Analysis, Milan, Italy, pp 53–58

Bezdek JC, Ehrlich R, Full W (1984) FCM: the fuzzy c-means clustering algorithm. Computers and Geosciences 10: 191–203

Binaghi E, Brivio PA, Ghezzi P, Rampini A (1999) A fuzzy set-based accuracy assessment of soft classification. Pattern Recognition Letters 20: 935–948

Bretschneider T, Kao O (1999) Image fusion in remote sensing, Report, Technical University of Clausthal

Brodley CE, Friedl MA, Strahler AH (1996) New approaches to classification in remote sensing using homogeneous and hybrid decision trees to map land cover. Proceedings of the IEEE International Geoscience and Remote Sensing Symposium 1, pp 532–534

Brown, LG (1992) A survey of image registration techniques. ACM Computing Surveys 24: 325–376

Bruzzone L, Conese C, Maselli F, Roli F (1997) Multi-source classification of complex rural areas by statistical and neural-network approaches. Photogrammetric Engineering and Remote Sensing 63(5): 523–533

Campbell, JB (2002) Introduction to remote sensing. Guilford Press, New York

Chi Z (1995) MLP classifiers: overtraining and solutions. Proceedings of the IEEE International Conference on Neural Networks 5, pp 2821–2824

Cipollini P, Corsini G, Diani M, Grasso R (2001) Retrieval of sea water optically active parameters from hyperspectral data by means of generalized radial basis function neural networks. IEEE Transactions on Geoscience and Remote Sensing 39(7): 1508–1524

Cohen J (1968) Weighted kappa: nominal scale agreement with provision for scaled disagreement or partial credit. Psychological Bulletin 70: 213–220

Congalton RG (1991) A review of assessing the accuracy of classifications of remotely sensed data. Remote Sensing of the Environment 37: 35–46

Dai X, Khorram S (1999) A feature-based image registration algorithm using improved chain-code representation combined with invariant moments. IEEE Transactions on Geoscience and Remote Sensing 37: 2351–2362

Folk M, McGrath RE, Yeager N (1999) HDF: an update and future directions. Proceedings of the IEEE International Geoscience and Remote Sensing Symposium 1, pp 273–275

Foody GM (1992) A fuzzy sets approach to the representation of vegetation continua from remotely sensed data: an example from lowland heath. Photogrammetric Engineering and Remote Sensing 58: 221–225

Foody GM (1995a) Cross-entropy for the evaluation of the accuracy of a fuzzy land cover classification with fuzzy ground data. ISPRS Journal of Photogrammetry and Remote Sensing 50: 2–12

Foody GM (1995b) Using prior knowledge in artificial neural network classification with a minimal training set. International Journal of Remote Sensing 16(2): 301–312

Foody GM (1998) Issues in training set selection and refinement for classification by a feed-forward neural network. Proceedings of the IEEE International Geoscience and Remote Sensing Symposium 1, pp 409–411

Foody GM, Arora MK (1996) Incorporating mixed pixels in the training, allocation and testing stages of supervised classifications. Pattern Recognition Letters 17(13): 1389–1398

Gersho A, Gray RM (1992) Vector quantization and signal compression. Kluwer Academic, Boston, MA

Gong P, Miller JR, Spanner M (1993) Forest canopy closure from classification and spectral mixing of scene components: multi-sensor evaluation of application to an open canopy. Proceedings of the IEEE International Geoscience and Remote Sensing Symposium 2, pp 747–749

Gonzalez RC, Woods RE (1992) Digital image processing. Addison-Wesley Publishing Company, Reading, Massachusetts

Gupta RP (2003) Remote sensing geology. Springer Verlag, Heidelberg

Hansen M, DeFries R, Townshend J, Sohlberg R (2000) Global land cover classification at 1km spatial resolution using a classification tree approach. International Journal of Remote Sensing. 21(6&7): 1331–1364

Heermann PD, Khazenie N (1992) Classification of multispectral remote sensing data using a back-propagation neural network. IEEE Transactions on Geoscience and Remote Sensing 30(1): 81–88

Huang X, Jensen JR (1997) A machine learning approach to automated construction of knowledge bases for image analysis expert systems that incorporate GIS data. Photogrammetric Engineering and Remote Sensing 63(10): 1185–1194

Jensen, JR (1996) Introductory digital image processing: a remote sensing perspective. Prentice-Hall, Upper Saddle River, NJ

Kavzoglu T, Mather PM (2003) The use of backpropagating artificial neural networks in land cover classification. International Journal of Remote Sensing 24(23): 4907-4938

Kimes DS (1991) Inferring vegetation characteristics using a knowledge-based system. Proceedings of the IEEE International Geoscience and Remote Sensing Symposium 4, pp 2299–2302

Lambin EF, Strahler AH (1994) Indicators of land-cover change for change-vector analysis in multitemporal space at coarse spatial scales. International Journal of Remote Sensing 15: 2099–2119

Lark RM (1995) Contribution of principal components to discrimination of classes of land cover in multispectral imagery. International Journal of Remote Sensing 6: 779–784

Liu X, Lathrop RG (2002) Urban change detection based on an artificial neural network. International Journal of Remote Sensing 23: 2513–2518

Ma Z, Redmond RL (1995) Tau coefficient for accuracy assessment of classification of remote sensing data. Photogrammetric Engineering and Remote Sensing 61: 435–439

Mas JF (1999) Monitoring land-cover changes: a comparison of change detection techniques. International Journal of Remote Sensing 20: 139–152

Maselli F, Conese C, Petkov L (1996) Use of probability entropy for the estimation and graphical representation of the accuracy of maximum likelihood classifications. ISPRS Journal of Photogrammetry and Remote Sensing 49(2): 13–20

Mather PM (1999) Computer processing of remotely-sensed images: an introduction. Wiley, Chichester

References

Melgani F, Bruzzone L (2002) Support vector machines for classification of hyperspectral remote-sensing images. Proceedings of the IEEE International Geoscience and Remote Sensing Symposium 1, pp 506–508

Mueller D, Hammerstorm D (1992) A neural network system component. Proceedings of the International Joint Conference on Neural Networks, pp 1258–1264.

Naesset E (1996) Conditional Tau coefficient for assessment of producer's accuracy of classified remotely sensed data. ISPRS Journal of Photogrammetry and Remote Sensing 51: 91–98

Rast M, Hook SJ, Elvidge CD, Alley RE (1991) An evaluation of techniques for the extraction of mineral absorption features from high spectral resolution remote sensing data. Photogrammetric Engineering and Remote Sensing 57(10): 1303–1309

Richards JA, Jia X (1999) Remote sensing digital image analysis: an introduction. Springer-Verlag, Berlin

Roberts RS (2000) Characterization of hyperspectral data using a genetic algorithm. Proceedings of the Thirty-Fourth Asilomar Conference on Signals, Systems and Computers 1, pp 169–173

Rosenfeld GH, Fitzpatrick-Lins K (1986) A coefficient of agreement as a measure of thematic classification accuracy. Photogrammetric Engineering and Remote Sensing 52: 223–227

Saha AK, Arora MK, Csaplovics E, Gupta RP (2004) Land cover classification using IRS 1C LISS and topographic data in a rugged terrain in Himalayas. Geo Carto International (Revision sent)

Schalkoff RJ (1997) Artificial neural networks. McGraw-Hill, New York

Settle JJ, Drake NA (1993) Linear mixing and the estimation of ground cover proportions. International Journal of Remote Sensing 14: 1159–1177

Simone G, Farina A, Morabito FC, Serpico SB, Bruzzone L (2002) Image fusion techniques for remote sensing applications. Information Fusion 3: 3–15

Smith GM, Milton EJ (1999) The use of the empirical line method to calibrate remotely sensed data to reflectance. International Journal of Remote Sensing 20: 2653–2662

Stehman SV (1997) Selecting and interpreting measures of thematic classification accuracy. Remote Sensing of the Environment 62: 77–89

Stehman SV (1999) Basic probability sampling designs for thematic map accuracy assessment. International Journal of Remote Sensing 20: 2423–2441

Story M, Congalton RG (1986) Accuracy assessment: a user's perspective. Remote Sensing of the Environment 52(3): 397–399

Swain PH, Davis SM (eds) (1978) Remote sensing: the quantitative approach, McGraw Hill, New York

Ullman RE (1999) HDF-EOS, NASA's standard data product distribution format for the Earth Observing System data information system. Proceedings of the IEEE International Geoscience and Remote Sensing Symposium 1, pp 276–278

Varshney PK, Chen H, Ramac LC, Uner M, Ferris D, Alford M (1999) Registration and fusion of infrared and millimeter wave images for concealed weapon detection. Proceedings of the IEEE International Conference on Image Processing 3, pp 532–536

Yu X, Hoff LE, Reed IS, Chen AM, Stotts LB (1997) Automatic target detection and recognition in multiband imagery: a unified ML detection and estimation approach. IEEE Transactions on Image Processing 6(1): 143–156

Zhang Z, Blum RS (1999) A categorization and study of multiscale-decomposition-based image fusion schemes. Proceedings of the IEEE, pp 1315–1328

Zitova B, Flusser J (2003) Image registration methods: a survey. Image and Vision Computing, pp 977–1000

**Part II
Theory**

CHAPTER 3

Mutual Information: A Similarity Measure for Intensity Based Image Registration

Hua-mei Chen

3.1
Introduction

Mutual information (MI) was independently proposed in 1995 by two groups of researchers (Maes and Collignon of Catholic University of Leuven (Collignon et al. 1995) and Wells and Viola of MIT (Viola and Wells 1995)) as a similarity measure for intensity based registration of images acquired from different types of sensors. Since its introduction, MI has been used widely for a variety of applications involving image registration. These include medical imaging (Holden et al. 2000; Maes et al. 1997; Studhilme et al. 1997; Wells et al. 1996), remote sensing (Chen et al. 2003ab), and computer vision (Chen and Varshney 2001a). The MI registration criterion states that an image pair is geometrically registered when the mutual information between the two images reaches its maximum. The strength of MI as a similarity measure lies in the fact that no assumptions are made regarding the nature of the relation between the intensity values of the image, as long as such a relationship exists. Thus, the MI criterion is very general and has been used in many image registration problems in a range of applications.

The basic concept of image registration using MI as the similarity measure may be explained with the help of Fig. 3.1. In this figure, F is referred to as the *floating image*, whose pixel coordinates are to be mapped to new coordinates on the *reference image R*, which are resampled according to the positions defined by the new coordinates. $\bar{x}$ represents the coordinates of a pixel (voxel in 3D case) in F and $\bar{y}$ represents the coordinates of a pixel in R. The transformation model is represented by T with associated parameters α. The dependence of T on α is indicated by the use of notation T_α. The joint histogram, which is used to compute the mutual information, I_{T_α}, between floating and reference images, is denoted by $h_{T_\alpha}(F, R)$. The subscript T_α indicates that h and I are dependent on T_α. The blocks in the shaded portion of Fig. 3.1 represent the global optimization process used to maximize mutual information.

In this chapter, we describe the concept of image registration using MI as a similarity measure by discussing each component of Fig. 3.1 in detail. The application of MI based image registration methodology to multisensor and multi-temporal remote sensing images will be discussed in Chap. 7. Two popular methods to estimate the joint histogram for the computation of MI

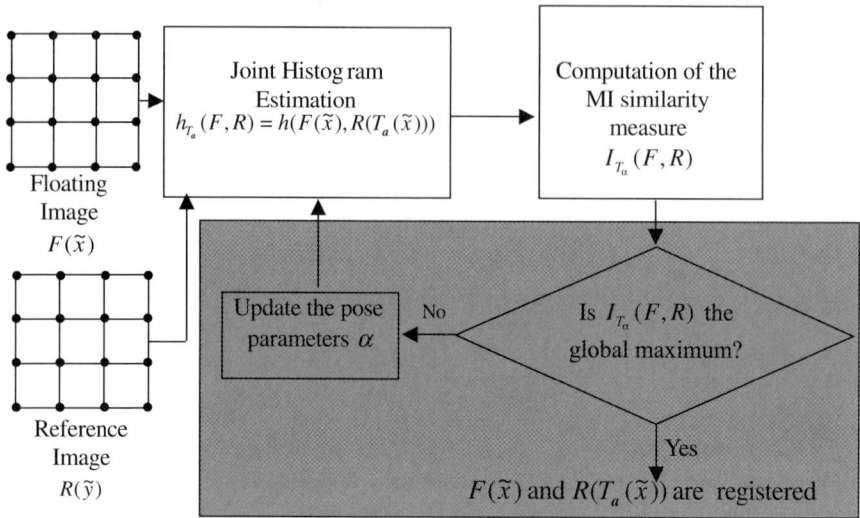

Fig. 3.1. General flow chart of intensity based image registration using mutual information as the similarity maesure

are explained. However, these methods, under certain conditions, suffer from a phenomenon called interpolation-induced artifacts that hampers the optimization process and may thereby reduce registration accuracy (Pluim et al. 2000). To overcome the problem of interpolation-induced artifacts, a new algorithm called the Generalized Partial Volume joint histogram Estimation (GPVE) algorithm (Chen and Varshney 2003) is also introduced.

3.2
Mutual Information Similarity Measure

Mutual information has its roots in information theory (Cover and Thomas 1991). It was developed to set fundamental limits on the performance of communication systems. However, it has made vital contributions to many different disciplines like physics, mathematics, economics, and computer science. In this section, we introduce the use of MI for image registration.

MI of two random variables A and B is defined by

$$I(A,B) = \sum_{a,b} P_{A,B}(a,b) \log \frac{P_{A,B}(a,b)}{P_A(a) P_B(b)}, \tag{3.1}$$

where $P_A(a)$ and $P_B(b)$ are the marginal probability mass functions, and $P_{A,B}(a,b)$ is the joint probability mass function. MI measures the degree of dependence of A and B by measuring the distance between the joint probability $P_{A,B}(a,b)$ and the probability associated with the case of complete independence $P_A(a)P_B(b)$, by means of the relative entropy or the Kullback-Leibler measure (Cover and Thomas 1991). MI is related to entropies by the relationship

$$I(A, B) = H(A) + H(B) - H(A, B), \tag{3.2}$$

where $H(A)$ and $H(B)$ are the entropies of A and B and $H(A, B)$ is their joint entropy. Considering A and B as two images, *floating image* (F) and *reference image* (R) respectively, the MI based image registration criterion states that the images shall be registered when $I(F, R)$ is maximum. The entropies and the joint entropy can be computed from,

$$H(F) = -\sum_f P_F(f) \log P_F(f), \tag{3.3}$$

$$H(R) = -\sum_r P_R(r) \log P_R(r), \tag{3.4}$$

$$H(F, R) = -\sum_{f,r} P_{F,R}(f, r) \log P_{F,R}(f, r), \tag{3.5}$$

where $P_F(f)$ and $P_R(r)$ are the marginal probability mass functions, and $P_{F,R}(f, r)$ is the joint probability mass function of the two images F and R. The probability mass functions can be obtained from,

$$P_{F,R}(f, r) = \frac{h(f, r)}{\sum_{f,r} h(f, r)}, \tag{3.6}$$

$$P_F(f) = \sum_r P_{F,R}(f, r), \tag{3.7}$$

$$P_R(r) = \sum_f P_{F,R}(f, r), \tag{3.8}$$

where h is the joint histogram of the image pair with each image being of size $(M \times N)$. It is a 2D matrix given by

$$h = \begin{bmatrix} h(0,0) & h(0,1) & \dots & h(0, N-1) \\ h(1,0) & h(1,1) & \dots & h(1, N-1) \\ \dots & \dots & & \dots \\ h(M-1, 0) & h(M-1, 1) & \dots & h(M-1, N-1) \end{bmatrix}. \tag{3.9}$$

The value $h(a, b)$, $a \in [0, M-1]$, $b \in [0, N-1]$, is the number of pixel pairs having intensity value a in the first image (i.e. F) and intensity value b in the second image (i.e. R). It can thus be seen from (3.2) to (3.8) that the joint histogram estimate is sufficient to determine the MI between two images.

To interpret (3.2) in the context of image registration with random variables F and R, let us first assume that both $H(F)$ and $H(R)$ are constant. Under this assumption, maximization of MI in (3.2) is equivalent to the minimization of

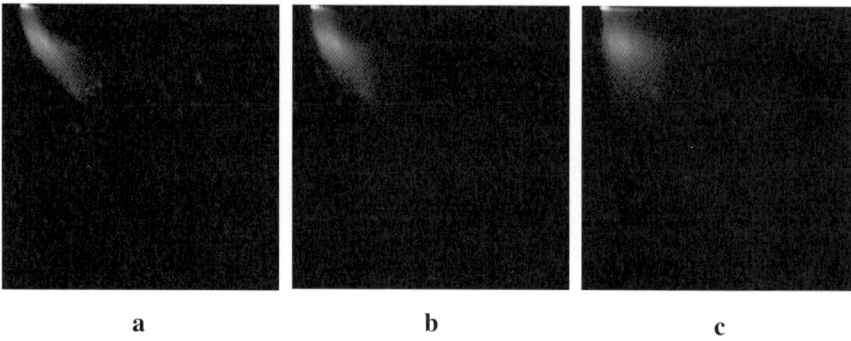

Fig. 3.2a–c. Joint histograms of a pair of Landsat TM band 3 and band 7 images. **a** Images are completely registered. **b** Images are shifted by one pixel vertically. **c** Images are shifted by three pixels vertically

joint entropy. Figure 3.2a-c show the joint histograms of a pair of Landsat TM band 3 and band 7 images. Figure 3.2a is obtained when the two images are registered. Figure 3.2b is obtained when the first image is shifted vertically by one pixel. Figure 3.2c shows the joint histogram when the first image is shifted vertically by three pixels. From Fig. 3.2, we can observe that when two images are registered, the joint histogram is very sharp whereas it gets more dispersed as the vertical displacement is increased. Therefore, it seems reasonable to devise a similarity measure that meaningfully represents the sharpness or degree of dispersion of the joint histogram. Joint entropy is one such metric that measures the *uncertainty* of the joint histogram. Thus, under the assumption of constant entropies $H(F)$ and $H(R)$, (3.2) may be used to interpret the mutual information criterion in that it tries to *minimize the uncertainty of the joint histogram of the two images to be registered.*

When the assumption of constant entropies $H(F)$ and $H(R)$ is not satisfied, we may employ the notion of conditional entropy to interpret the MI criterion. We rewrite MI in terms of conditional entropy as,

$$I(F, R) = H(F) - H(F|R)$$
$$= H(R) - H(R|F), \tag{3.10}$$

where the conditional entropies $H(F|R)$ and $H(R|F)$ are defined as:

$$H(F|R) = -\sum_{f,r} P_{F,R}(f,r) \log P_{F|R}(f|r), \tag{3.11}$$

$$H(R|F) = -\sum_{f,r} P_{F,R}(f,r) \log P_{R|F}(r|f). \tag{3.12}$$

Conditional entropy $H(F|R)$ represents the entropy (uncertainty) of F when R is known. Similarly, $H(R|F)$ is the entropy (uncertainty) of R given F. Therefore, mutual information can be realized as *the reduction in the uncertainty (entropy) of one image by the knowledge of the other image.* In other words, the MI criterion implies that when two images are registered, one gains the most knowledge about one image by observing the other one. For example, based on (3.10), the uncertainty of F without the knowledge of R is $H(F)$ and its uncertainty when R is known is $H(F|R)$. Therefore, the reduction in the uncertainty $H(F) - H(F|R)$ defines the MI between F and R and its maximization implies the most reduction in the uncertainty. Since its introduction, MI has been accepted as one of the most accurate and robust similarity measures for image registration.

3.3
Joint Histogram Estimation Methods

From (3.2) to (3.8), it is clear that the main task involved in determining the mutual information between two images is to estimate the joint histogram of the two images. Existing joint histogram estimation schemes may be categorized into two groups, depending on whether or not an intermediate image resampling procedure for the reference image is required.

3.3.1
Two-Step Joint Histogram Estimation

Intuitively, a joint histogram can be estimated using a two-step procedure. Assume that $\tilde{x}_i$ represents the coordinates of the ith pixel of the floating image and $\tilde{y}_i$ represents the transformed coordinates of the pixel (i.e. $\tilde{y}_i = T_\alpha(\tilde{x}_i)$). In general, $\tilde{y}_i$ may not coincide with the coordinates of any grid point of the reference image. Therefore, during the first step, one needs to evaluate the intensity values of the reference image at the transformed grid point positions through interpolation. Often, linear interpolation is employed. Other intensity interpolation algorithms like cubic convolution interpolation (Keys 1981) and cubic spline interpolation (Unser et al. 1993) may also be adopted. In general, the interpolated values are not integers. Thus, one may resort to rounding off the interpolated values to the nearest integer. After obtaining this intermediate resampled reference image, the second step is to obtain the joint histogram by updating the corresponding entry by one. This updating procedure is expressed mathematically as

$$h_\alpha\left(F\left(\tilde{x}_i\right), round\left(R\left(T_\alpha\left(\tilde{x}_i\right)\right)\right)\right) = h_\alpha\left(F\left(\tilde{x}_i\right), round\left(R\left(T_\alpha\left(\tilde{x}_i\right)\right)\right)\right) + 1 \tag{3.13}$$

for every i such that $T_\alpha(\tilde{x}_i) \in \widehat{Y}$. In (3.13), $round(\cdot)$ represents the rounding off operation and $\widehat{Y}$ is the continuous domain of the reference image.

The general drawback of this two-step procedure is that the resulting MI registration function is usually not very smooth due to the rounding off operation. Smoothness of the MI registration function facilitates the subsequent optimization process. This is illustrated in Fig. 3.3a–c where a and b represent

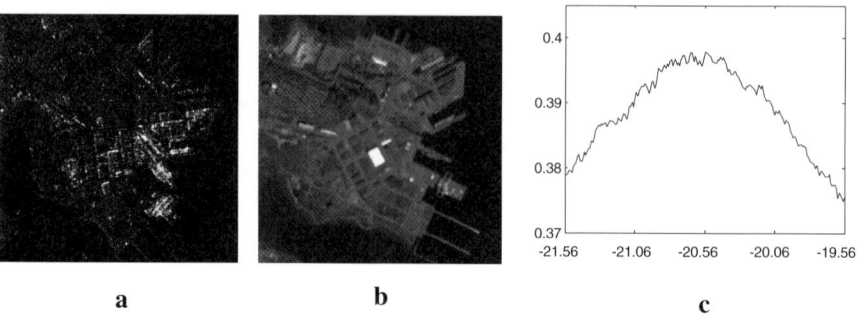

Fig. 3.3. a Radarsat SAR image, b IRS PAN image image and c MI registration function using linear interpolation. In c, rotation angle is presented in degrees

a pair of Radarsat Syntehtic Aperture Radar (SAR) and IRS PAN images and c shows the MI registration function using linear interpolation by varying the rotation angle from $-21.56°$ to $-19.56°$ after registration while keeping the values of the other transformation parameters fixed. From Fig. 3.3c, we can observe that the MI registration function obtained using linear interpolation is not smooth. To overcome this problem, Maes et al. (1997) devised a one-step joint histogram estimation method that produces a very smooth MI curve and is described next.

3.3.2
One-Step Joint Histogram Estimation

This method directly estimates the joint histogram without resorting to the interpolation of intensity values at the transformed grid points on the reference image. As a result, rounding off operation involved in the two-step procedure is avoided and the resulting MI registration function becomes very smooth. This method is called partial volume interpolation (PVI) as it was originally devised for 3D brain image registration problems. Its 2D implementation is now described.

In Fig. 3.4, $\tilde{n}_1$, $\tilde{n}_2$, $\tilde{n}_3$, and $\tilde{n}_4$ are the grid points of the reference image that are closest to the transformed point $T_\alpha(\tilde{x})$. These points define a cell. $T_\alpha(\tilde{x})$

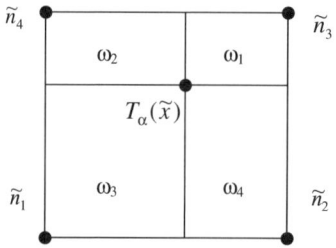

Fig. 3.4. Graphical illustration of partial volume interpolation (PVI) in two dimensions

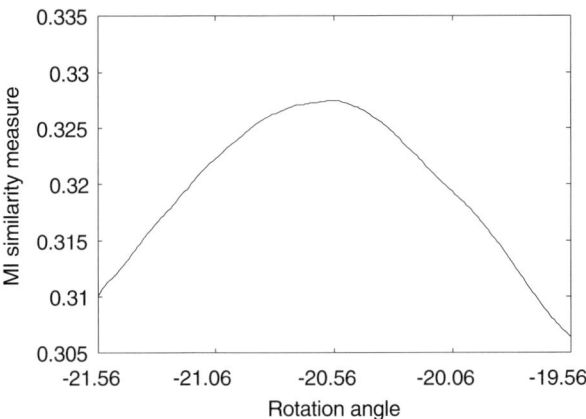

Fig. 3.5. MI registration function obtained using the same pair of images as shown in Fig. 3.3 but employing 2D implementation of the PVI algorithm

splits the cell $(\tilde{n}_1, \tilde{n}_2, \tilde{n}_3, \tilde{n}_4)$ into four sub-cells having areas $\omega_1, \omega_2, \omega_3$, and ω_4 with the constraint $\sum_i \omega_i(T_\alpha(\tilde{x})) = 1$. The joint histogram is then obtained by updating four entries defined by $(F(\tilde{x}), R(\tilde{n}_i))$, $i = 1, \ldots, 4$, as follows:

$$\forall i : h_\alpha\left(F(\tilde{x}), R(\tilde{n}_i)\right) = h_\alpha\left(F(x), R(\tilde{n}_i)\right) + \omega_i . \tag{3.14}$$

The method results in a very smooth MI registration function, and does not introduce any extra intensity values. That is, the indices of all the non-zero entries in the estimated joint histogram are completely defined by the set of intensity values in the floating image and the set of intensity values in the reference image (see (3.9), where each element is specified by its 2D index $[0 \sim M - 1, 0 \sim N - 1]$). In the two-step procedure described previously, new intensity values, which are not in the original set of intensity values in the reference image, may be introduced because of interpolation. In addition, in the two-step procedure, if a more sophisticated interpolation method like cubic convolution interpolation or cubic B-spline interpolation is employed, the interpolated values may go beyond the range of the intensity values used (e.g. $0 \sim 255$ in general). In this case, extra effort may be required to obtain a meaningful estimate of the joint histogram by keeping the interpolated values between 0 and 255. Fig. 3.5 shows the registration function using the same image data as shown in Fig. 3.3 but the 2D implementation of PVI is employed. Clearly, in this case, the MI registration function is much smoother.

3.4
Interpolation Induced Artifacts

Very often, when a new similarity measure is adopted for image registration, it is a common practice to use a pair of identical images to test its efficacy.

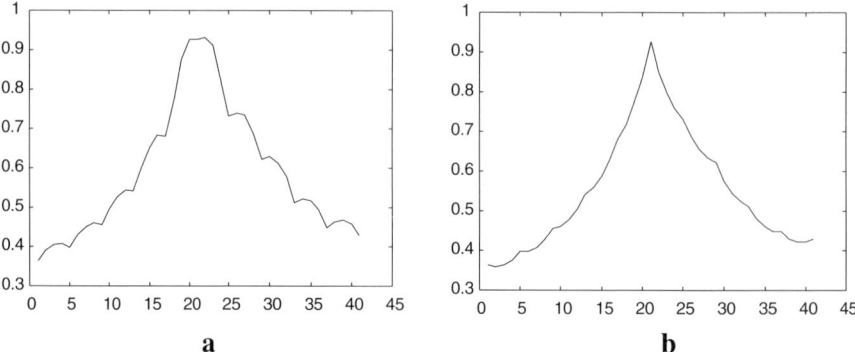

Fig. 3.6a,b. Typical interpolation-induced artifact patterns for a MI based registration function. The concave pattern shown in **a** is a common pattern from a two-step histogram estimation procedure and the convex pattern as shown in **b** results from partial volume interpolation estimation. In both cases, the vertical axis is the MI based measure, and the horizontal axis can either be vertical or horizontal displacement

Usually, the similarity measure is plotted as a function of one of the transformation parameters to observe whether the optimum is reached or not when the two images are registered. However, adopting this procedure in MI based registration may confuse many people due to the occurrence of periodic patterns (as shown in Fig. 3.6) where the MI measure is plotted as a function of either vertical displacement or horizontal displacement at *sub-pixel* level. In fact, the occurrence of periodic patterns is a phenomenon referred to as interpolation-induced artifacts (Pluim et al. 2000). Figures 3.6a and 3.6b show typical artifact patterns obtained from the two-step and one step procedures for joint histogram estimation respectively. It is stated in Pluim et al. (2000) that when two images have equal sample spacing in one or more dimensions, existing joint histogram estimation algorithms like the PVI and the linear interpolation may result in certain types of artifact patterns in an MI based registration function. This is due to the fact that many of the grid lines (or grid planes for 3D image volumes) in this case may be aligned along that dimension under certain geometric transformations. For example, if two images have the same sample spacing along the horizontal axis, then any integer-valued horizontal displacement between the two images results in the alignments of all vertical grid lines within the overlap area. Therefore, fewer interpolations are needed to estimate the joint histogram of these two images than in the case in which none of the grid lines are aligned. In linear interpolation, the averaging (blurring) operation is believed to be the cause of the artifacts. As a consequence, the artifact pattern is more pronounced if the images contain noise. This is because the averaging effect of linear interpolation is made clearer in a noisy image than a noise-free image. Figure 3.7a–d shows the plots of the MI registration function for a pair of Landsat TM images (band 1 and band 3) having displacements in vertical direction. Different amounts of Gaussian noise were added to both images to produce these results. Clearly, noise plays an

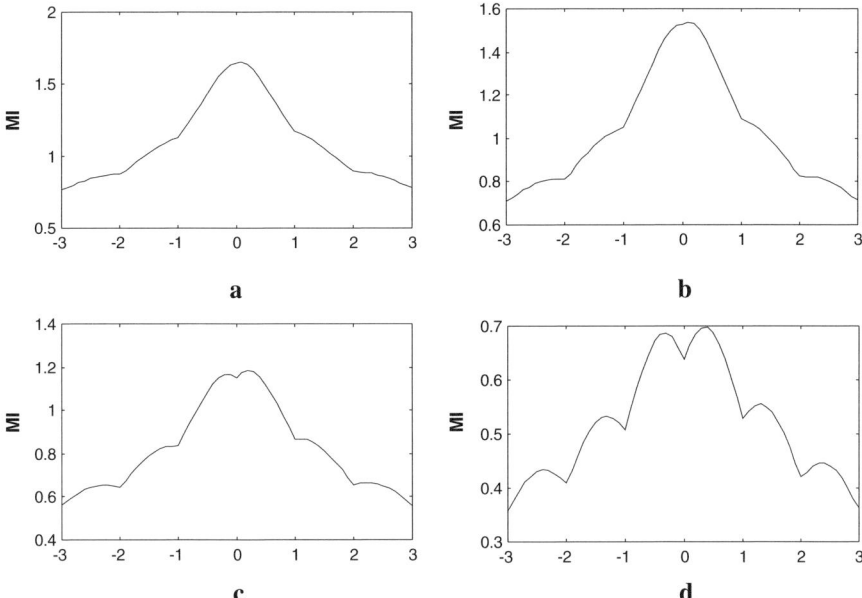

Fig. 3.7a–d. MI artifacts for a pair of Landsat TM band 1 and band 3 images along the vertical displacement (x axis) in pixels. **a** No extra noise is added. **b** Gaussian noise with variance equal to 1 is added. **c** Gaussian noise with variance equal to 3 is added. **d** Gaussian noise with variance equal to 7 is added

important role in the formation of the artifact pattern. As the noise in the images is increased, the artifact patterns become more pronounced.

To gain some additional insight into the formation of the artifacts resulting from linear interpolation, let us consider the following. From (3.2), we know that three components are involved in the determination of MI. These are entropy of the floating image, entropy of the reference image, and the joint entropy of the floating and reference images. For simplicity, we can make the floating image always within the scope of the reference image. In this manner, the overlap for the floating image remains the same and the corresponding entropy $H(F)$ remains a constant. Thus, only the entropy of the reference image and the joint entropy vary as the transformation parameters change. Figures 3.8 and 3.9 show the effect of noise on these two components when linear interpolation is employed for joint histogram estimation. From these figures, we can see that when the vertical displacement is non-integer, both the entropy of the reference image and the joint entropy decrease as a result of the blurring effect of linear interpolation. Moreover, this decrease is more manifested for joint entropy than marginal entropy. Therefore, the pattern of MI artifacts shown in Fig. 3.7 is basically the inverse of the pattern of the joint entropy shown in Fig. 3.9.

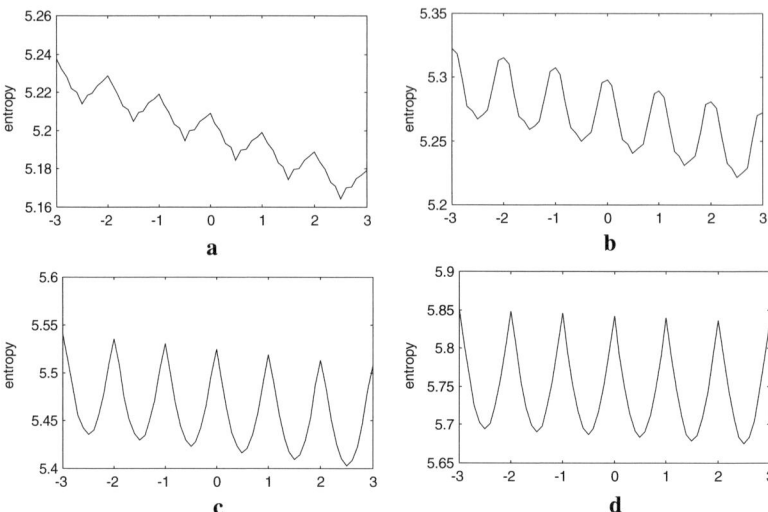

Fig. 3.8a–d. Entropy of the reference image. **a** No extra noise is added. **b** Gaussian noise with variance equal to 1 is added. **c** Gaussian noise with variance equal to 3 is added. **d** Gaussian noise with variance equal to 7 is added. In each case, y axis represents entropy and x axis the displacement in pixels

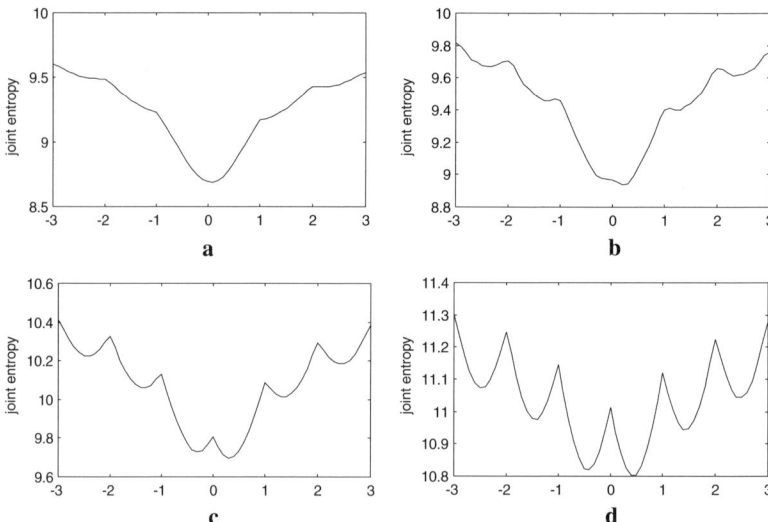

Fig. 3.9a–d. Joint entropy of the floating and reference images. **a** No extra noise is added. **b** Gaussian noise with variance equal to 1 is added. **c** Gaussian noise with variance equal to 3 is added. **d** Gaussian noise with variance equal to 7 is added. In each case, x axis represents the displacement in pixels

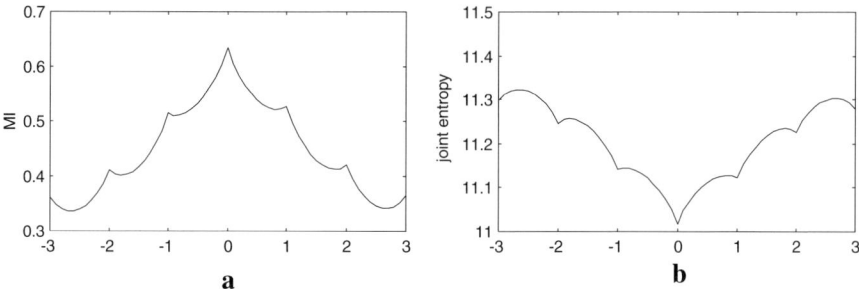

Fig. 3.10a,b. Artifact patterns resulting from the PVI algorithm. **a** Artifact pattern in the MI registration function. **b** Artifact pattern in the joint entropy function. In each case, x axis represents the displacement in pixels

Next, let us turn to the artifacts resulting from the PVI algorithm. The mechanism that causes the artifacts in the PVI algorithm is the splitting of a joint histogram entry into a few smaller entries, or equivalently, a probability into a few smaller probabilities (Pluim et al. 2000). The result of this probability split is an increase in entropy. Again, from (3.2) we know that there are three components that determine the mutual information. Generally, the change in the joint entropy of the two images caused by the probability split behaves in a similar but more severe manner as the entropies of the two individual images. Therefore, the artifact pattern in the MI registration function resulting from PVI is dominated by the artifact pattern in the joint entropy, as is the case in linear interpolation. Figure 3.10 shows the artifact patterns in the MI registration function and the joint entropy resulting from PVI using the same image pair used to generate Fig. 3.7d, 3.8d and 3.9d. Notice that the artifact pattern in the joint entropy (Fig. 3.10b) is similar to that in the MI measure (Fig. 3.10a) but is inverted. This is because of the negative sign in (3.2). For more information about the mechanisms resulting in artifact patterns from linear and partial volume interpolations, interested readers are referred to Pluim et al. (2000). A new joint histogram estimation algorithm that can reduce the interpolation artifact problem effectively is discussed next.

3.5
Generalized Partial Volume Estimation of Joint Histograms

Severity of interpolation artifacts was first discussed in Pluim et al. (2000). In an effort to reduce the artifacts, resampling of one of the two images (floating or reference) in such a way that none of the sample spacing along any of the x, y and z axes has equal distance, was suggested. However, the resampling procedure is expected to decrease registration accuracy due to extra rounding off operations in the resampling procedure (see Sect. 3.3.2). As an alternative procedure to reduce the artifacts, a two-step joint histogram estimation procedure involving a more sophisticated image interpolation algorithm with less blurring effect

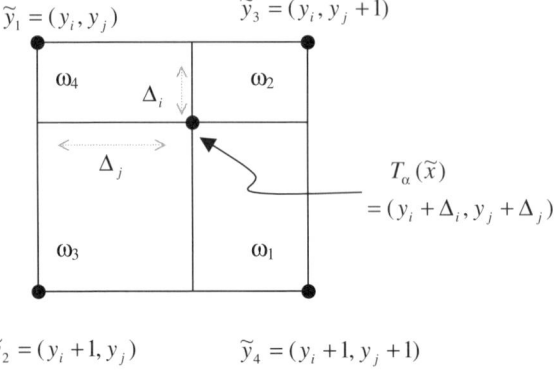

Fig. 3.11. Another graphical illustration of PVI in two dimensions. compare this figure with Fig. 3.4

may be employed. However, Chen and Varshney (2002) clearly showed that such an attempt was not successful as even after applying cubic convolution and cubic B-spline interpolations, the artifacts were still present. Although, there are many other image interpolation algorithms (Lehmann et al. 1999), the chances of devising a two-step artifact-free joint histogram estimation scheme appear remote.

Alternatively, a one-step joint histogram estimation scheme that will reduce the problem of artifacts may be devised. As discussed previously, the occurrence of artifacts in the PVI algorithm is due to the introduction of additional joint histogram dispersion. Thus, if this additional joint histogram dispersion can be reduced, artifacts can be reduced. In this section, we introduce such a scheme called Generalized Partial Volume joint histogram Estimation (GPVE) (Chen and Varshney 2003). It turns out that PVI is a special case of GPVE scheme.

Before introducing the GPVE algorithm, let us first rewrite the PVI algorithm, proposed by Maes et al. (1997), for the 2D case. With reference to Fig. 3.11, let F and R be the floating image and the reference image respectively, and T_α be the transformation characterized by the parameter set α that is applied to the grid points of F. Assume that T_α maps the grid point (x_i, x_j) in image F onto the point $(y_i + \Delta_i, y_j + \Delta_j)$ in the image R, where (y_i, y_j) is a grid point in R and $0 \leq \Delta_i, \Delta_j < 1$. $\tilde{y}_1, \tilde{y}_2, \tilde{y}_3,$ and $\tilde{y}_4$ are the grid points of the reference image R that are closest to the transformed grid point $T_\alpha(\tilde{x})$ that splits the cell defined by grid points $\tilde{y}_1, \tilde{y}_2, \tilde{y}_3,$ and $\tilde{y}_4$ into four sub-cells having areas $\omega_1, \omega_2, \omega_3,$ and ω_4 with the constraint $\sum_i \omega_i (T_\alpha(\tilde{x})) = 1$. Now, let us express the original PVI algorithm described in (3.13) in terms of a kernel function f.

Let f be a triangular function defined by

$$f(t) = \begin{cases} 1 - t & \text{if } 0 \leq t \leq 1 \\ 1 + t & \text{if } -1 \leq t < 0 \\ 0 & \text{otherwise} \end{cases}, \quad (3.15)$$

then (3.14) can be written as,

$$h\left(F\left(x_i, x_j\right), R\left(y_i + p, y_j + q\right)\right) + = f\left(p - \Delta_i\right) f\left(q - \Delta_j\right) \quad \forall p, q \in Z, \quad (3.16)$$

where Z is the set of all integers. Notice that in (3.16), the increments are all zero except when $p, q \in \{0, 1\}$. In fact, when $p = 0$ and $q = 0$, the increment is ω_1, when $p = 0$ and $q = 1$, the increment is ω_2, when $p = 0$ and $q = 1$, the increment is ω_3 and when $p = 1$ and $q = 1$, the increment is ω_4. Next the GPVE algorithm for the 2D case is described in terms of a more general kernel function.

Let f be a real valued function that satisfies the following two conditions:

1) $f(x) \geq 0$, where x is a real number, (3.17)

2) $\sum_{n=-\infty}^{\infty} f(n + \Delta) = 1$, where n is an integer, $0 \leq \Delta < 1$, (3.18)

then for each grid point $\bar{x} = (x_1, x_2) \in X$ in the image F, the joint histogram h is updated in the following manner:

$$h\left(F\left(x_i, x_j\right), R\left(y_i + p, y_j + q\right)\right) + = f\left(p - \Delta_i\right) f\left(q - \Delta_j\right) \quad \forall p, q \in Z, \quad (3.19)$$

where f is referred to as the kernel function of GPVE and Z is the set of all integers. The first condition on f ensures that the joint histogram entries are non-negative while the second condition makes the sum of the updated amounts equal to one for each corresponding pair of points (x_i, x_j) in F and $(y_i + \Delta_i, y_j + \Delta_j)$ in R.

From this generalization, it can be seen that the PVI algorithm proposed by Maes et.al (1997) is a special case when f is a triangular function defined by (3.15). In GPVE, a family of functions called B-splines may be used as the kernel function f as it satisfies both the conditions in (3.17) and (3.18) and furthermore, it has finite support (the domain within which the function value is not zero). It is also interesting to note that the PVI algorithm corresponds to using a first order B-spline as the kernel function, which is a triangular function. Figure 3.12 shows the shapes of first, second, and third order B-splines. Notice that they have different supports. The size of the support determines

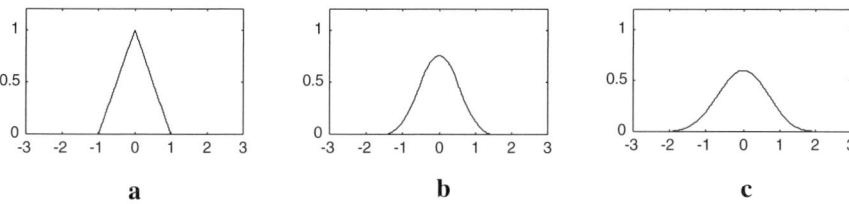

Fig. 3.12a–c. B-spline functions of different orders. **a** First order. **b** Second order. **c** Third order

how many neighboring grid points will be involved in the joint histogram updating procedure. For more details on B-spline functions, interested readers are referred to Unser et al. (1993a,b).

In the GPVE scheme, the kernel functions can be of different types along different directions. That is, we can rewrite (3.19) as

$$h(F(x_i, x_j), R(y_i + p, y_j + q)) + = f_1(p - \Delta_i)f_2(q - \Delta_j) \quad \forall p, q \in Z, \qquad (3.20)$$

where f_1 and f_2, can be different kernels. For example, if we know that the artifacts are to appear in the y-direction only, we can choose f_1 as the 1st order B-spline function but may choose f_2 as the 3rd order B-spline function.

Fig. 3.13 shows the grids in R (shown as "○") that are involved in updating the joint histogram in the 2D case using the 1st, 2nd, and 3rd order B-splines as the kernel functions. In each case, the transformed grid point of F appears at the center of each plot. Figure 3.13a shows the case when the transformed grid point of F is coincident with a grid point of R and Fig. 3.13b shows the case when the transformed grid point of F does not coincide with a grid point in R and is surrounded by four grid points in R. We observe that one to four entries of the joint histogram are involved in updating each pixel in F if the PVI algorithm (or the 1st order GPVE) is used. This is evident from Fig. 3.13a,b. In Fig. 3.13a, only the central pixel is involved in updating whereas in Fig. 3.13b all the four grid points surrounding the point marked by "∗" are involved in updating. Similarly, nine and four grid points will be involved in updating in Fig. 3.13a and Fig. 3.13b respectively when 2nd order GPVE is employed. The number of grid points involved in updating is determined by the size of the support of the kernel function, which is shown as the shaded area in Fig. 3.13a,b. Each side of the shaded region is four times the sample spacing for the case of 3rd order GPVE. In this case, 9 to 16 grid points are involved in updating as seen in Fig. 3.13. The ratios of the maximum number to minimum number of updated entries are 4, 2.25, and 1.78 when using the 1st, 2nd, and 3rd order GPVE respectively. The *reduction in the values of these ratios* gives GPVE the ability to reduce the artifacts. Intuitively, the ratio needs to be one to remove the artifacts completely because different numbers of updated entries introduce different amounts of dispersion in the joint histogram, and therefore influence the mutual information measure differently. However, in

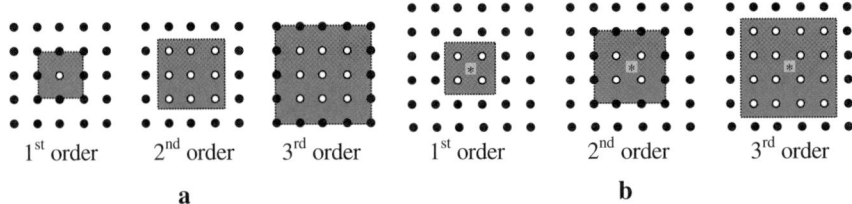

Fig. 3.13a,b. Grid points corresponding to R that are involved in updating the joint histogram in the 2D case. **a** When the transformed grid point is coincident with a grid point in R. **b** When the transformed grid point is surrounded by grid points in R

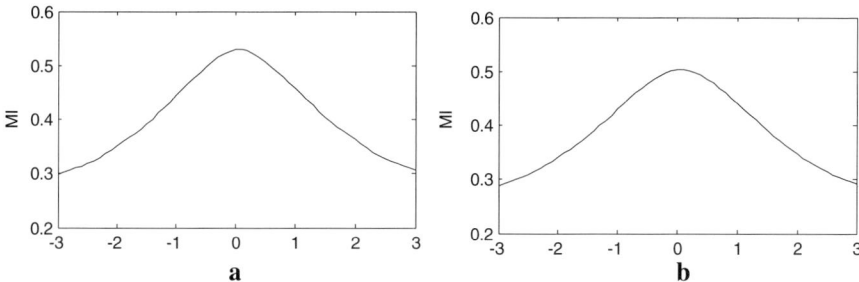

Fig. 3.14a,b. MI similarity measure plotted as a function of vertical displacement along *x*-axis. **a** 2nd order GPVE is employed. **b** 3rd order GPVE is employed

many cases, artifacts can hardly be seen when either the 2nd or the 3rd order GPVE is used. Figure 3.14a,b shows the MI similarity measure as a function of vertical displacement using the same image data used to produce Fig. 3.10a. Figure 3.14a is obtained using the 2nd order GPVE algorithm, and Fig. 3.14b results from the 3rd order GPVE algorithm. Clearly, artifacts can be hardly seen in either case. It is shown in Chen and Varshney (2003) that, for medical brain MR to CT image registration application, the use of higher order GPVE algorithm not only reduces artifacts visually, but also improves registration accuracy when it is affected by the artifact pattern resulting from the PVI method.

3.6
Optimization Issues in the Maximization of MI

According to the mutual information criterion for image registration, one needs to find the pose parameter set that results in the *global* maximum of the registration function as shown in Fig. 3.1 earlier. The existing local optimization algorithms may result in just a local maximum rather than a global maximum while the existing global optimization algorithms are very time consuming and lack an effective termination criterion. Therefore, almost all the intensity based registration algorithms that claim to be automatic run the risk of producing inaccurate registration results. Thus, to develop a reliable fully automated registration algorithm, a robust global optimizer is desirable.

In general, existing global optimization algorithms such as genetic algorithms (Michalewicz 1996) and simulated annealing (Farsaii and Sablauer 1998) are considered to be robust only when the process runs long enough such that it converges to the desired global optimum. In other words, if an optimization algorithm terminates too early, it is very likely that the global optimum has not been reached yet. On the other hand, if a strict termination criterion is adopted, despite having achieved the global optimum, the search will not end until the termination criterion is satisfied resulting in poor efficiency. One way to build a robust yet efficient global optimizer is to determine whether a local optimum of a function, once found, is actually the global opti-

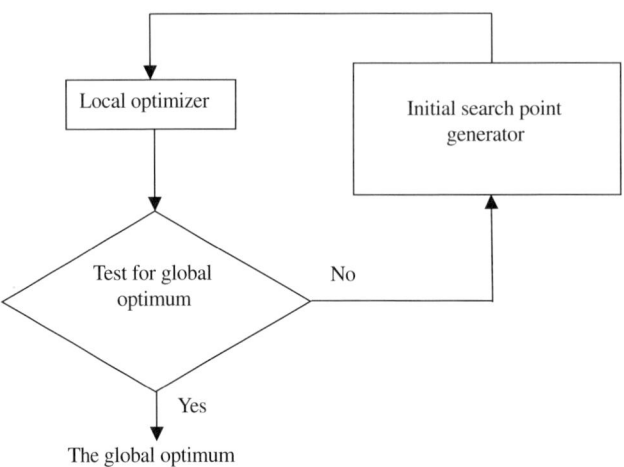

Fig. 3.15. A robust and efficient global optimization scheme

mum. If there is a way to distinguish the global optimum of a function from its local optimum, then a robust yet efficient global optimizer is achievable. Figure 3.15 shows such a global optimization scheme. It is efficient because a local optimizer is employed to accelerate convergence, and once the global optimum is reached, the search process terminates immediately. It is also robust because the process will not terminate until the global optimum is reached.

Generally, it is very difficult, if not impossible, to determine whether or not the optimum found for a function is global without evaluating the function completely. However, for the global image registration problem (a single transformation is sufficient for registering entire images), there exists a heuristic test to determine whether the optimum determined is global or not (Chen 2001b). The heuristic test algorithm may be described with the help of Fig. 3.16a that shows a function with several local maxima. The goal is to find the position of the global maximum. As mentioned earlier, to identify the global optimum of a function without completely evaluating the whole function is very hard, if not impossible. However, if a second function like the one (thick line) shown in Fig. 3.16b is available, then it is possible to identify the global maximum of the function represented by the thin line. If we observe Fig. 3.16b carefully, we can see the unique relationship of the two functions: *their global maxima appear at the same location.* Using this property, it is expedient to identify the global maximum of the function represented by the thin line. Fig. 3.16c illustrates this approach. For example, if we use a local optimizer and find Point 1 in Fig. 3.16c as an optimum and would like to know whether it is a global maximum or just a local maximum, we can use the position of Point 1 as the initial position and use a local optimizer to find a local maximum of the second function shown by the thick line. In this case, the local optimizer will result in Point 3 as the local maximum. Now we can compare the positions of Point 1 and Point 3. Since they are different in this case, we can conclude that Point 1 is just a local maximum

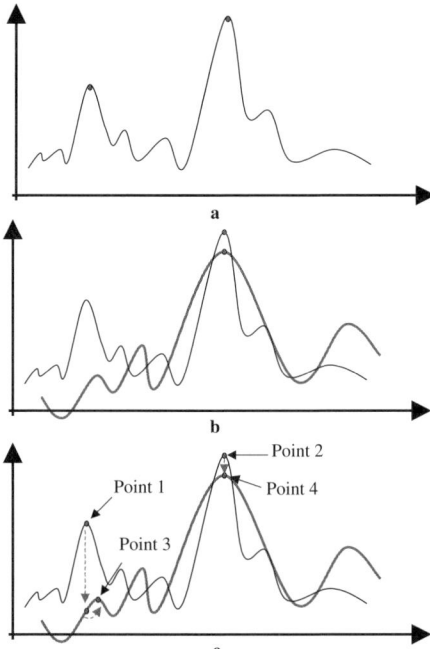

Fig. 3.16a–c. An illustration of the heuristic test to identify the global maximum. For a colored version of this figure, see the end of the book

rather than the global maximum. In the same manner, we can identify Point 2 as the global maximum of the function shown as the thin line curve in Fig. 3.16c. From this example, we see that it is possible to identify the global optimum of a function if *multiple uncorrelated functions whose global optima occur at the same location in the parameter space are available*. The question now is: how do we obtain the multiple uncorrelated functions? In general, it is difficult. However, it is possible to find such functions for image registration problems. Let us consider the example shown in Figs. 3.17 and 3.18. Figure 3.17a,b shows a pair of Radarsat SAR and IRS PAN images to be registered as shown earlier in Fig. 3.3a,b. Fig. 3.17c shows the MI similarity measure as a function of rotation angle after registration. If we partition the SAR image into two sub-images as shown in Fig. 3.18a,b and use each of them as the floating image, we attain two more MI registration functions as shown in Fig. 3.18c,d. Observing Fig. 3.17c together with Fig. 3.18c,d, we find that their global maxima occur at the same position (i.e. at about −22°).

The basic idea of this approach is the following: if two images are geometrically aligned through a global transformation T, then any corresponding portions of the two images (sub-images) are also geometrically aligned through T.

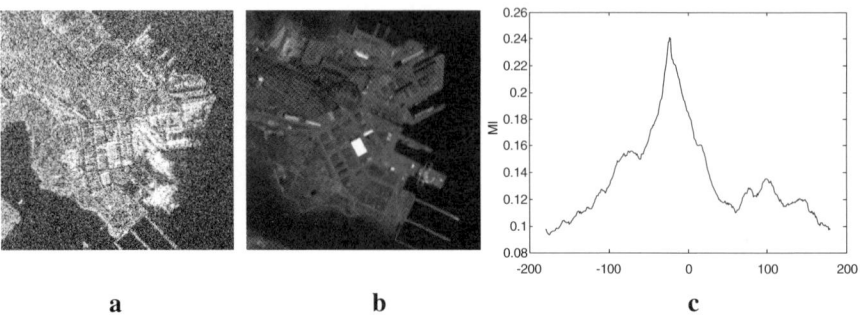

Fig. 3.17a–c. A pair of **a** Radarsat SAR (histogram equalized for display purposes), **b** IRS PAN images, **c** the MI registration function along the rotation axis. Rotation angle along absica is shown in degrees

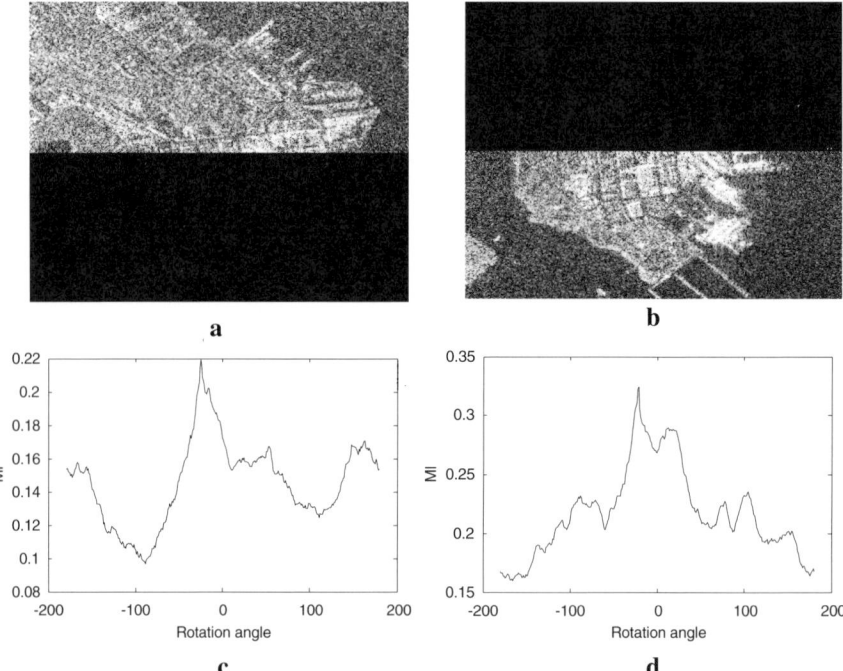

Fig. 3.18a–d. a and **b** Partitioned SAR images. **c** and **d** the corresponding MI registration functions along the rotation axis. In **c** and **d**, both rotation angles are represented in degrees

Mathematically, this can be expressed as follows:

$$\text{Let } \alpha^* = \arg\max_{\alpha}\left(I\left(F, R_{T_\alpha}\right)\right) \text{ and } \alpha_i^* = \arg\max_{\alpha}\left(I\left(F^i, R_{T_\alpha}\right)\right)$$
$$\text{then } \alpha^* = \alpha_i^* \text{ for every } i, \tag{3.21}$$

where F^i is a subimage of F with domain $X^i \in X$. Notice that the inverse of this statement is not true in general; however, we can use it as a test for the global optimum of a registration function. In practice, the probability of false alarm (the probability that an identified global optimum is just a local optimum) is very low (Chen and Varshney 2001b), especially when the dimension of the search space is high. Based on this idea, the global maximum of an MI based registration function can be distinguished from local maxima and a robust yet efficient global optimizer as the one shown in Fig. 3.15 can be successfully designed.

3.7
Summary

Most of the existing similarity measures for intensity based image registration problems depend on certain specific relationships between the intensities of the images to be registered. For example, mean squared difference assumes an identical relationship and cross correlation assumes a linear relationship. If these assumptions are not strictly satisfied, registration accuracy may be affected significantly. One example to illustrate this will be given in Chap. 7. In this chapter, we have introduced the use of mutual information as a similarity measure for image registration. It does not assume specific relationship between the intensity values of the images involved and hence, it is a very general similarity measure having been adopted for many different registration applications involving different types of imaging sensors.

Tasks involved in MI based registration include joint histogram estimation and global optimization of the MI similarity measure. These tasks have been discussed in detail in this chapter. A phenomenon called interpolation-induced artifacts was also discussed. A new joint histogram estimation scheme called generalized partial volume joint histogram estimation was described. It eliminates or reduces the severity of this phenomenon. To build a robust yet efficient global optimizer, a heuristic test to distinguish the global optimum from the local ones was presented. Image registration examples using the techniques introduced in this chapter will be provided in Chap. 7.

References

Chen H (2002) Mutual information based image registration with applications. PhD dissertation, Syracuse University

Chen H, Varshney PK (2000a) A pyramid approach for multimodality image registration based on mutual information. Proceedings of 3rd international conference on information fusion 1, pp MoD3 9–15

Chen H, Varshney PK (2002) Registration of multimodal brain images: some experimental results. Proceedings of SPIE Conference on Sensor Fusion: Architectures, Algorithms, and Applications 6, Orlando, FL, 4731, pp 122–133

Chen H, Varshney PK (2001a) Automatic two stage IR and MMW image registration algorithm for concealed weapon detection. IEE Proceedings, vision, image and signal processing 148(4): 209–216

Chen H, Varshney PK (2001b) A cooperative search algorithm for mutual information based image registration. Proceedings of SPIE Conference on Sensor Fusion: Architectures, Algorithms, and Applications 5, Orlando, FL, 4385 117–128, 2001.

Chen H, Varshney PK (2003) Mutual information based CT-MR brain image registration using generalized partial volume joint histogram estimation. IEEE Transactions on medical imaging 22(9): 1111–1119

Chen H, Varshney PK, Arora MK (2003a) Mutual information based image registration for remote sensing data. International Journal of Remote Sensing 24(18): 3701–3706

Chen H, Varshney PK, Arora MK (2003b) Automated registration of multi-temporal remote sensing images using mutual information. IEEE Transactions on Geoscience and Remote Sensing 41(11): 2445–2454

Collignon A, Maes F, Delaere D, Vandermeulen D, Suetens P, Marchal G (1995) Automated multimodality medical image registration using information theory. Proceedings of 14th International Conference on Information Processing in Medical Imaging (IPMI'95), Ile de Berder, France, pp. 263–274.

Cover TM, Thomas JA (1991) Elements of information theory, John Wiley and Sons, New York.

Farsaii B, Sablauer A (1998) Global cost optimization in image registration using Simulated Annealing. Proceedings of SPIE conference on mathematical modeling and estimation techniques in computer vision 3457, San Diego, California, pp 117–125

Holden M, Hill DLG, Denton ERE, Jarosz JM, Cox TCS, Rohlfing T, Goodey J, Hawkes DJ (2000) Voxel similarity measures for 3-D serial MR brain image registration. IEEE Transactions on Medical Imaging 19(2): 94–102.

Keys RG (1981) Cubic convolution interpolation for digital image processing. IEEE Transactions on Acoustics, Speech and Signal Processing 29(6): 1153–1160

Lehmann TM, Gonner C, Spitzer K (1999) Survey: interpolation methods in medical image processing. IEEE Transactions on Medical Imaging 18: 1049–1075

Maes F, Collignon A, Vandermeulen D, Marchal G, Suetens P (1997) Multimodality image registration by maximization of mutual information. IEEE Transactions on Medical Imaging 16(2): 187–197

Michalewicz Z (1996) Genetic algorithms + data structures = evolution programs, 3rd and extended edition. Springer, New York.

Pluim JPW, Maintz IBA, Viergever MA (2000) Interpolation artifacts in mutual information-based image registration. Computer vision and image understanding 77: 211–232

Studholme C, Hill DLG, Hawkes DJ (1997) Automated three-dimensional registration of magnetic resonance and positron emission tomography brain images by multiresolution optimization of voxel similarity measures. Medical Physics 24(1): 25–35

Unser M, Aldroubi A, Eden M (1993) B-spline signal processing: Part I-theory. IEEE Transactions on Signal Processing 41(2): 821–833

Unser M, Aldroubi A, Eden M (1993) B-spline signal processing: Part II-efficient design. IEEE Transactions on Signal Processing 41(2): 834–848

Viola P, Wells III WM (1995) Alignment by maximization of mutual information. Proceedings of 5th International Conference on Computer Vision, Cambridge, MA, pp. 16–23

Wells WM, Viola P, Atsumi H, Nakajima S (1996) Multi-modal volume registration by maximization of mutual information. Medical Image Analysis 1(1): 35–51

West J, et al (1997) Comparison and evaluation of retrospective intermodality brain image registration techniques. Journal of Computer Assisted Tomography 21(4): 554–566

CHAPTER 4
Independent Component Analysis

Stefan A. Robila

4.1
Introduction

Independent component analysis (ICA) is a multivariate data analysis method that, given a linear mixture of statistical independent sources, recovers these components by producing an unmixing matrix. Stemming from a more general problem called blind source separation (BSS), ICA has become increasingly popular in recent years with several excellent books (e.g. Cichocki and Amari 2002; Haykin 2000; Hyvärinen et al. 2001) and a large number of papers being published. Its attractiveness is explained by its relative simplicity as well as from a large number of application areas (Haykin 2000). For example, ICA has been successfully employed in sound separation (Lee 1998), financial forecasting (Back and Weingend 1997), biomedical data processing (Lee 1998), image filtering (Cichocki and Amari 2002), and remote sensing (Tu et al. 2001). While in many of these applications, the problems do not exactly fit the required setting, ICA based algorithms have been shown to be robust enough to produce accurate solutions. In fact, it is this robustness that has fueled the theoretical advances in this area (Haykin 2000).

In view of such success, and given the fact that hyperspectral data can be modeled as multivariate data, it is natural to examine how ICA can be used to process hyperspectral images. This chapter provides an overview of ICA and its application to the processing of hyperspectral images. The ICA concept is introduced in Sect. 4.2 followed by the description of some of the most popular algorithms. Finally, in the last section, a few applications are presented.

4.2
Concept of ICA

Consider a cocktail party in a room in which several persons are talking to each other. When they talk at the same time, their voices cannot be distinguished even with the help of several microphones. This is because the recorded sound signals will be mixtures of the *source* sound signals (voices). The exact manner in which the voices have been mixed is unknown and may depend on the distance from the microphone and other factors. We are, thus, faced with the problem of recovering unknown signals that were mixed in an unknown manner (Lee 1998). Our goal is to distinguish the conversation made by each person present in the room. This problem can formally be defined as follows.

Consider an m-dimensional random vector s, and a matrix A of size $n \times m$. The problem is to recover this pair (s, A) from the available n-dimensional observation vector x defined as:

$$x = As .\tag{4.1}$$

Generally, this problem is called *Blind Source Separation* (BSS). Each component of s corresponds to a source (thus there are m sources). The term blind indicates that little or no information on the mixing matrix A, or the source signals is available (Hyvärinen et al. 2001). The number of possible solutions for this problem can be infinite, i.e., for a given x there is an infinite number of pairs (s, A) that satisfy (4.1).

Consider now the situation when the *components* of the source s are assumed to be statistically *independent* meaning thereby that the probability density function of s, $p(s)$ can be expressed as:

$$p(s) = \prod_{i=1}^{m} p(s_i) .\tag{4.2}$$

In this case, the problem is called the independent component problem, and a solution to this problem is called an *Independent Component Analysis* (ICA) solution (Common 1994). In addition to the independence assumption, in order to provide a unique solution, ICA has the following restrictions (Hyvärinen et al. 2001):

1. All the components of s have non-Gaussian distribution

2. The number of sources is smaller or equal to the number of observations ($m \leq n$)

3. Only low noise is permitted

The above conditions are required to ensure the existence and the uniqueness of the ICA solution. The second restriction states that there should be enough observations available in order to be able to recover the sources. It is similar to the condition in linear algebra, where the number of equations needs to be at least equal to the number of variables. If this condition is not satisfied, the solution obtained for ICA is not unique (Lee 1998). For simplicity, in most applications, it is assumed that $m = n$ and that the mixing matrix A is invertible. In this case, once the matrix A is computed, its inverse can be used to retrieve the independent components. Thus, in the rest of this chapter, the matrix A will be assumed to be square. Extension of ICA to non-square matrix cases has been proposed in the literature (Common 1994; Girolami 2000; Chang et al. 2002).

The restriction regarding non-Gaussian components also contributes to the uniqueness of the solution. If non-Gaussian components are involved, the ICA solution consists of the unique independent non-Gaussian components and of mixtures of independent Gaussian components (Hyvärinen et al. 2001). However, when only one of the components is Gaussian, the ICA solution in terms of independent components can still be correctly obtained. Since in many signal and image processing applications, the noise is considered to be Gaussian, relaxing restriction i) to include a Gaussian component allows greater applicability of ICA.

A very simple method for selecting the non-Gaussian variables is based on *kurtosis*. Kurtosis is the fourth order central moment and, for a random variable u, is defined in the normalized form as (Common 1994):

$$K(u) = \frac{E\left\{(u - E\{u\})^4\right\}}{\left(E\left\{(u - E\{u\})^2\right\}\right)^2} - 3 \ . \tag{4.3}$$

In general, for a random variable u, $E\{f(u)\}$ denotes the expected value of $f(u)$ where $f(\cdot)$ is a function of u. With this notation, $E\{u\}$ is the expected value of u (or average of u for the discrete case).

For Gaussian distributed random variables, the kurtosis value is zero. The random variables with positive kurtosis values are called *super-Gaussian* (or *leptokurtic*). Their probability density functions have larger peak value at the mean and longer tails when compared with the Gaussian probability density function. The random variables with negative kurtosis values are called *sub-Gaussian* (or *platykurtic*). They are characterized by flatter probability density functions than that of the normal variables. Figure 4.1 presents examples of both super-Gaussian and sub-Gaussian probability density functions. The super-Gaussian variable is modeled using the Laplacian distribution and the sub-Gaussian is derived from a uniform distribution for the plotted interval.

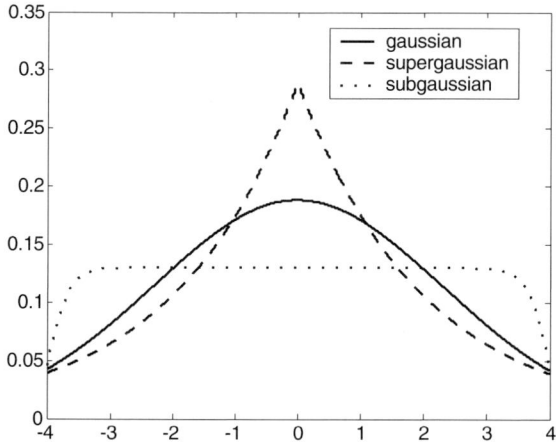

Fig. 4.1. Examples of super-Gaussian and sub-Gaussian probability density functions. The Gaussian distribution is provided for comparison

Kurtosis is highly sensitive to outliers. A small number of outliers can significantly affect its value. For example, Fig. 4.2a shows the histogram of a Gaussian random variable with zero mean and standard deviation one generated from 100,000 observations. Figure 4.2b shows the histogram of the same random variable perturbed by the addition of 500 observations generated by a Gaussian random variable with mean value as five and standard deviation as one. The perturbation is almost unnoticeable in the histogram even when enlarged (Fig. 4.2d). However, this small perturbation was sufficient to change the kurtosis value from close to zero (0.0022) (before) to (2.5639) (after) showing its sensitivity to outliers.

The kurtosis for super-Gaussian variables can have large positive values (in theory up to infinity) whereas it is limited to −2 for sub-Gaussian variables (Hyvärinen et al. 2001). There is a possibility for a random variable to have the kurtosis equal to zero without being Gaussian. However, this occurrence is rare.

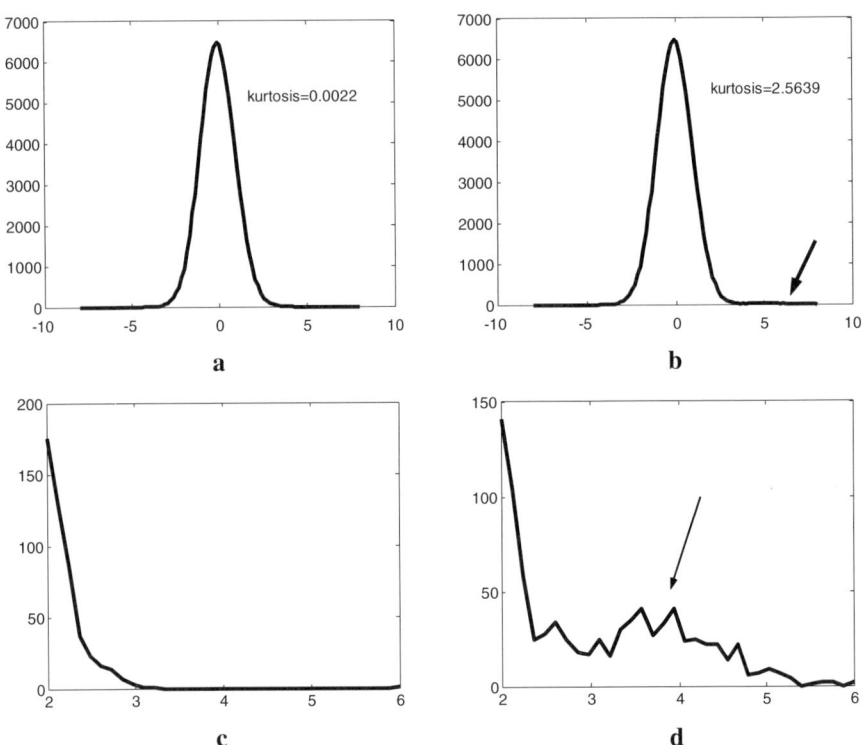

Fig. 4.2. Example of kurtosis instability. **a** Gaussian distributed random variable (mean = 0, standard deviation = 1 and number of observations = 100,000) **b** A perturbation of 500 observations is added on the right tail **c, d** Area enlarged of the two graphs (**a** and **b** respectively) showing that the perturbation is almost unnoticeable in the histogram

Independent Component Analysis

As mentioned earlier, when some of the resulting components are non-Gaussian in nature, kurtosis can be used to identify the non-Gaussian independent components from the ICA solution. Since a linear mixture of independent Gaussian components is also Gaussian, its kurtosis will be zero. Selecting the components with non-zero kurtosis values is equivalent to identifying the non-Gaussian independent components.

If the assumptions for ICA are satisfied, a relatively unique ICA solution can be obtained. The term "relative" indicates that multiplying a component with a scalar or permuting the order of the components preserves independence. Therefore, the solution is, in fact, identifiable up to scaling and multiplication (Hyvärinen et al. 2001; Lee 1998).

In practice, it is unlikely that the observed data are characterized as a linear mixture of components that are independent in that they satisfy (4.2). This may be due to the fact that the components are not independent, that there is noise or interference present and that the data available are of discrete nature. Even in these situations, the ICA algorithm provides a solution where the components are "as independent as possible", i.e., dependence among them is as small as possible.

4.3
ICA Algorithms

We proceed to solve the ICA problem stated in the previous section. The first step is to decorrelate and sphere the observed data, thereby eliminating the first and second order statistics (i.e. mean and standard deviation). The next step is to determine the matrix A such that its inverse allows the recovery of the independent components. A is usually obtained through gradient based techniques that optimize a certain cost function $c(\cdot)$:

$$\Delta A = \frac{\partial c(A, x)}{\partial A} . \quad (4.4)$$

The cost function is designed such that its optimization corresponds to achieving independence. One direct approach for the solution of the ICA problem is to consider $c(\cdot)$ as the mutual information of the components of $A^{-1}x$. This method is described in detail in Sect. 4.3.2. A second approach is based on the observation that projecting the data x into components, which are decorrelated and are as non-Gaussian as possible, also leads to independence (see Sect. 4.3.3) Both methods can be shown to be equivalent and choice of one over the other, or modification of either of them depends on the specific problem or the preference of the user.

4.3.1
Preprocessing using PCA

Whitening (decorrelation and sphering) of the data is usually performed before the application of ICA as a mean to reduce the effect of first and second order statistics, and to speed up the convergence process. A random vector is said to be *white* if the mean of each component is zero and the covariance matrix is the identity matrix. Among the various choices for preprocessing techniques, principal component analysis (PCA) (Hyvärinen et al. 2001; Chang et al. 2002) is widely used. This method has the additional advantage of eliminating dependence up to the second order between the components (i.e. the components are decorrelated).

PCA attempts to solve the following problem. For the multidimensional random vector x find a linear transform W so that the components obtained are uncorrelated (Richards and Jia 1999):

$$y = Wx \tag{4.5}$$

such that

$$\sum\nolimits_y = E\left\{(y - E\{y\})(y - E\{y\})^T\right\} \tag{4.6}$$

is diagonal. The vector of expected values $E\{y\}$ and the covariance matrix Σ_y can be expressed in terms of the vector of expected values and the covariance matrix for x:

$$E\{y\} = E\{Wx\} = WE\{x\}, \tag{4.7}$$

$$\sum\nolimits_y = W \sum\nolimits_x W^T . \tag{4.8}$$

If A_x is the matrix denoting the normalized eigenvectors for the covariance matrix Σ_x, we have (Rencher 1995):

$$C_x = A_x^T \sum\nolimits_x A_x , \tag{4.9}$$

where C_x is the corresponding diagonal eigenvalue matrix. Since in (4.6), Σ_y is required to be diagonal, we note that $W = A_x^T$ leads to the PCA solution:

$$y = A_x^T x . \tag{4.10}$$

The components of y are called principal components and the eigenvector matrix A_x is called the principal component transform (PCT). The eigenvalues of the covariance matrix for x correspond to the variances of the principal components. When these eigenvalues are arranged in a decreasing order (along with the corresponding permutation of the eigenvectors), we get the components of y sorted in the decreasing order of their variance. High variance is usually associated with high signal to noise ratio. Therefore, it is useful to consider the highest variance components for further processing as they are expected to contain most of the information. For this reason, PCA has been considered as a possible tool for data compression or feature extraction (by dropping the lowest variance components).

It is also interesting to note that ICA is frequently compared with PCA. The similarity between the two originates from their goals – that is they both try to

Independent Component Analysis

minimize the dependence between the components of the data. In the case of PCA, dependence minimization is achieved when the covariance between the components is zero, whereas ICA considers the components to be independent when the probability density function of the vector can be expressed as the product of the probability density functions of the components (see (4.2)). When the data is Gaussian, decorrelation is equivalent to independence, and thus PCA can be considered to be equivalent to ICA. However, when the data is non-Gaussian, PCA does not fully achieve independence. This is because, in this case, dependence in the data needs to be characterized through third and fourth order statistical measures, which is not the case with PCA that depends on second order statistics only.

Nevertheless, decorrelation prior to ICA processing allows ICA algorithms to achieve faster convergence. In addition, the data is also sphered. This transform is given by (Hyvärinen et al. 2001):

$$x' = C_x^{-\frac{1}{2}} A_x^T (x - E\{x\}) . \tag{4.11}$$

Once the data is transformed, we are ready to proceed with the ICA solution.

4.3.2
Information Minimization Solution for ICA

For an m dimensional random vector u, the mutual information of its components is defined as:

$$I(u_1, \ldots, u_m) = E\left\{\log \frac{p(u)}{\prod_{i=1}^{m} p(u_i)}\right\} . \tag{4.12}$$

The mutual information is the most natural way of assessing statistical independence of random variables (Hyvärinen et al. 2001). When the components are independent, the ratio inside the logarithm in (4.12) reduces to one, and thus the mutual information becomes zero. It should be pointed out that $I(\cdot)$ is lower bounded by zero. It can also be shown that when the mutual information is zero, the components of u are independent (Lee 1998). This property of the mutual information is used here to achieve independence of components.

The mutual information can also be defined based on the entropy. For a random variable u, the entropy of u is defined as:

$$H(u) = -E\{\log(p(u))\} . \tag{4.13}$$

Similarly, for a random vector u, the joint entropy of u is defined as:

$$H(u) = -E\{\log(p(u))\} . \tag{4.14}$$

In this case, the mutual information of the components of u can be expressed as the difference between the sum of marginal (individual) entropies and the joint entropy:

$$I(u_1,\ldots,u_m) = \sum_{i=1}^{m} H(u_i) - H(u) . \tag{4.15}$$

The solution of the ICA problem can be obtained by minimizing the mutual information and making it as close to zero as possible. Thus, we define the minimal mutual information (MMI) problem as follows. For an n dimensional random vector x, find a pair (u, W), where u is an m dimensional random vector, and W is an $m \times n$ matrix, such that:

$$u = Wx \tag{4.16}$$

and

$$I(u_1,\ldots,u_m) = \min \{I(v_1,\ldots,v_m) | v = Vx\} , \tag{4.17}$$

where V is a $m \times n$ matrix.

The MMI problem can be used to obtain the solution for the ICA problem. From (4.1), if (s, A) is the solution to the ICA problem, under the assumption that A is invertible, we have:

$$s = A^{-1}x , \tag{4.18}$$

where the components of s are independent. The MMI problem defined by (4.16) is identical to the ICA problem if s and A^{-1} are set equal to u and W, respectively. Since the components of s are independent, the mutual information is zero, i.e.:

$$I(s_1,\ldots,s_m) = 0 . \tag{4.19}$$

In this case, s satisfies the second condition for the MMI solution (4.17). Hence, the ICA solution (s, A^{-1}) can be obtained by solving the MMI problem. The equivalence between the ICA and the MMI problems allows the development of a practical procedure to solve the ICA problem. In other words, the ICA solution is found by minimizing the mutual information. If MI reaches the minimum achievable value of zero, complete independence of the components is achieved. Otherwise, the components are as independent as possible. This will occur in situations where complete independence is not achievable (Hyvärinen et al. 2001; Lee 1998).

Let us now develop an algorithm for the solution of the MMI problem. The goal is to determine the gradient of the mutual information with respect to the elements of W. Once the gradient is computed, it is used as the iterative step for updating the elements of W in the gradient-based optimization algorithm:

$$W = W + \Delta W = W - \frac{\partial I(u_1,\ldots,u_m)}{\partial W} . \tag{4.20}$$

Independent Component Analysis

In this case, the role of the cost function $c(\cdot)$ in (4.4) is played by the mutual information I. In order to compute the gradient, expand the mutual information as (Papoulis 1991):

$$I(u_1, \ldots, u_m) = -H(\mathbf{u}) + \sum_{i=1}^{m} H(u_i)$$

$$= E\{\log(p(\mathbf{u}))\} - \sum_{i=1}^{m} E\{\log(p(u_i))\}. \tag{4.21}$$

Since $\mathbf{u}$ is a function of $\mathbf{x}$ when both have the same number of components (i.e. $m = n$), we can express $p(\mathbf{u})$ using $p(\mathbf{x})$ and the Jacobian matrix $J(\mathbf{x})$ of $\mathbf{u}$ with respect to $\mathbf{x}$:

$$p(\mathbf{u}) = \frac{p(\mathbf{x})}{\det J(\mathbf{x})}, \tag{4.22}$$

where 'det' represents the determinant and

$$J(\mathbf{x}) = \begin{pmatrix} \frac{\partial u_1}{\partial x_1} & \cdots & \frac{\partial u_1}{\partial x_n} \\ \vdots & & \vdots \\ \frac{\partial u_n}{\partial x_1} & \cdots & \frac{\partial u_n}{\partial x_n} \end{pmatrix} = \begin{pmatrix} w_{11} & \cdots & w_{1n} \\ \vdots & & \vdots \\ w_{n1} & \cdots & w_{nn} \end{pmatrix} = \mathbf{W}. \tag{4.23}$$

Using (4.22) and (4.23), express the first term of the mutual information given in (4.21) using $\mathbf{x}$ and $\mathbf{W}$:

$$I(u_1, \ldots, u_n) = E\{\log(p(\mathbf{x}))\} - \log(\det \mathbf{W}) - \sum_{i=1}^{n} E\{\log(p(u_i))\}. \tag{4.24}$$

The gradient of $I(\cdot)$ with respect to $\mathbf{W}$ can be expressed as:

$$\frac{\partial I(u_1, \ldots, u_n)}{\partial \mathbf{W}} = \frac{\partial E\{\log(p(\mathbf{x}))\}}{\partial \mathbf{W}} - \frac{\partial \log(\det \mathbf{W})}{\partial \mathbf{W}} - \frac{\partial \sum_{i=1}^{n} E\{\log(p(u_i))\}}{\partial \mathbf{W}}$$

$$= -\frac{\partial \log(\det \mathbf{W})}{\partial \mathbf{W}} - \sum_{i=1}^{n} \frac{\partial E\{\log(p(u_i))\}}{\partial \mathbf{W}}. \tag{4.25}$$

Since the first term $E\{\log(p(\mathbf{x}))\}$ does not involve $\mathbf{W}$, we will analyze the two remaining terms separately. The first term becomes (Lee 1998):

$$\frac{\partial \log(\det(\mathbf{W}))}{\partial \mathbf{W}} = \frac{1}{\det \mathbf{W}} \frac{\partial \det \mathbf{W}}{\partial \mathbf{W}} = \frac{1}{\det(\mathbf{W})} (adj(\mathbf{W}))^T = (\mathbf{W}^{-1})^T, \tag{4.26}$$

where adj represents the adjoint of the matrix.

The second term in (4.25) can be further developed:

$$\sum_{i=1}^{n} \frac{\partial E\{\log(p(u_i))\}}{\partial W} = \sum_{i=1}^{n} E\left\{\frac{\partial \log(p(u_i))}{\partial W}\right\} = \sum_{i=1}^{n} E\left\{\frac{1}{p(u_i)} \frac{\partial p(u_i)}{\partial W}\right\}$$

$$= \sum_{i=1}^{n} E\left\{\frac{1}{p(u_i)} \frac{\partial p(u_i)}{\partial u_i} \frac{\partial u_i}{\partial W}\right\}. \qquad (4.27)$$

The derivative of u_i with respect to the (i,j) element of w, (i.e. $w_{j,k}$) can be computed as follows:

$$\frac{\partial u_i}{\partial w_{j,k}} = \frac{\partial \left(\sum_{q=1}^{n} w_{i,q} x_q\right)}{\partial w_{j,k}} = \sum_{q=1}^{n} \frac{\partial (w_{i,q} x_q)}{\partial w_{j,k}} = \begin{cases} x_k & \text{if } i = j \\ 0 & \text{if } i \neq j \end{cases}. \qquad (4.28)$$

Now, define a family of nonlinear functions $g_i(\cdot)$ such that they approximate the probability density function for each u_i. Using the above result, the second term in (4.25) becomes:

$$\sum_{i=1}^{n} \frac{\partial E\{\log(p(u_i))\}}{\partial W} \approx \sum_{i=1}^{n} E\left\{\frac{1}{g_i(u_i)} \frac{\partial g_i(u_i)}{\partial u_i} \frac{\partial u_i}{\partial W}\right\} = E\left\{\frac{1}{g(u)} \frac{\partial g(u)}{\partial u} x^T\right\}, \qquad (4.29)$$

where $g(u)$ denotes $(g_1(u_1), \ldots, g_n(u_n))$. Finally, compute an approximation of the gradient of $I(\cdot)$ with respect to W and obtain the update step:

$$\Delta W \approx -\frac{\partial I(u)}{\partial W} = (W^{-1})^T + E\left\{\left(\frac{1}{g(u)} \frac{\partial g(u)}{\partial u}\right) \cdot x^T\right\}. \qquad (4.30)$$

In the form given in (4.30), a matrix inversion is required. This may result in slow processing. To speed up, the update step is multiplied by the 'natural gradient' $W^T W$ (Lee 1998):

$$\Delta W \approx -\frac{\partial I(u)}{\partial W} W^T W = (W^{-1})^T W^T W + E\left\{\left(\frac{1}{g(u)} \frac{\partial g(u)}{\partial u}\right) x^T W^T\right\} W$$

$$= W + E\{h(u) u^T\} W, \qquad (4.31)$$

where

$$h(u) = \frac{1}{g(u)} \frac{\partial g(u)}{\partial u}.$$

It can be proved that multiplication with the natural gradient not only preserves the direction of the gradient but also speeds up the convergence process (Lee 1998). The MMI algorithm for ICA will repeatedly perform an update of the matrix W:

$$W = W + k\Delta W, \qquad (4.32)$$

where the update step is given by (4.31) and k is an update coefficient (< 1) that controls the convergence speed.

The above update procedure does not restrict the search space to vectors of uncorrelated components. This contradicts our earlier statement that ICA is an extension of decorrelation methods (like PCA). However, if the components of u are uncorrelated, since x is also uncorrelated (after PCA processing) we have:

$$\sum_u = W \sum_x W^T = WIW^T = WW^T. \qquad (4.33)$$

Therefore, for the resulting components to be decorrelated, the matrix W needs to be orthogonal. It makes sense to restrict the optimization space only to orthogonal matrices, by adding an additional step in which W is orthogonalized after each iteration:

$$W = \left(WW^T\right)^{-\frac{1}{2}} W. \qquad (4.34)$$

In Hyvärinen et al. (2001), a direct solution to the orthogonalization is given by:

$$W \leftarrow \left(A_W C_W^{-\frac{1}{2}} A_W^T\right) W, \qquad (4.35)$$

where C_W and A_W are the eigenvalue and eigenvector matrices respectively for WW^T.

Results available in the literature indicate that full independence is not always achieved by the ICA solution obtained by the above algorithm even under ideal conditions (when the components are independent and there is no noise) (Hyvärinen et al. 2001; Lee 1998). This is because, similar to all other gradient-based algorithms, the MMI algorithm may converge to a local minimum. To reduce the likelihood of this, remedial methods (for example, the use of variable update coefficients) can be employed. This way, the likelihood of achieving full independence is increased (Hyvärinen et al. 2001).

Suitable selection of the nonlinear functions $g_i(\cdot)$ is essential to obtain an accurate solution from the algorithm since they need to approximate the probability density functions of the components of u. In Cochocki and Amari (2002), it was suggested to model the probability density functions of the components by a weighted sum of parametric logistic functions. However, the estimation of the weights proved to be computationally expensive and the accuracy of the results did not always improve (Lee 1998). On the other hand, experimental results have indicated that even with a single fixed nonlinearity, the

MMI algorithm converges very close to an ICA solution. This is very important since the use of a single nonlinearity implies that the estimation of the probability density functions does not have to be very precise. The success of using only one function can be explained by the fact that a probability density function mismatch may still lead to a solution that is sufficient for our purposes because its corresponding independent sources are scaled versions of the initial independent sources. It is, however, important to indicate that not all functions are equal. For example, a super-Gaussian-like function may not model a sub-Gaussian source correctly. Therefore, remedial approaches that involve prediction of the nature of the sources along with a switch between sub-Gaussian and super-Gaussian functions have been proposed (Lee 1998).

Algorithms, other than the MMI algorithm have also been proposed. One such algorithm is the *Information Maximization ICA (Infomax)* algorithm presented in Bell and Sejnowski (1995). While the update formula is identical to the one generated through MMI, the setting for the Infomax algorithm is slightly different. Following the transformation through W, the random vector u is further processed through a set of nonlinear functions $f_i(\cdot)$ (see Fig. 4.3). If these functions are invertible, the independence of the components of y is equivalent to the independence of the components of u.

To understand how the nonlinearity can be used, it is interesting to look at the mutual information for the components of y (Bell and Sejnowski 1995):

$$I(y_1,\ldots,y_n) = -H(y) + \sum_{i=1}^{n} H(y_i) = -H(y) + \sum_{i=1}^{n} E\{\log(p(y_i))\}$$

$$= -H(y) - \sum_{i=1}^{n} E\left\{\log\left(p(u_i) \bigg/ \frac{\partial f_i(u_i)}{\partial u_i}\right)\right\} \qquad (4.36)$$

since, according to Popoulis (1991), for a random variable v and a differentiable function $h(\cdot)$:

$$p(h(v)) = \left(p(v) \bigg/ \frac{\partial h(v)}{\partial v}\right). \qquad (4.37)$$

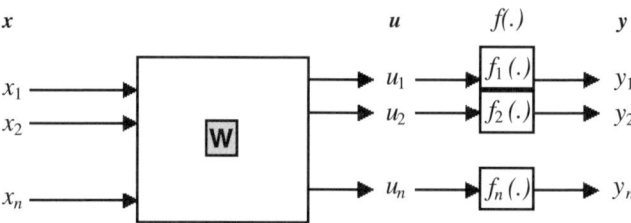

Fig. 4.3. Infomax algorithm for ICA solution. Following the multiplication with W, the data is further transformed through nonlinear functions. Independence of the components of u is equivalent to the independence of the components of y. Infomax algorithm proceeds to modify W such that components of y become independent

If, for all i, the derivative of $f_i(u_i)$ with respect to u_i matches the probability density function of u_i, i.e.

$$p(u_i) = \frac{\partial f_i(u_i)}{\partial u_i}, \qquad (4.38)$$

then all the marginal entropies become zero and

$$I(y_1, \ldots, y_n) = -H(y). \qquad (4.39)$$

Thus, maximization of the joint entropy becomes equivalent to the minimization of the mutual information. If the nonlinearities do not match the probability density functions, then the algorithm may introduce errors in the estimation of the independent components.

The gradient-based iterative step for the Infomax algorithm (Bell and Sejnowski 1995) is given as

$$\Delta W \approx \frac{\partial H(y)}{\partial W} W^T W = W + E\left\{\left(\frac{\frac{\partial^2 f(u)}{\partial^2 u}}{\frac{\partial f(u)}{\partial u}}\right)\right\} u^T W, \qquad (4.40)$$

which is identical to the one in (4.31) with the observation that $g_i(\cdot)$ are the derivatives of $f_i(\cdot)$.

4.3.3
ICA Solution through Non-Gaussianity Maximization

A different approach for the solution of the ICA problem is to find a linear transform such that the resulting components are uncorrelated and as non-Gaussian as possible (Hyvärinen et al. 2001).

According to the central limit theorem (Hyvärinen et al. 2001), the distribution of any linear mixture of non-Gaussian independent components tends to be closer to the Gaussian distribution than the distributions of the initial components. In the effort to recover the independent components from the observed vector x, we are trying to find the transform W such that:

$$u = Wx = WAs \qquad (4.41)$$

and that the components of u are as independent as possible. In (4.41), s is the vector of original independent components, and $x = As$ is the vector of observed components.

W and A are linear transforms, so the components of u will be more Gaussian than the components of s. Since we assume that the components of s are non-Gaussian and independent, the maximum non-Gaussianity for u is achieved when the components of u are equal to the components of s, i.e. when:

$$W = A^{-1}. \qquad (4.42)$$

To find the independent components, we proceed to determine the projection vector w_1 such that $w_1^T x$ is as non-Gaussian as possible. We then search for a second projection vector w_2 such that $w_2^T x$ is as non-Gaussian as possible, making sure that the projection is orthogonal to the one already found. The process continues until all the projection vectors $(w_1, w_2, w_3, \ldots, w_m)$ are found. It can be shown (Hyvärinen et al. 2001) that, if we consider:

$$u = Wx = WAs, \qquad (4.43)$$

where $W = (w_1, w_2, w_3, \ldots, w_m)^T$, the components of u are estimates of the original independent components (s).

There are several methods for quantifying the non-Gaussianity. One of them is based on the negentropy, (i.e. the distance between the probability density function of the component and the Gaussian probability density function) (Common 1994). For a random vector u the negentropy is defined as:

$$J(u) = H(u_G) - H(u), \qquad (4.44)$$

where u_G denotes a Gaussian variable with the same covariance as u.

While negentropy is a good measure for non-Gaussianity, it is difficult to use it in practice due to the need for an accurate approximation for the probability density function. Instead, approximations of the entropy have been suggested. One approximation is based on general nonquadratic functions and the other one is based on skewness and kurtosis. In both cases, once the approximate formula for negentropy is computed a gradient-based algorithm is devised.

In the first case, the negentropy for a random variable u is approximated as (Hyvärinen et al. 2001):

$$J(u) = \left[E\{G(u)\} - E\{G(u_G)\} \right]^2, \qquad (4.45)$$

where $G(\cdot)$ is a general nonquadratic function. It is interesting to note that again, as in the case of information maximization, the accuracy of the results relies heavily on the choice of the function. Additionally, in this case, the independent components are sequentially generated, making sure that they remain uncorrelated.

Based on (4.44) we proceed to generate the expression for the update step (Hyvärinen et al. 2001):

$$\begin{aligned}
\Delta w = \frac{\partial J(u)}{\partial w} &= \left[E\{G(u)\} - E\{G(u_G)\} \right] E \left\{ \frac{\partial G(u)}{\partial w} \right\} \\
&= \left[E\{G(u)\} - E\{G(u_G)\} \right] E \left\{ \frac{\partial u}{\partial w} \frac{\partial G(u)}{\partial u} \right\} \\
&= \left[E\{G(u)\} - E\{G(u_G)\} \right] E \left\{ xg\left(w^T x\right) \right\}, \qquad (4.46)
\end{aligned}$$

followed by the normalization ($w = w/||w||$). In the above expression, $g(\cdot)$ represents the derivative of $G(\cdot)$.

In the second case, the negentropy is approximated using skewness and kurtosis:

$$J(x) = \frac{1}{12} E\left\{x^3\right\} + \frac{1}{48} K(x)^2, \qquad (4.47)$$

where $E\left\{x^3\right\}$ estimates the skewness and $K(\cdot)$ denotes the kurtosis. If the data is initially whitened, $E\left\{x^3\right\} = 0$, (4.47) becomes:

$$J(x) = K(x)^2. \qquad (4.48)$$

In most of the ICA algorithms that are based on non-Gaussianity maximization, the components of u in (4.42) are generated in a sequential order with each component constrained to be orthogonal to the ones previously generated. Other algorithms that allow simultaneous generation of the components have also been proposed (Hyvärinen et al. 2001; Chang et al. 2002).

To summarize, ICA is an active area of research and new solutions are being continuously developed. Therefore, only the most common approaches to solving the ICA problem have been presented here. For additional approaches, see Cichocki and Amari (2002); Haykin (2000); Hyvärinen et al. (2001); Lee 1998).

4.4
Application of ICA to Hyperspectral Imagery

A limited amount of research on the use of ICA to process hyperspectral data has been reported in the literature (e. g. Tu et al. 2001; Chang et al. 2002; Chen and Zhang 1999; Robila and Varshney 2002; Bayliss et al. 1997; Robila et al. 2000; Parra et al. 2000 Healy and Kuan 2002). There are many reasons for the scarcity of research in this important area. First, both ICA and hyperspectral imaging have evolved mostly in the last few years and the research connecting them is still at the incipient stage. Second, the size of the hyperspectral data to be processed coupled with the iterative nature of ICA leads to very large computational times, reducing the appeal of the method. Third, the initial research has focused on direct application of ICA without considering how the hyperspectral data is modeled, how the resulting components are interpreted and whether modifications of the original algorithms are required (Tu et al. 2001; Chen and Zhang 1999; Bayliss et al. 1997). Only recently, ICA has been linked with the linear mixing model used in hyperspectral image processing where it has been shown to provide an approximation of the solution to the linear unmixing problem (Chang et al. 2002).

Most of the research investigating the applicability of ICA to hyperspectral imagery has been focused on the maximization of non-Gaussianity (i. e. kurtosis) of each component. This approach is similar to projection pursuit (Hyvärinen et al. 2001) and provides a good approximation of the ICA solution only when the most non-Gaussian components correspond to the independent components of interest. Unfortunately, hyperspectral data are sometimes affected by impulse noise (Chang et al. 2002) that results in components with very high kurtosis. With respect to this, an algorithm using minimization of mutual information may prove to be more robust than the ones based on non-Gaussianity (Hyvärinen et al. 2001).

The success of the application of ICA to hyperspectral imagery relies on how well a hyperspectral image cube can be modeled to fit the ICA requirements. We discuss two different models, one is based on feature extraction to reduce the dimensionality of the dataset and is inspired by PCA. The other is based on the concept of linear mixture model frequently used to determine the class proportions of mixed pixels or perform fuzzy classification (see Sect. 2.6.4 in Chap. 2) in the image.

4.4.1
Feature Extraction Based Model

Let us construct a vector space of size equal to the number of spectral bands. A pixel vector in the hyperspectral image is a point in such a space, with each coordinate given by the corresponding intensity value (Richards and Jia 1999). The visualization of the image cube plotted in the vector space provides useful information related to the data. In the two-dimensional case, a grouping of the points along a line segment signifies that the bands are highly correlated. This is usually the case with adjacent bands in a hyperspectral image cube since the adjacency of corresponding spectral bandwidths does not yield a major change in the reflectance of the objects (see Fig. 4.4a). On the other hand, for spectral bands with spectral wavelength ranges that are very distant, the plot of corresponding bands will have scattered points. This is the situation shown in Fig. 4.4b where the two bands belong to different spectral ranges (visible

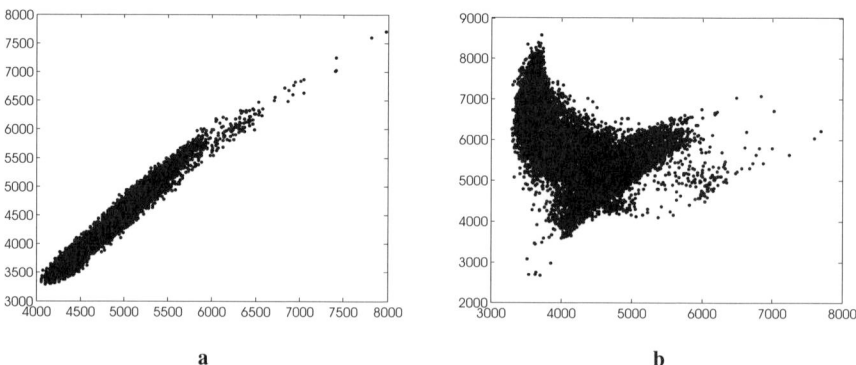

Fig. 4.4a,b. Model of a hyperspectral image as a random vector. **a** Two dimensional plot of the pixel vectors for adjacent bands (blue and green). **b** Two-dimensional plot for the distant bands (green and near infrared). The range for pixel values is due to the calibration process that transformed the raw data to compensate for illumination conditions, angle of view, etc.

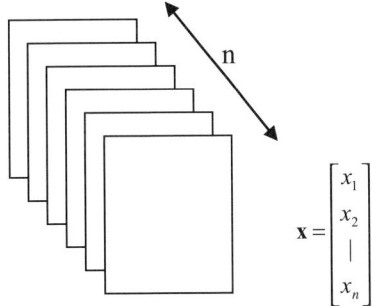

Fig. 4.5. Model of an *n*-band multispectral/hyperspectral data set as a random vector. Each band is associated with a component of the random vector. Each realization of the random vector corresponds to a pixel vector from the image cube

and near infrared). Same conclusions may be drawn by inspecting data in more than two dimensions. However, the visualization becomes difficult, if not impossible, once the dimensionality exceeds three.

Thus, a hyperspectral image cube can be associated with a random vector with each pixel vector corresponding to an observation. The number of observations is, therefore, equal to the size (in pixels) of the image. Individually, for each component (or *feature*) of the random vector, the observations are the intensity values of the pixels in the associated spectral band (see Fig. 4.5). This model can then be used for extracting features containing most of the information by applying ICA, as is done in PCA (Richards and Jia 1999).

4.4.2
Linear Mixture Model Based Model

A second model can be drawn from the linear mixture model (LMM). In LMM, each pixel vector is assumed to be a linear combination of a finite set of endmembers. For a specific endmember, the contribution (abundance) corresponding to each pixel vector in the hyperspectral image can be seen as an image band itself. The relationship between LMM and ICA is achieved by considering the columns in the ICA mixing matrix to represent endmembers as described in the linear mixture model, and each independent component as the abundance of an endmember (Chang et al. 2002).

A precise equivalence between LMM and ICA is not possible. In LMM, the abundances of the endmembers are not required to be independent. However, since in LMM we are looking for the most representative pixel vectors, it seems natural to modify the model by assuming that the abundance of one endmember in a specific pixel does not provide any information regarding the abundance of other endmembers for that pixel. Another major difference is the presence of noise in the LMM, which is not considered in ICA. However, we may include noise in ICA as Gaussian noise with its abundances assumed to be independent of one of the endmembers. This satisfies one of the restrictions of the ICA model that the model is allowed to have one of the components as Gaussian.

In the linear mixture model, for each image pixel, the corresponding endmember abundances need to be positive and should sum to one. Neither of the conditions may be satisfied by the ICA generated components. The positivity of the abundances can be achieved by simple transformations that preserve independence. For example, for each independent component v, the following transformation can be used:

$$v = \frac{v - \min_v}{\max_v - \min_v}, \qquad (4.49)$$

where $\min_v$ and $\max_v$ correspond to the minimum and maximum values of the random variable v.

The additivity constraint is more difficult to satisfy. A solution suggested in (Healy and Kuan 2002) lets ICA recover all but one of the endmembers. The abundance of the remaining endmember can be determined by subtracting the other abundances from a constant. Unfortunately, in this case, the ICA model is no longer valid since we allow the existence of a component that is not independent from the others. Alternatively, it has been suggested that the criterion be modified by requiring the sum of abundances to be less than one. This is justified by the fact that ground inclination as well as change in viewing angles lead to differences in scaling across all bands (Parra et al. 2000).

The two models presented above have close relationships if we consider the endmember abundances from LMM to correspond to the features from the feature based model. From this analogy, we can also associate each ICA derived independent component with a class present in the image. We may also note that since the models accounting for classes are not always pure (in the sense that they are formed by more than one material or mixed), there may be more than one feature associated with a class, one for each of the composing materials.

4.4.3
An ICA algorithm for Hyperspectral Image Processing

In the following, we use a practical experiment to illustrate the appropriateness of the two models. The ICA algorithm used as example in this section is exactly the one described in Sect. 4.3 (see Fig. 4.6). It starts by preprocessing the data with PCA and is followed by the application of the iterative minimum mutual information (MMI) algorithm. Note that in the current format, the algorithm will generate the same number of components as the original number of spectral bands. This will slow down its computational speed.

From the theoretical issues of ICA presented in the previous sections, it follows that minimization of mutual information leads to the same solution as the maximization of non-Gaussianity, in the case of square mixing matrices. Therefore, the algorithm leads to results similar to the ones obtained by other proposed methods that perform kurtosis maximization preserving the decor-

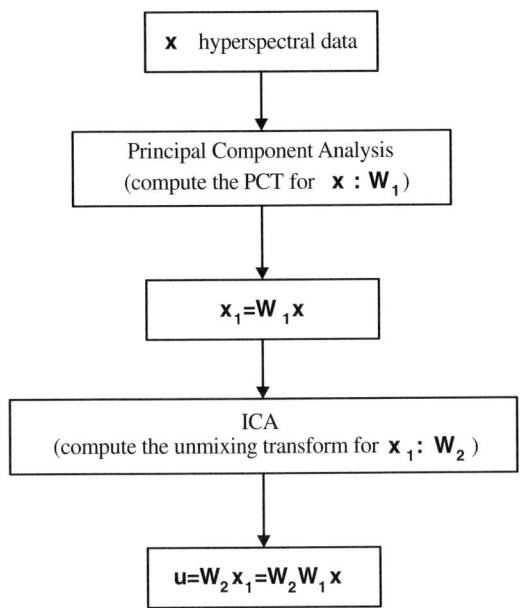

Fig. 4.6. ICA Algorithm for processing hyperspectral image

relation of the components (Chang et al. 2002) or the equivalent sequential projection pursuit method (Chiang et al. 2001).

The ICA algorithm was run on an AVIRIS image available at the website http://dynamo.ecn.purdue.edu/~biehl/Multispec/. The image was acquired in June 1992 over the Indian Pine Test Site in NW Indiana (see Fig. 4.7). This image has been used as a test image for conducting several experiments reported in different chapters in this book. It consists of 145 × 145 pixels in 220 contiguous spectral bands, at 10 nm intervals in the spectral region from 0.40 to 2.45 µm, at a spatial resolution of 20 m. Four of the 224 AVIRIS bands contained no data or zero values. The advantage of using this dataset is the availability of the reference data image (also called the ground truth) for accuracy assessment purposes (see Fig. 4.8). This makes the dataset an excellent source for conducting experimental studies, and, therefore, this image has been used in many other studies (e.g. Gualtieri et al. 1999; Gualtieri and Cromp 1998; Melgani and Bruzzone 2002; Tadjudin and Landgrebe 1998a and 1998b) also.

Two-thirds of the scene is covered with agricultural land while one-third is forest and others. Two major dual lane highways, a smaller road, a rail line, low density housing, and other building structures can also be seen in the scene. The scene represents 16 classes as defined in Fig. 4.8.

In this experiment, from the original bands, 34 bands were discarded (due to sensor malfunctioning, water absorption, and artifacts not related to the scene). The remaining 186 bands were first whitened using PCA and then processed through ICA. A visual inspection of the results reveals that the

Fig. 4.7. One band of the AVIRIS image with four locations indicated for further analysis of pixel vectors (*a*) road, (*b*) soybean, (*c*) grass, (*d*) hay

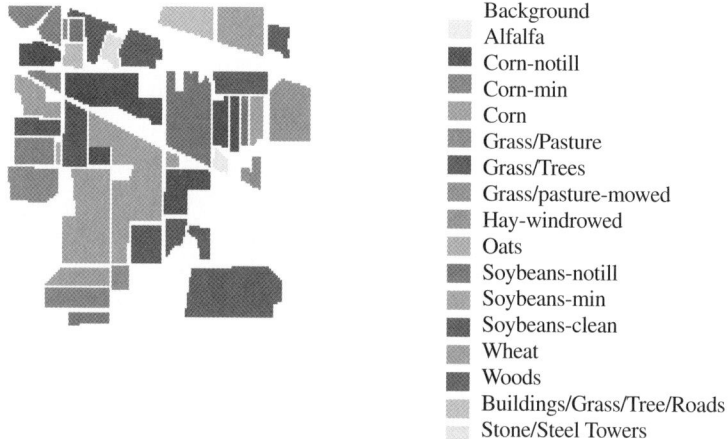

Fig. 4.8. Reference data (ground truth) corresponding to AVIRIS image in Fig. 4.7. For a colored version of this figure, see the end of the book

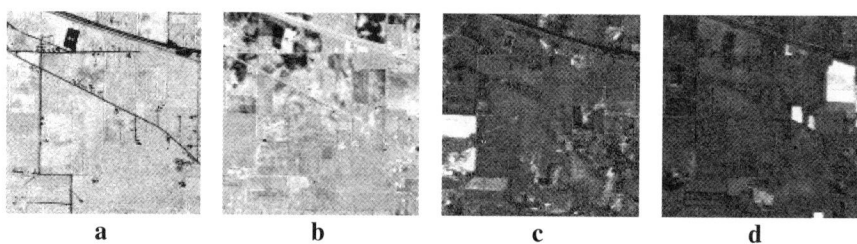

Fig. 4.9a–d. Resulting ICA components. Notice that the features correspond to information related to various classes in the scene (such as **a** road, **b** soybean, **c** grass, **d** hay) are concentrated in different components

information related to the classes has concentrated in only few of the bands. Figure 4.9a–d display four of the resulting ICA components.

Next, we analyzed four different pixel locations in the image (indicated by arrows in Fig. 4.7a), corresponding to four of the sixteen classes road, soybean, grass and hay. For each location, we have plotted the pixel vectors for the data produced by ICA. For comparison, the pixel vectors from the PCA-produced data have also been plotted (Fig. 4.10). There are significant differences between the plots. In PCA, the pixel vectors seem to lack a clear organization, i. e., none of the features seem to have information related to the type of class. In contrast, in ICA, we notice information belonging to one class has been concentrated only in one or very few components.

Further visual inspection reveals that most of the pixel vectors from the image cube display close to zero values and only one or very few peaks. This behavior resembles that of the vectors of abundances for the pixel location. However, since the current algorithm is designed to recover the same number of endmembers as the number of bands, most of these endmembers will not contribute to the image and will be associated with noise.

We note that the range of the abundances varies widely and is not restricted to the interval [0,1], as required by the linear mixture model. Scalar multiplication, as well as shifting may solve this problem. Unfortunately, there is no clear indication as to how the scaling may be done when dealing with components that show up reversed. This aspect was also mentioned in general as affecting ICA (Hyvärinen et al. 2001). The fact that the LMM conditions cannot always be satisfied turns out to be a limitation of ICA as a linear unmixing tool.

4.5
Summary

In this chapter, we have provided theoretical background of ICA and shown its utility as an important data processing tool for hyperspectral imaging. From the point of view of the linear mixture model, the ability of ICA to recover an approximation of the endmember abundances is the key indicator that ICA is a powerful tool for hyperspectral image processing. However, the independent components are identifiable only up to scaling and shifting with scalars. This indicates that, in fact, the endmembers as well as their abundances can not be determined exactly. Moreover, the number of endmembers to be determined is fixed by the number of spectral bands available. Since there are usually hundreds of spectral bands, it is improbable that there will be that many distinct endmembers. This is also indicated by the result of the experiments where we obtained only a few significant components, the rest being mostly noise. Given this fact, one may try to find only the components that contain useful information. Unfortunately, there is no clear method for selecting the bands that contain useful information.

The feature-based model provides a better understanding of the usefulness of ICA in the context of hyperspectral imagery. If we consider ICA as a feature extraction method, the resulting features will be statistically independent. This

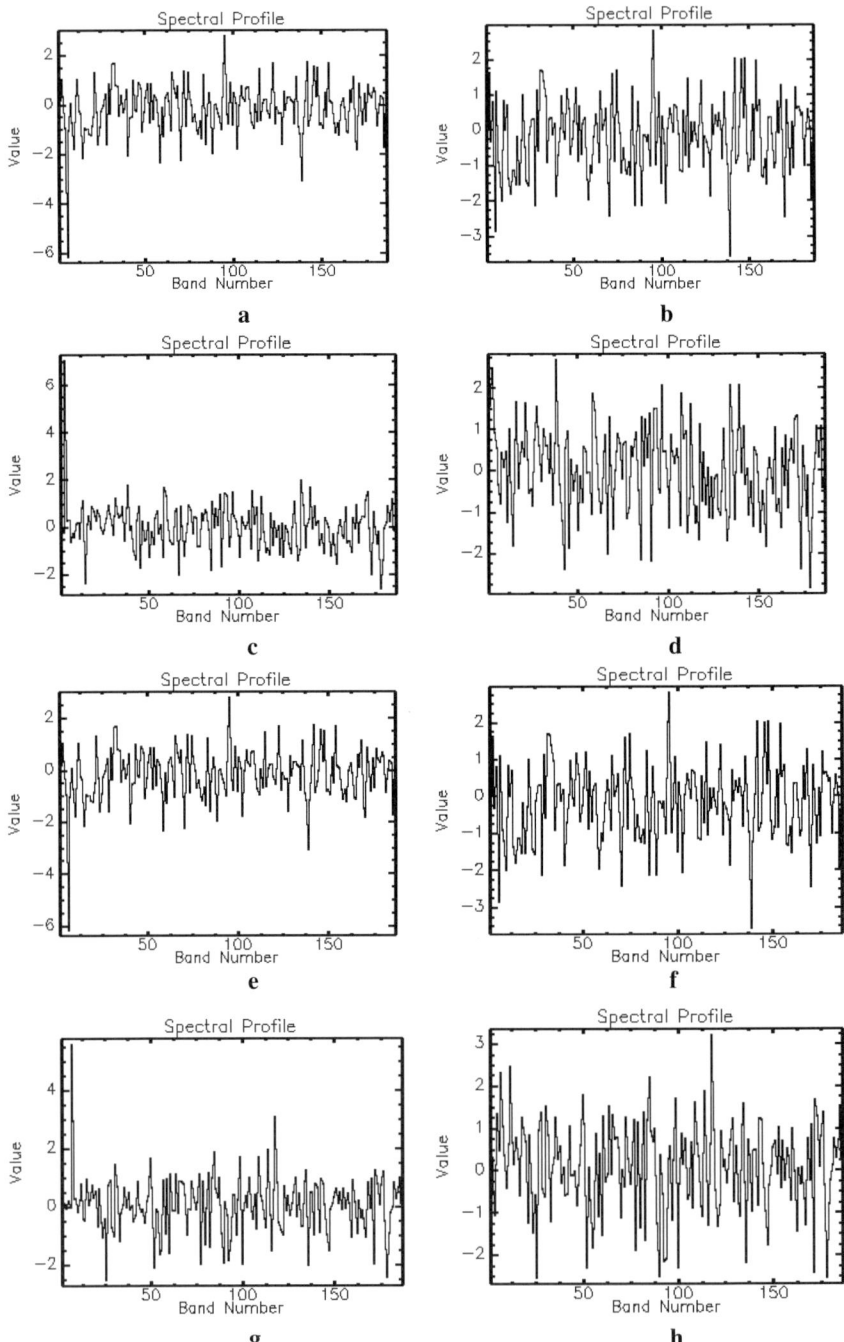

Fig. 4.10a–h. Pixel vector plots for classes road, soybean, grass and hay derived from ICA (**a, c, e,** and **g**) and PCA (**b, d, f,** and **h**)

provides a stronger separation than simple decorrelation and allows a better processing for further class identification (either classification or target detection). The ICA can be viewed, therefore, as an improvement over PCA as an unsupervised feature extraction tool. A more extensive discussion on ICA based feature extraction is presented in Chap. 8.

We also note that apart from the direct application of ICA to hyperspectral imagery, other approaches have been used. A very promising new method, associating mixtures of independent components to the classes, and thus allowing unsupervised classification is presented in Chap. 9.

References

Back AD, Weingend AS (1997) A first application of independent component analysis in evaluating structure from stock returns. International Journal of Neural Systems 8(4): 473–484

Bayliss J, Gualtieri JA, Cromp RF (1997) Analyzing hyperspectral data with independent component analysis. Proceedings of SPIE 26th AIPR Workshop: Exploiting New Image Sources and Sensors, 3240, pp 133–143

Bell A, Sejnowski TJ (1995) An information-maximization approach to blind separation and blind deconvolution. Neural Computation 7: 1129–1159

Chang C-I, Chiang S-S, Smith JA, Ginsberg IW (2002) Linear spectral random mixture analysis for hyperspectral imagery. IEEE Transactions on Geoscience and Remote Sensing 40(2): 375–392.

Chen CH, Zhang X (1999) Independent component analysis for remote sensing. Proceedings of SPIE Image and Signal Processing Conference for Remote Sensing V, 3871, pp 150–157

Chiang S-S, Chang C-I, Ginsberg IW (2001) Unsupervised target detection in hyperspectral images using projection pursuit. IEEE Transactions on Geoscience and Remote Sensing 39(7): 1380–1391.

Cichocki A, Amari S-I (2002) Adaptive blind signal and image processing – learning algorithms and applications. John Wiley and Sons, New York

Common P (1994) Independent component analysis, a new concept?. Signal Processing 36: 287–314.

Girolami M (2000) Advances in independent component analysis. Springer, London, New York

Gualtieri JA, Cromp RF (1998) Support vector machines for hyperspectral remote sensing classification. Proceeding of the SPIE, 27th AIPR Workshop, Advances in Computer Assisted Recognition, 3584, pp 221–232

Gualtieri JA, Chettri SR, Cromp RF., Johnson LF (1999). Support vector machine classifiers as applied to AVIRIS data. Summaries of the Eighth JPL Airborne Earth Science Workshop

Haykin S (Ed) (2000) Unsupervised adaptive filtering: blind source separation 1. John Wiley and Sons, New York

Healey G, Kuan C (2002) Using source separation methods for end-member selection. Proceedings of SPIE Aerosense, 4725, pp 10–17

Hyvärinen A, Karhunen J, Oja E (2001) Independent component analysis. John Wiley and Sons, New York

Lee TW (1998) Independent component analysis: theory and applications. Kluwer Academic Publishers, Boston

Melgani F, Bruzzone L (2002) Support vector machines for classification of hyperspectral remote-sensing images. International Geoscience and Remote Sensing Symposium, IGARSS'02, CD

Papoulis A (1991) Probability, random variables, and stochastic processes, McGraw-Hill, New York

Parra L, Spence CD, Sajda P, Ziehe A, Muller K-R (2000) Unimixing hyperspectral data, Advances in Neural Information Processing Systems 12: 942–948

Rencher AC (1995) Methods of multivariate analysis. John Wiley and Sons, New York

Richards JA, Jia X (1999) Remote sensing digital image analysis: an introduction. Springer-Verlag, Berlin

Robila SA, Varshney PK (2002) Target detection in hyperspectral images based on independent component analysis. Proceedings of SPIE Automatic Target Recognition XII, 4726, pp 173–182

Robila SA, Haaland P, Achalakul T, Taylor S (2000) Exploring independent component analysis for remote sensing. Proceedings of the Workshop on Multi/Hyperspectral Sensors, Measurements, Modeling and Simulation, Redstone Arsenal, Alabama, U.S. Army Aviation and Missile Command, CD

Tadjudin S, Landgrebe D (1998a) Classification of High Dimensional Data with Limited Training Samples. Ph.D. thesis, School of Electrical Engineering and Computer Science, Purdue University.

Tadjudin S, Landgrebe D (1998b) Covariance estimation for limited training samples. International Geoscience and Remote Sensing Symposium (http://dynamo.ecn.purdue.edu/~landgreb/LOOCPAMI-web.pdf)

Tu TM, Huang PS, Chen PY (2001) Blind separation of spectral signatures in hyperspectral imagery. IEEE Proceedings on Vision, Image and Signal Processing 148(4): 217–226

CHAPTER 5
Support Vector Machines

Mahesh Pal, Pakorn Watanachaturaporn

5.1 Introduction

Support Vector Machines (SVMs) are a relatively new generation of techniques for classification and regression problems. These are based on *Statistical Learning Theory* having its origins in *Machine Learning*, which is defined by Kohavi and Foster (1998) as,

> ...*Machine Learning* is the field of scientific study that concentrates on induction algorithms and on other algorithms that can be said to "learn."

The process of learning involves identification of a model based on the training data that are used to make predictions about unknown data sets. Two possible learning approaches can be employed (see Sect. 2.6 of Chap. 2): supervised and unsupervised. Both the learning approaches may be grouped into two categories: parametric and non-parametric learning algorithms. Parametric learning algorithms such as the supervised maximum likelihood classifier (MLC) require that the data be based on some pre-defined statistical model. For example, the MLC assumes that the data follow Gaussian distribution. The performance of a parametric learning algorithm, thus, depends on how well the data match the pre-defined model and on the accurate estimation of model parameters. Moreover, parametric learning algorithms generally suffer from the problem called the *curse of dimensionality*, also known as the *Hughes Phenomenon* (Hughes 1968), especially under the situation when data dimensions are very high such as the hyperspectral data from remote sensing sensors. The Hughes phenomenon states that the ratio of the number of pixels with known class identity (i.e. the training pixels) and the number of bands must be maintained at or above some minimum value to achieve statistical confidence. In hyperspectral datasets, we have hundreds of bands, and it is difficult to have a sufficient number of training pixels. Also, parametric learning algorithms may have difficulty in classifying data at different measurement scales and units. To overcome the limitations of parametric learning algorithms, several non-parametric algorithms are in vogue for classification of remote sensing data. These include nearest neighbor (Friedman 1994; Hastie and Tibshirani

1996; Mather 1999), decision tree (Richards and Jia 1999; Pal 2002) and neural network (Haykin 1999; Tso and Mather 2001) algorithms. The latter two fall in the category of machine learning algorithms. Neural network classifiers are sometimes touted as substitutes for the conventional MLC algorithm in the remote sensing community. The preferred neural network classifier is the feed-forward multi-layer perceptron learnt with a back-propagation algorithm (see Sect. 2.6.3 of Chap. 2). However, even though neural networks have been successful in classifying complex data sets, they are slow during the training phase. A number of studies have also reported that neural network classifiers have problems in setting various parameters during training. Moreover, these may also have limitations in classifying hyperspectral datasets since the complexity of the network architecture increases manifolds. Nearest-neighbor algorithms are sensitive to the presence of irrelevant parameters in the dataset such as noise in a remote sensing image. In case of decision tree classifiers, as the dimensionality of the data increases, class structure becomes dependent on a combination of features thereby making it difficult for the classifier to perform well (Pal 2002).

Recently, the Support Vector Machine (SVM), another machine learning algorithm, has been proposed that may overcome the limitations of aforementioned non-parametric algorithms. SVMs, first introduced by Boser et al. (1992) and discussed in more detail by Vapnik (1995, 1998), have their roots in *statistical learning theory* (Vapnik 1999) whose goal is to create a mathematical framework for learning from input training samples with known identity and predict the outcome of data points with unknown identity. This results in two important theories. The first theory is called empirical risk minimization (ERM) where the aim is to minimize the learning or training error. The second theory is called structural risk minimization (SRM), which is aimed at minimizing the upper bound on the expected error over the whole dataset. SVMs are based on the SRM theory while neural networks are based on ERM theory.

An SVM is basically a linear learning machine based on the principle of optimal separation of classes. The aim is to find a *linear separating hyperplane* that separates classes of interest. The hyperplane is a plane in a multidimensional space and is also called *a decision surface* or *an optimal separating hyperplane* or *an optimal margin hyperplane*. The linear separating hyperplane is placed between classes in such a way that it satisfies two conditions. First, all the data vectors that belong to the same class are placed on the same side of the hyperplane. Second, the distance or *margin* between the closest data vectors in both the classes is maximized (Vapnik and Chervonenkis 1974; Vapnik 1982). In other words, the optimum hyperplane is the one that provides the maximum margin between the two classes. For each class, the data vectors forming the boundary of classes are located on supporting hyperplanes – the term used in the theory of convex sets. Thus, these data vectors are called the *Support Vectors* (Schölkopf 1997). It is noteworthy that the data vectors located along the class boundary are the most significant ones for SVMs.

Many times, a linear separating hyperplane is not able to classify input data without error. Under such circumstances, the data are transformed to a higher dimensional space using a non-linear transformation that spreads the data apart such that a linear separating hyperplane may be found. But, due to very large dimensionality of the feature space, it is not practical to compute the inner product of two transformed data vectors (see Sect. 5.3.3). This may, however, be achieved by using a *kernel trick* instead of explicitly computing transformations in the feature space. Kernel trick is played by substituting a kernel function in place of the inner product of two transformed data vectors. The use of kernel trick reduces the computational effort by a significant amount.

In this chapter, several theoretical aspects of support vector machines will be described. In Sect. 5.2, statistical learning theory is briefly reviewed. Section 5.3 describes the construction of SVMs for the binary classification problem for three different cases: linearly separable case, linearly non-separable case, and non-linear case. Section 5.4 explains the extension of the binary classification problem to the multiclass problem. A discussion on various optimization methods used in the construction of SVMs is presented in Sect. 5.5. A summary of the chapter is presented in Sect. 5.6. Though, all the discussion here is directed towards the classification problem, it is equally applicable for solving regression problems.

5.2
Statistical Learning Theory

Statistical Learning Theory, developed by Vapnik (1982, 1995, 1998), views a supervised classification problem as an input-output relationship. In the case of a two-class (i.e. binary) classification problem, an algorithm learns from a given set of k training samples, $(x_1, y_1), \ldots, (x_k, y_k)$, $x_i \in \mathbb{R}^N$, $y_i \in \{-1, +1\}$, which are drawn from a fixed but unknown cumulative (probability) distribution function $P(x, y)$, where x is an N-dimensional observed data vector and y_i is a class label. $\mathbb{R}$ is the set of all real numbers. A decision rule or classification rule is represented by, $\{f_\alpha(x) : \alpha \in \Lambda\}$, $f_\alpha : \mathbb{R}^N \to \{-1, +1\}$, where Λ is the set of parameters used in the decision rule (Osuna et al. 1997). For example, in a multilayer neural network, Λ is the set of weights of the network.

The aim of classification is to assign the class label y, based on the training samples x, and a decision rule f_α that provides the smallest possible error over the independent data samples or the smallest possible *expected risk* defined as

$$R(\alpha) = \int L\left(y, f_\alpha(x)\right) \, \mathrm{d}P\left(x, y\right) . \tag{5.1}$$

The function f_α is called the *hypothesis*. The set $\{f_\alpha(x) : \alpha \in \Lambda\}$ is called the *hypothesis space* (Osuna et al. 1997), and $L(y, f_\alpha(x))$ is the loss or discrepancy between the response y of the supervisor or teacher to a given input x and the response $f_\alpha(x)$ provided by the learning machine. In other words, the expected risk is a measure of the performance of a decision rule that assigns the class label y to an input data vector x. However, evaluation of the expected risk is difficult, since the cumulative distribution function $P(x, y)$ is unknown and thus one may not be able to evaluate the integral in (5.1). The only known information is contained in the training samples. Therefore, a stochastic approximation of integral in (5.1) is desired that can be computed empirically by a finite sum given by

$$R_{emp}(\alpha) = \frac{1}{k}\sum_{i=1}^{k} L\left(y_i, f_\alpha(x_i)\right) \tag{5.2}$$

and, thus, is known as the *empirical risk*. The value $R_{emp}(\alpha)$ is a fixed number for a given α and a particular training data set. Next we discuss the minimization of risk functions.

5.2.1
Empirical Risk Minimization

The empirical risk is different from the expected risk in two ways (Haykin 1999):

1. it does not depend on the unknown cumulative distribution function
2. it can be minimized with respect to the parameter α

Based on the *law of large numbers* (Gray and Davisson 1986), the empirical mean of a random variable converges to its expected value if the size of the training samples is infinitely large. This remark justifies the use of the empirical risk $R_{emp}(\alpha)$ instead of the risk function $R(\alpha)$. However, convergence of the empirical mean of the random variable to its expected value does not imply that the value α that minimizes the empirical risk will also minimize the risk function $R(\alpha)$. If convergence of the minimum of the empirical risk to the minimum of the expected risk does not occur, this principle of empirical risk minimization is said to be *inconsistent*. In this case, even though the empirical risk is minimized, the expected risk may be high. In other words, a small error rate of a learning machine on the training samples does not necessarily guarantee high *generalization* ability (i.e. the ability to work well on unseen data). This situation is commonly referred to as *overfitting*. Vapnik and Chervonenkis (1971, 1991) have shown that *consistency* occurs if and only if convergence in probability of the empirical risk to the expected risk is substituted by *uniform convergence in probability*. Note that convergence in probability of $R(\alpha)$ means that for any $\varepsilon > 0$ and for any $\eta > 0$, there exists a number $k_0 = k_0\left(\varepsilon, \eta\right)$ such that for any $k > k_0$, the inequality $R\left(\alpha_k\right) - R\left(\alpha_0\right) < \varepsilon$ holds true with a probability of at least $1 - \eta$ (Vapnik 1999). ε is a small number close to zero, and η is referred to as the *level of significance* – similar to the α value in statistics. Uniform convergence in probability is defined as

$$\lim_{k\to\infty} \text{Prob}\left(\sup_{\alpha\in\Lambda} |R(\alpha) - R_{emp}(\alpha)| > \varepsilon\right) \to 0, \quad \forall \varepsilon, \tag{5.3}$$

where 'sup A' is the supremum of a nonempty set A and is defined as the smallest scalar x such that $x \geq y$ for all $y \in A$. Uniform convergence is a necessary and sufficient condition for the consistency of the principle of empirical risk minimization.

Vapnik and Chervonenkis (1971, 1979) also showed that the necessary and sufficient condition for consistency amounts to the fitness of the *VC-dimension* to the hypothesis space. The VC-dimension (named after its originators Vapnik and Chervonenkis) is a measure of the *capacity* of a set of classification functions or the complexity of the hypothesis space. The VC-dimension, generally denoted by h, is an integer that represents the largest number of data points that can be separated by a set of functions f_α in all possible ways. For example, for a binary classification problem, the VC-dimension is the maximum number of points, which can be separated into two classes without error in all possible 2^k ways. The proof of consistency of the empirical risk minimization (ERM) can be found in Vapnik (1995, 1998).

The theory of uniform convergence in probability also provides a bound on the deviation of empirical risk from the expected risk given by

$$R(\alpha) \leq R_{\text{emp}}(\alpha) + \varphi\left(\frac{h}{k}, \frac{\log(\eta)}{k}\right), \tag{5.4}$$

where k is the number of training samples, and the *confidence term* φ is defined as

$$\varphi\left(\frac{h}{k}, \frac{\log(\eta)}{k}\right) = \sqrt{\frac{h\left(\log\frac{2k}{h} + 1\right) - \log(\eta/4)}{k}}. \tag{5.5}$$

The parameter h is the VC-dimension of a set of classifiers, and the bound in (5.4) holds for any $\alpha \in \Lambda$, and $k > h$ with a probability of at least $1 - \eta$ such that $0 \leq \eta \leq 1$. From bound in (5.4), a good generalization performance (i.e. the smallest expected risk $R(\alpha)$) can be obtained when both the empirical risk and the ratio between the VC-dimension and the number of training samples are small. With a fixed number of training samples, the empirical risk is usually a decreasing function of the VC-dimension while the confidence term is an increasing function. This means that there exists an *optimal* value of the VC-dimension that can give the smallest expected risk. Therefore, to obtain accurate classification, the choice of an appropriate VC-dimension is also crucial.

5.2.2
Structural Risk Minimization

For the selection of an appropriate VC-dimension for a given set of functions, Vapnik (1982) proposed the principle of SRM that is based on the fact that the minimization of the expected risk is possible by simultaneous minimization of the two terms in (5.4). First is the empirical risk and second is the confidence term, which depends on the VC-dimension of the set of functions. SRM minimizes the expected risk function with respect to both the empirical risk and the VC-dimension. To achieve this aim, a nested structure of hypothesis space is introduced by dividing the entire class of functions into nested subsets

$$H_1 \subset H_2 \subset \ldots \subset H_n \subset \ldots . \tag{5.6}$$

The symbol $\subset$ indicates "is contained in." Each hypothesis space has the property that $h(n) \leq h(n+1)$ where $h(n)$ is the VC-dimension of the set H_n. This implies that the VC-dimension of each hypothesis space is finite. The principle of SRM can be mathematically represented as

$$\min_{H_n} \left(R_{\text{emp}}(\alpha) + \phi \left(\frac{h}{k}, \frac{\log(\eta)}{k} \right) \right). \tag{5.7}$$

Even though the SRM is mathematically defined, its implementation may not be easy due to the difficulty in computing the VC-dimension for each H_n. Only a few models are known for the computation of the VC-dimension (Osuna et al. 1997). Thus, the SRM procedure is to minimize the empirical risk for each hypothesis space so as to identify the hypothesis space with the smallest expected risk. The hypothesis with the smallest expected risk is the best compromise between the empirical risk (i.e. an approximation of the training data) and the confidence term (i.e. a measure of the complexity of the approximating function). The support vector machine (SVM) algorithm is an implementation that is able to minimize the empirical risk as well as a bound on the VC-dimension. We discuss the implementation of the SVM in the next section.

5.3
Design of Support Vector Machines

An SVM is based on SRM to achieve the goal of minimizing the bound on the VC-dimension and the empirical risk at the same time. An SVM is constructed by finding a *linear separating hyperplane* to separate classes of interest. The linear separating hyperplane is placed between classes such that the data belonging to the same class are placed on the same side of the hyperplane and the distance between the closest data vectors in both the classes is maximized. In this case, called the *linearly separable case*, the empirical risk is set to zero, and the bound on the VC-dimension is minimized by maximizing the distance between the closest data vectors of class 1 and class 2. When the classes in the dataset are mixed (i.e. erroneous or noisy data), these cannot be separated by a linear separating hyperplane. This case is known as the *linearly nonseparable case*. Bennett and Mangasarian (1992) and Cortes and Vapnik (1995) introduced slack variables and a regularization parameter to compensate for the noisy data. Thus, in a linearly non-separable case, the empirical risk is controlled by the slack variables and the regularization parameter. The VC-dimension is minimized in the same fashion as in the linearly separable case. Often, a linear separating hyperplane is not able to classify input data (either noiseless or noisy), but a non-linear separating hyperplane can. This has been referred to as the *nonlinear case* (Boser et al. 1992). In this case, the input data are transformed into a higher dimensional space that spreads the data out such that a linearly separable hyperplane can be obtained. They also suggested the use of kernel functions as transformation functions to reduce the computational cost. The design of SVMs for the three cases is described next.

5.3.1
Linearly Separable Case

Linearly separable case is the simplest of all to design a support vector machine. Consider a binary classification problem under the assumption that data can be separated into two classes using a *linear separating hyperplane* (Fig. 5.1). Consider k training samples obtained from the two classes, represented by $(x_1, y_1), \ldots, (x_k, y_k)$, where $x_i \in \mathbb{R}^N$ is an N-dimensional observed data vector with each sample belonging to either of the two classes labeled by $y \in \{-1, +1\}$. These training samples are said to be linearly separable if there exists an N-dimensional vector w that determines the orientation of a discriminating plane and a scalar b that determines the offset of this plane from the origin such that

$$w \cdot x_i + b \geq +1 \quad \text{for all } y = +1, \tag{5.8}$$
$$w \cdot x_i + b \leq -1 \quad \text{for all } y = -1. \tag{5.9}$$

Points lying on the optimal hyperplane satisfy the equation $wx_i + b = 0$. The

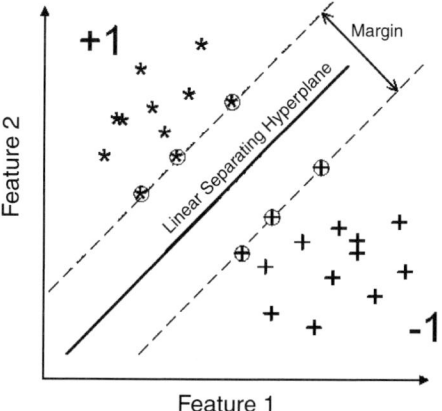

Fig. 5.1. A linear separating hyperplane for the linearly separable data sets. *Dashed lines* pass through the support vectors

inequalities in (5.8) and (5.9) can be combined into a single inequality as

$$y_i \left(\mathbf{w} \cdot \mathbf{x}_i + b \right) - 1 \geq 0 . \tag{5.10}$$

The decision rule for the linearly separable case can be defined by a set of classifiers (or decision functions) as

$$f_{\mathbf{w},b} = \text{sign} \left(\mathbf{w} \cdot \mathbf{x} + b \right) , \tag{5.11}$$

where sign(·) is the signum function. It returns +1 if the element is greater than or equal to zero, and returns −1 if it is less than zero.

The distance $D(\mathbf{x}; \mathbf{w}, b)$, or *margin of separation* or *margin*, for a point $\mathbf{x}$ from the hyperplane defined by both $\mathbf{w}$ and b is given by

$$D \left(\mathbf{x}; \mathbf{w}, b \right) = \frac{|\mathbf{w} \cdot \mathbf{x} + b|}{\|\mathbf{w}\|_2} , \tag{5.12}$$

where $|\cdot|$ is the absolute function, and $\|\cdot\|_2$ is the 2-norm.

Let γ be the value of the margin between two separating planes. To maximize the margin, we express the value of γ as

$$\gamma = \frac{\mathbf{w} \cdot \mathbf{x} + b + 1}{\|\mathbf{w}\|_2} - \frac{\mathbf{w} \cdot \mathbf{x} + b - 1}{\|\mathbf{w}\|_2} = \frac{2}{\|\mathbf{w}\|_2} . \tag{5.13}$$

The maximization of (5.13) is equivalent to the minimization of the 2-norm $\|\mathbf{w}\|_2 / 2$. Thus, the objective function $\Phi(\mathbf{w})$ may be written as

$$\Phi(\mathbf{w}) = \frac{1}{2} \mathbf{w}^T \mathbf{w} . \tag{5.14}$$

The scale factor 1/2 is used for computational convenience only.

A constrained optimization problem can be constructed with the goal to minimize the objective function in (5.14) under the constraints given in (5.10). This constrained optimization problem is called a *primal problem* that has two properties; the objective function is a convex function of $\mathbf{w}$ and the constraints are linear.

Equation (5.14) can be solved using standard Quadratic Programming (QP) optimization techniques. The QP optimization technique to solve (5.14) under the constraints in (5.10) can be implemented by replacing the inequalities in a simpler form by transforming the problem into a dual space representation using Lagrange multipliers (Leunberger 1984). The method of Lagrange multipliers transforms the *primal problem* to its corresponding *dual problem*. The primal Lagrangian is given by

$$L \left(\mathbf{w}, b, \lambda \right) = \frac{1}{2} \|\mathbf{w}\|^2 - \sum_{i=1}^{k} \lambda_i y_i \left(\mathbf{w} \cdot \mathbf{x}_i + b \right) + \sum_{i=1}^{k} \lambda_i , \tag{5.15}$$

where $\lambda_i \geq 0$ are the unknown Lagrange multipliers.

The solution to the corresponding *dual problem* is obtained by minimizing the primal Lagrangian with respect to w and b and maximizing with respect to λ:

$$\min_{w,b} \max_{\lambda} L(w, b, \lambda) , \qquad (5.16)$$

subject to the constraints

$$y_i (w \cdot x_i + b) - 1 \geq 0 \qquad (5.17)$$

and

$$\lambda_i \geq 0 \quad \text{for } i = 1, \ldots, k . \qquad (5.18)$$

By differentiating (5.15) with respect to w and b and equating them to zero, the following equations are obtained:

$$\frac{\partial L(w, b, \lambda)}{\partial w} = 0 , \qquad (5.19)$$

$$\frac{\partial L(w, b, \lambda)}{\partial b} = 0 . \qquad (5.20)$$

After differentiating and rearranging (5.19) and (5.20), the optimality conditions become

$$w = \sum_{i=1}^{k} \lambda_i y_i x_i , \qquad (5.21)$$

$$\sum_{i=1}^{k} \lambda_i y_i = 0 . \qquad (5.22)$$

From (5.21), the weight vector w is obtained from the Lagrange multipliers corresponding to the k training samples.

Substituting (5.21) and (5.22) into (5.15), the dual optimization problem becomes

$$\max_{\lambda} L(w, b, \lambda) = \sum_{i=1}^{k} \lambda_i - \frac{1}{2} \sum_{i=1}^{k} \sum_{j=1}^{k} \lambda_i \lambda_j y_i y_j (x_i \cdot x_j) , \qquad (5.23)$$

subject to the constraints

$$\sum_{i} \lambda_i y_i = 0 \qquad (5.24)$$

and

$$\lambda_i \geq 0 \quad \text{for } i = 1, \ldots, k . \qquad (5.25)$$

Thus, the solution of the optimization problem given in (5.23) is obtained in terms of the Lagrange multipliers λ_i. According to the Karush-Kuhn-Tucker (KKT) optimality condition (Cristianini and Shawe-Taylor 2000), some of the multipliers will be zero. The multipliers that have the nonzero values are called the *support vectors*. It is also noteworthy that the solution *w* obtained from (5.21) is unique and independent of the optimizer used. However, the values of the Lagrange multipliers obtained from different optimizers are not necessarily unique. This also implies that the support vectors and the number of support vectors are also not unique and may vary depending upon the optimizer used.

The results from an optimizer, also called an *optimal solution*, is a set $\lambda^o = (\lambda_1^o, \ldots, \lambda_k^o)$ that is substituted in (5.21) to get

$$w^o = \sum_i y_i \lambda_i^o x_i \ . \tag{5.26}$$

The intercept b^o is determined from

$$b^o = \frac{1}{2} [w^o x_{+1}^o + w^o x_{-1}^o] \ , \tag{5.27}$$

where x_{+1}^o and x_{-1}^o are the support vectors of class labels +1 and −1 respectively.

The following decision rule is then applied to classify the data vector into two classes +1 and −1:

$$f(x) = \text{sign} \left(\sum_{\text{support vectors}} y_i \lambda_i^o (x_i \cdot x) + b^o \right) . \tag{5.28}$$

As mentioned before, an SVM is based on SRM whose goal is to minimize the bound on the VC-dimension and the empirical risk at the same time. In the linearly separable case, the empirical risk is automatically minimized as the data belonging to a class are placed on the same side (i.e. no training error). To control the VC dimension, a nested structure of classifiers each with a different VC-dimension is created (see Sect. 5.2.2). Based on the fact that an SVM is constructed by optimizing the Euclidean norm of the weight vector *w*, its VC-dimension can best be described by the following theorem (proof given in Vapnik (1999)),

Theorem 5.1 *Let vectors $x \in X$ belong to a sphere of radius R. Then the set of Δ-margin separating hyperplanes has the VC dimension h bounded by the inequality*

$$h \leq \min \left(\left\lceil \frac{R^2}{\Delta^2} \right\rceil, N \right) + 1$$

where $\lceil \varepsilon \rceil$ denotes the ceiling function that round the element ε to the nearest integer greater than or equal to the value ε, Δ is the margin of separation, and N is the dimensionality of the input space.

Support Vector Machines

This theorem shows that the VC-dimension of the set of hyperplanes is equal to $N+1$. However, it can be less than $N+1$ if the margin of separation is large.

In case of an SVM, a SRM nested structure of hypothesis space can be described in terms of the separating hyperplanes as $A_1 \subset A_2 \subset \ldots \subset A_n \subset \ldots$ with

$$A_k = \{w \cdot x + b : \|w\|^2 \leq c_k\}, \quad k = 1, 2, \ldots, \tag{5.29}$$

where c_k is a constant. The nested structure may be reformulated in terms of the bound on the VC-dimension as

$$A_k = \left\{ \left\lceil \frac{R^2}{\Delta^2} \right\rceil + 1 : \Delta^2 \geq a_k \right\}, \quad k = 1, 2, \ldots, \tag{5.30}$$

where a_k is a constant.

After a nested structure is created, the expected risk is then minimized by minimizing the empirical risk and the confidence term simultaneously. The empirical risk is automatically minimized by setting it to zero from the requirement that data belonging to the same class are placed on the same side without error. Since the empirical risk of every hypothesis space is set to zero, the expected risk depends on the confidence term alone. However, the confidence term depends on the VC-dimension and the number of training samples. Let the number of training samples be a constant value, then the confidence term relies on the VC-dimension only. The smallest value of VC-dimension is determined from the largest value of the margin (from Theorem 5.1). The margin is equal to $2/\|w\|_2$ (i.e. the smallest value of the vector w also produces the smallest value of VC-dimension). Therefore, the classifier that produces the smallest expected risk is identified as the one that minimizes the vector w. This clearly shows that the construction of an SVM completely complies with the principle of SRM.

5.3.2
Linearly Non-Separable Case

It is true that the linearly separable case is an ideal case to understand the concept of support vector machines. All data are assumed to be separable into two classes with a linear separating hyperplane. However, in practice, this assumption is rarely met due to noise or mixture of classes during the selection of training data. In other words, it is not possible in practice to create a linear separating hyperplane to separate classes of interest without any misclassification error for a given training data set (see Fig. 5.2). Thus, the classes are not linearly separable. This problem can be tackled by using a *soft margin classifier* introduced by Bennett and Mangasarian (1992) and Cortes and Vapnik (1995). Soft margin classification relaxes the requirement that every data point belonging to the same class must be located on the same side of a linear separating hyperplane. It introduces slack variables $\xi_i \geq 0, i = 1, \ldots, l$, to take into account the noise or error in the dataset due to misclassification.

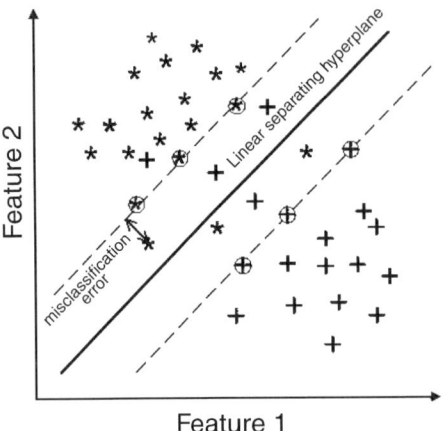

Fig. 5.2. Illustration of the linearly non-separable case

The constraint in (5.10) is thus relaxed to

$$y_i \left(w \cdot x_i + b \right) - 1 + \xi_i \geq 0 . \tag{5.31}$$

In the objective function, a new term called the penalty value $0 < C < \infty$ is added. The penalty value is a form of regularization parameter and defines the trade-off between the number of noisy training samples and the classifier complexity. It is usually selected by trial and error.

It can be shown that the optimization problem for the linearly non-separable case becomes

$$\min_{w,b,\xi_1,\dots,\xi_k} \left[\frac{1}{2} \|w\|^2 + C \sum_{i=1}^{k} \xi_i \right], \tag{5.32}$$

subject to the constraints

$$y_i \left(w \cdot x_i + b \right) - 1 + \xi_i \geq 0 \tag{5.33}$$

and

$$\xi_i \geq 0 \quad \text{for } i = 1,\dots,k . \tag{5.34}$$

The penalty value may have a significant effect on the performance of the resulting support vector machines (see experimental results in Chap. 10). From (5.32), it can also be seen that when $C \to 0$, the minimization problem is not affected by the misclassifications even though $\xi_i > 0$. The linear separating hyperplane will be located at the midpoint of the two classes with the largest possible separation. When $C > 0$, the minimization problem is affected by ξ_i. When $C \to \infty$, the values of ξ_i approach zero and the minimization problem reduces to the linearly separable case.

From the point of view of statistical learning theory, it can be shown that the minimization of the VC-dimension is achieved by minimizing the first term of (5.32) similar to the linearly separable case. Unlike the linearly separable case, minimization of the empirical risk in this case is achieved by minimizing the second term of (5.32). Thus, here also, the principle of structural risk minimization is satisfied.

The constrained optimization problem given by (5.32) to (5.34) is again solved using Lagrange multipliers. The primal Lagrangian for this case can be written as

$$L(w, b, \lambda, \mu) = \frac{1}{2} \|w\|^2 + C \sum_i \xi_i - \sum_i \lambda_i \{y_i (w \cdot x_i + b) - 1 + \xi_i\} - \sum_i \mu_i \xi_i, \quad (5.35)$$

where $\lambda_i \geq 0$ and $\mu_i \geq 0$ are the Lagrange multipliers. The terms μ_i are introduced to enforce positivity of ξ_i. The constraints for the optimization of (5.35) are

$$y_i (w \cdot x_i + b) - 1 + \xi_i \geq 0, \quad (5.36)$$

$$\xi_i \geq 0, \quad (5.37)$$

$$\lambda_i \geq 0, \quad (5.38)$$

$$\mu_i \geq 0, \quad (5.39)$$

$$\lambda_i \{y_i (w \cdot x_i + b) - 1 + \xi_i\} = 0, \quad (5.40)$$

$$\mu_i \xi_i = 0, \quad (5.41)$$

where $i = 1, \ldots, k$. Equations (5.36) and (5.37) are the constraints as given by (5.33) and (5.34), (5.38) and (5.39) are obtained from the definition of the Lagrangian method, and the necessary conditions for the Lagrangian method are represented by (5.40) and (5.41).

By differentiating (5.35) with respect to w, b, and ξ_i and setting the derivatives to zero, we obtain

$$\frac{\partial L(w, b, \lambda, \mu, \xi)}{\partial w} = w - \sum_i \lambda_i y_i x_i = 0, \quad (5.42)$$

$$\frac{\partial L(w, b, \lambda, \mu, \xi)}{\partial b} = \sum_i \lambda_i y_i = 0, \quad (5.43)$$

$$\frac{\partial L(w, b, \lambda, \mu, \xi)}{\partial \xi_i} = C - \lambda_i - \mu_i = 0. \quad (5.44)$$

Substituting (5.42), (5.43) and (5.44) into (5.35), the dual optimization problem is obtained as

$$\max_\lambda L(w, b, \lambda) = \sum_{i=1}^{k} \lambda_i - \frac{1}{2} \sum_{i=1}^{k} \sum_{j=1}^{k} \lambda_i \lambda_j y_i y_j (x_i \cdot x_j), \quad (5.45)$$

subject to the constraints

$$\sum_{i=1}^{k} \lambda_i y_i = 0 \qquad (5.46)$$

and

$$C \geq \lambda_i \geq 0 \quad \text{for } i = 1,\ldots,k. \qquad (5.47)$$

It can thus be seen that the objective function of the dual optimization problem for the linearly non-separable case is the same as that of the linearly separable case except that the Lagrange multipliers are bounded by the penalty value C.

After obtaining the solution of (5.45), w and b can be found in the same manner as explained in (5.26) and (5.27) earlier. The decision rule is also the same as defined in (5.28).

5.3.3
Non-Linear Support Vector Machines

SVM seeks to find a linear separating hyperplane that can separate the classes. There are instances where a linear hyperplane cannot separate classes without misclassification. However, those classes can be separated by a nonlinear separating hyperplane. In fact, most of the real-life problems are non-linear in nature (Minsky and Papert 1969). In this case, data are mapped to a higher dimensional feature space with a nonlinear transformation function. In the higher dimensional space, data are spread out, and a linear separating hyperplane can be constructed. For example, two classes in the input space of Fig. 5.3 may not be separated by a linear hyperplane but a nonlinear hyperplane can make them separable. This concept is based on *Cover's theorem* on the *separability of patterns* (Cover 1965).

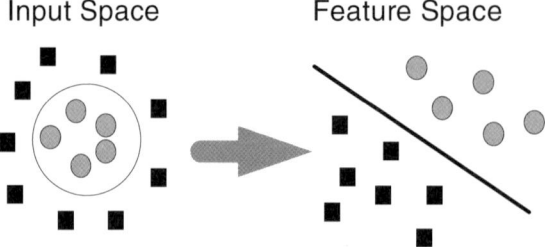

Fig. 5.3. Non linear case. Mapping nonlinear data to a higher dimensional feature space where a linear separating hyperplane can be found

The concept of creating a linear separating hyperplane in a higher dimensional feature space can be incorporated in SVMs by rewriting the dual optimization problem defined in (5.23) and (5.45) as

$$\max_{\lambda} L\left(w, b, \lambda\right) = \sum_{i=1}^{k} \lambda_i - \frac{1}{2} \sum_{i=1}^{k} \sum_{j=1}^{k} \lambda_i \lambda_j y_i y_j \left(x_i \cdot x_j\right). \qquad (5.48)$$

Let a nonlinear transformation function ϕ map the data into a higher dimensional space. In other words, $\phi(x)$ represents the data x in the higher dimensional space. The dual optimization problem, then, for a nonlinear case may be expressed as

$$\max_{\lambda} L\left(w, b, \lambda\right) = \sum_{i=1}^{k} \lambda_i - \frac{1}{2} \sum_{i=1}^{k} \sum_{j=1}^{k} \lambda_i \lambda_j y_i y_j \phi\left(x_i\right) \cdot \phi\left(x_j\right). \qquad (5.49)$$

A *kernel function* is substituted for the dot product of the transformed vectors. The explicit form of the transformation function ϕ is not necessarily known. Further, the use of the kernel function is less computationally intensive. The formulation of the kernel function from the dot product is a special case of *Mercer's theorem* (Mercer 1909; Courant and Hilbert 1970; Schölkopf 1997; Haykin 1999; Cristianini and Shawe-Taylor 2000; Schölkopf and Smola 2002). Suppose there exists a kernel function K such that

$$K\left(x_i, x_j\right) \equiv \phi\left(x_i\right) \cdot \phi\left(x_j\right). \qquad (5.50)$$

The dual optimization problem for a nonlinear case can then be expressed as

$$\max_{\lambda} L\left(w, b, \lambda\right) = \sum_{i=1}^{k} \lambda_i - \frac{1}{2} \sum_{i=1}^{k} \sum_{j=1}^{k} \lambda_i \lambda_j y_i y_j K\left(x_i, x_j\right), \qquad (5.51)$$

subject to the constraints

$$\sum_{i=1}^{k} \lambda_i y_i = 0 \qquad (5.52)$$

and

$$C \geq \lambda_i \geq 0 \quad \text{for } i = 1, \ldots, k. \qquad (5.53)$$

In a manner similar to the other two cases, the dual optimization problem can be solved by using Lagrange multipliers that maximizes (5.51) under the constraints (5.52) and (5.53). The decision function can be expressed as

$$f(x) = \text{sign}\left(\sum_{\text{support vectors}} y_i \lambda_i^o K\left(x_i, x\right) + b^o\right). \qquad (5.54)$$

Table 5.1. Examples of some kernel functions

Kernel Function	Definition	Parameters
Linear	$x \cdot x_j$	
Polynomial with degree d	$(x \cdot x_i + 1)^d$	d is a positive integer
Radial Basis Function	$\exp\left(-\frac{\|x-x_i\|^2}{2\sigma}\right)$	σ is a user defined value
Sigmoid	$\tanh\left(\kappa\left(x \cdot x_i\right) + \Theta\right)$	κ and Θ are user defined values

In fact, every condition of the linearly separable case can be extended to the nonlinear case with a suitable kernel function. The kernel function for a linearly separable case will simply be a dot product of two data vectors, $K\left(x_i, x_j\right) = x_i \cdot x_j$. Other examples of well-known kernel functions are provided in Table 5.1.

The selection of a suitable kernel function is essential for a particular problem. For example, the performance of the simple dot product linear kernel function may deteriorate when decision boundaries between the classes are non-linear. The performance of sigmoid, polynomial and radial basis kernel functions may depend on the selection of appropriate values of the user-defined parameters, which may vary from one dataset to another. An experimental investigation on the choice of kernel functions for the classification of multi and hyperspectral datasets has been provided in Chap. 10.

5.4
SVMs for Multiclass Classification

Originally, SVMs were developed to perform binary classification, where the class labels can either be +1 or −1. However, applications of binary classification are very limited. The classification of a dataset into more than two classes, called *multiclass classification*, is of more practical relevance and has numerous applications. For example, in a digit recognition application, the system has to classify ten classes of interest from 0 to 9. In remote sensing, land cover classification is generally a multiclass problem. A number of methods to generate multiclass SVMs from binary SVMs have been proposed by researchers. It is still a continuing research topic. In this section, we provide a brief description of some multiclass methods that may be employed to perform classification using SVMs.

5.4.1
One Against the Rest Classification

This method is also called *winner-take-all* classification. Suppose the dataset is to be classified into M classes. Therefore, M binary SVM classifiers may be created where each classifier is trained to distinguish one class from the remaining $M - 1$ classes. For example, class one binary classifier is designed to discriminate between class one data vectors and the data vectors of the remaining classes. Other SVM classifiers are constructed in the same manner. During the testing or application phase, data vectors are classified by finding the margin from the linear separating hyperplane (i. e. (5.28) or (5.54) without the *sign* function):

$$g^j(x) = \sum_{i=1}^{m} y_i \lambda_i^j K(x, x_i) + b^j \quad \text{for } j = 1, \ldots, M, \tag{5.55}$$

where m is the number of support vectors and M is the number of classes. A data vector is assigned the class label of the SVM classifier that produces the maximum output.

However, if the outputs corresponding to two or more classes are very close to each other, those points are labeled as *unclassified*, and a subjective decision may have to be taken by the analyst. Otherwise, a *reject decision* (Schölkopf and Smola 2002) may also be applied using a *threshold* to decide the class label. For example, considering the two largest outputs resulting from (5.55), the difference between the two outputs must be larger than the reject decision threshold to assign a class label.

This multiclass method has an advantage in the sense that the number of binary classifiers to construct equals the number of classes. However, there are some drawbacks. First, during the training phase, the memory requirement is very high and is proportional to the square of the total number of training samples. This may cause problems for large training data sets and may lead to computer memory problems. Second, suppose there are M classes and each has an equal number of training samples. During the training phase, the ratio of training samples of one class to the rest of the classes will be 1 : $(M - 1)$. This ratio, therefore, shows that training sample sizes will be unbalanced. Because of these limitations, the pairwise multiclass method has been proposed.

5.4.2
Pairwise Classification

In this method, SVM classifiers for all possible pairs of classes are created (Knerr et al. 1990; Friedman, 1996; Hastie and Tibshirani 1998; Kreßel 1999). Therefore, for M classes, there will be $\frac{1}{2}M(M - 1)$ binary classifiers. The output from each classifier in the form of a class label is obtained. The class label that occurs the most is assigned to that point in the data vector. In case of a tie, a tie-breaking strategy may be adopted. A common tie-breaking strategy is to randomly select one of the class labels that are tied.

The number of classifiers created by this method is generally much larger than the previous method. However, the number of training data vectors required for each classifier is much smaller. The ratio of training data vector size for one class against another is also 1 : 1. Therefore, this method is considered more symmetric than the one-against-the-rest method. Moreover, the memory required to create the kernel matrix $K(x_i, x_j)$ is much smaller. However, the main disadvantage of this method is the increase in the number of classifiers as the number of classes increases. For example, for a ten class problem, 45 classifiers will be created.

5.4.3
Classification based on Decision Directed Acyclic Graph and Decision Tree Structure

Platt et al. (2000) proposed a multiclass classification method called Directed Acyclic Graph SVM (DAGSVM) based on the Decision Directed Acyclic Graph (DDAG) structure that has a tree-like structure (see Fig. 5.4). The DDAG method in essence is similar to pairwise classification in that, for an M class classification problem, the number of binary classifiers is equal to $\frac{1}{2}M(M-1)$ and each classifier is trained to classify two classes of interest. Each classifier is treated as a node in the graph structure. Nodes in DDAG are organized in a triangle with the single root node at the top and increasing thereafter in an increment of one in each layer until the last layer that will have M nodes. The i-node in layer $j < M$ is connected to the ith and $(i+1)^{\text{th}}$ nodes in the $(j+1)^{\text{th}}$ layer.

The DDAG evaluates an input vector x starting at the root node and moves to the next layer based on the output values. For instance, it exits to the left edge if the output from the binary classifier is negative, and it exits to the right edge if the output from the binary classifier is positive. The binary classifier of the next node is then evaluated. The path followed is called the *evaluation path*. The DDAG method basically eliminates one class from the list at each layer. The list initially contains all the classes. Each node evaluates the first class against the last class in the list. For example, the root node evaluates class 1 against class M. The evaluation at each layer results in one class out of the two classes, and the other class is eliminated from the list. The process then tests the first and the last class in the new list. The procedure is terminated when only one class remains in the list. The class label associated with the input data will be the class label of the node in the final layer of the evaluation path or the class remaining in the list. Although the number of binary classifiers is still the same as in the pairwise classification method, the inputs are evaluated $M-1$ times instead of $\frac{1}{2}M(M-1)$ times.

A classification based on a decision tree structure has also been proposed by Takahashi and Abe (2002). Each node of the decision tree structure is a binary classifier that separates either one class or some number of classes from the remaining classes. The number of binary classifiers needed by this approach is equal to $M-1$. However, there are numerous ways to build the decision

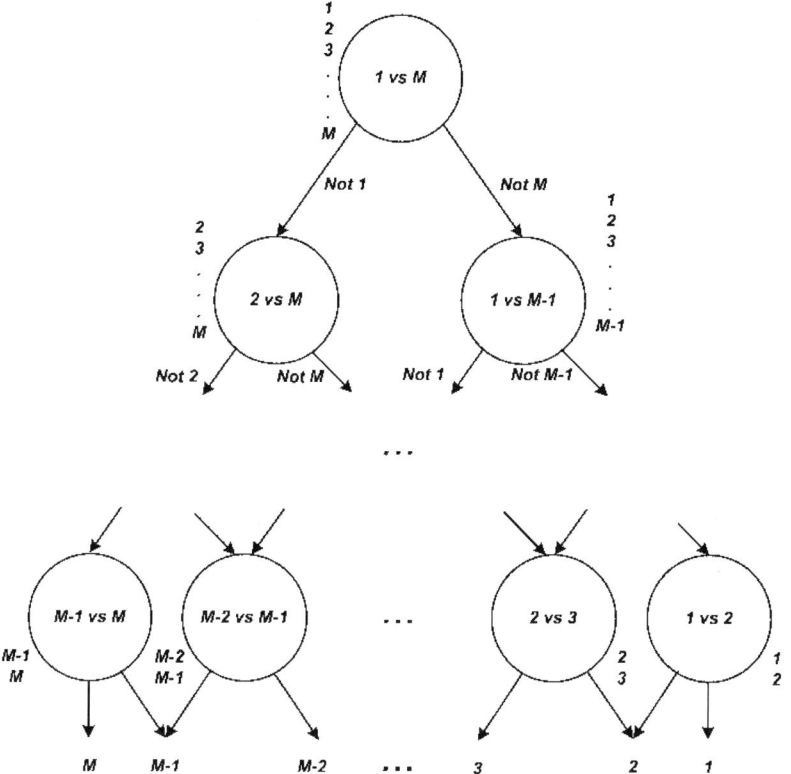

Fig. 5.4. The directed acyclic graph SVM (DAGSVM)

tree. The question is how to determine its structure. Use of Euclidean distance or the Mahalanobis distance as a criterion has been proposed to determine the decision tree structure. The top node is the most important classifier. The better the classifier at the higher node, the better will be the overall classification accuracy. Therefore, the higher nodes are designed to classify a class or some classes that has/have the farthest distance from the remaining classes.

The input vector x is evaluated starting from the top of the decision tree. The sign of the value of the decision function determines the path of the input vector. The process is repeated until a leaf node is reached. The class label corresponding to the final leaf node is associated with the input vector. It clearly shows that the number of binary classifiers for this method is less than the other aforementioned methods since the input is evaluated at most $M - 1$ times. However, it is important to mention that the construction of the decision tree is critical to the overall classification performance.

5.4.4
Multiclass Objective Function

Instead of creating many binary classifiers to determine the class labels, this method attempts to directly solve a multiclass problem (Weston and Watkins 1998, 1999; Lee et al. 2001; Crammer and Singer 2001; Schölkopf and Smola 2002). This is achieved by modifying the binary class objective function and adding a constraint to it for every class. The modified objective function allows simultaneous computation of multiclass classification and is given by Weston and Watkins (1998):

$$\min_{w,b,\xi} \left[\frac{1}{2} \sum_{i=1}^{M} \|w\|^2 + C \sum_{i=1}^{k} \sum_{r \neq y_i} \xi_i^r \right], \qquad (5.56)$$

subject to the constraints

$$w_{y_i} \cdot x_i + b_{y_i} \geq w_r \cdot x_i + b_r + 2 - \xi_i^r \quad \text{for } i = 1, \ldots, k \qquad (5.57)$$

and

$$\xi_i^r \geq 0 \quad \text{for } i = 1, \ldots, k, \qquad (5.58)$$

where $y_i \in \{1, \ldots, M\}$ are the multiclass labels of the data vectors and $r \in \{1, \ldots, M\} \setminus y_i$ are multiclass labels excluding y_i.

Lee et al. (2001) and Schölkopf and Smola (2002) showed that the results from this method and the one-against-the-rest method are similar. However, in this method, the optimization algorithm has to consider all the support vectors at the same time. Therefore, although it may be able to handle massive data sets but the memory requirement and thus, the computational time may be very high.

Thus, the choice of a multiclass method depends on the problem at hand. A user should consider the accuracy requirements, the computational time, the resources available and the nature of the problem. For example, the multiclass objective function approach may not be suitable for a problem that contains a large number of training samples and classes due to the requirement of large memory and extremely long computational time.

5.5
Optimization Methods

One of the key processing steps in the development of SVM algorithms is to employ an optimization method to find the support vectors. A variety of optimization methods may be used. Typically, the conventional SVMs have used an optimizer based on quadratic programming (QP) or linear programming (LP) methods to solve the optimization problem. Most of the QP algorithms are based on a conjugate gradient, quasi-Newton or a prime-dual interior-point method, which are implemented through a number of optimization packages such as CPLEX (CPLEX Optimization Inc. 1992), MINOS (Murtagh and Saunders 1987), and LOQO (Vanderbei 1997). The major disadvantage of QP algorithms is the storage requirement of the kernel data matrix in the memory. When the size of the kernel data matrix is large, it requires huge memory, which may not always be available. Therefore, these algorithms may be suitable for small datasets and may have limitations in processing large datasets such as hyperspectral images. Alternate optimization methods may therefore have to be considered.

Bennett and Campbell (2000) and Campbell (2002) have suggested an optimization method that sequentially updates the Lagrange multipliers based on the Hildreth's optimization technique called the *Kernel Adatron (KA)* algorithm (Campbell and Cristianini 1998; Luenberger 1984). This method is relatively easy to implement. However, it is not as fast as most of the QP and LP optimization algorithms, especially for small datasets.

Other alternatives that work on a smaller or a subset of the whole dataset are *chunking* and *decomposition* methods. These methods are based on the assumption that the number of support vectors is quite small in comparison to the total number of training samples. Instead of sequentially updating the Lagrange multipliers, these methods update the Lagrange multipliers in parallel since they update many parameters in each iteration unlike other methods that update one parameter at a time. In the chunking method, QP is used to optimize an initial arbitrary subset of data. The support vectors found by the optimizer from this subset are kept while other data points are discarded. A new subset is then selected based on these support vectors and the additional data. The process is then repeated until the margin is maximized. However, the chunking method fails when the training data set is too large or most of the training data points become support vectors. The decomposition method, on the other hand, works on the fixed size working dataset rather than an arbitrary size as in the chunking method. A QP optimizer updates the Lagrange multipliers on the fixed size working dataset only; the other Lagrange multipliers that are not in the working set are kept fixed. To summarize, the chunking and decomposition methods use a QP or LP optimizer to solve the problem by considering many small datasets rather than a single huge dataset. The *Sequential Minimal Optimization (SMO)* algorithm (Platt 1999) is a special case of the decomposition method when the size of working dataset is fixed at two such that an analytical solution is derived in very few numerical operations. This discards the use of QP or LP optimizer. This method needs more number of iterations but requires a small number of operations thus resulting in an increase in optimization speed for very large data sets.

Mangasarian and Musicant (1998) introduced the Successive Overrelaxation method (SOR) for SVM, which combines SVM with SOR, and uses linear programming to solve the optimization problem. This is done by introducing an extra term in the objective function (5.14) and (5.32), which results in the elimination of the equality constraint (5.24), (5.43) and (5.52). The optimization problem requires the adjustment of only one point (one Lagrange multiplier) at a time rather than two points during each iteration as in the SMO algorithm.

This algorithm can process large datasets without large memory requirements. Hsu and Lin (2002) found that these algorithms work quite well and the results are similar to any standard QP optimization method. Further, Mangasarian and Musicant (2000a) proposed the use of mathematical programming to solve the optimization problem. An 'active set' strategy is used to generate a fast algorithm that consists of solving a finite number of linear equations of the order of the dimensionality of the original input space at each step. This method consists of maximizing the margin between two separating hyperplanes with respect to both *w* and *b* and using a squared 2-norm of slack variables in place of the 1-norm defined in (5.30). Thus, the active set optimization algorithm requires no specialized quadratic or linear programming software, but a linear solver to solve an $N + 1$ by $N + 1$ matrix, where N is the dimensionality of the input data space.

Mangasarian and Musicant (2000b) also proposed the Lagrangian SVM (LSVM) that reformulates the constrained optimization problem as an unconstrained optimization problem. The problem is solved through an optimizer based on the system of linear equalities. Ferris and Munson (2000a, 2000b) proposed interior point and semi-smooth support vector machines. The interior point SVM uses proximal point modification to the underlying algorithm, the Sherman-Morrison-Woodbury formula, and the Schur complement to solve a linear system. In the semi-smooth support vector machine, Ferris and Munson reformulated the optimality conditions as a semi-smooth system using the Fischer-Burmeister function, applied as a damped Newton method, and exploited the Sherman-Morrison-Woodbury formula to efficiently solve the problem. Both methods can solve linear classification problems proficiently. These optimization methods are just some examples and are still being pursued in ongoing research. Some of the methods will also be investigated in Chap. 10.

5.6
Summary

In this chapter, we have introduced the basic concepts of support vector machines (SVMs) for classification problems. SVMs originate from the structural risk minimization concept of statistical learning theory. The basic mathematical background of SVM was discussed by formulating a binary classification problem. All the three cases – linearly separable and linearly non-separable cases, and the nonlinear case – were considered. In the nonlinear case, data are required to be mapped to a higher dimensional feature space through a kernel function suitably incorporated in the objective function. The binary classification problem was then extended to multiclass classification and the associated methods were briefly discussed. A discussion on various optimization methods was also provided. This methodology will be employed for the classification of a set of multi- and hyperspectral data in Chap. 10.

References

Bennett KP, Campbell C (2000) Support vector machines: hype or hallelujah. Special Interest Group on Knowledge Discovery in Data Mining Explorations 2(2): 1–13

Bennett KP, Mangasarian OL (1992) Robust linear programming discrimination of two linearly inseparable sets. Optimization Methods and Software 1: 23–34

Boser H, Guyon IM, Vapnik VN (1992) A training algorithm for optimal margin classifiers. Proceedings of the 5th Annual ACM Workshop on Computational Learning Theory, Pittsburgh, PA, pp 144–152

Campbell C (2002) Kernel methods: a survey of current techniques. Neurocomputing 48: 63–84

Campbell C, Cristianini N (1998) Simple training algorithms for support vector machines. Technical Report, Bristol University (http://lara.bris.ac.uk/cig)

Cortes C, Vapnik VN (1995) Support vector networks. Machine Learning 20: 273–297

Courant R, Hilbert D (1970) Methods of mathematical Physics, I and II. Wiley Interscience, New York

Cover TM (1965) Geometrical and statistical properties of systems of linear inequalities with applications in pattern recognition. IEEE Transaction on Electronic Computers EC-14: 326–334.

CPLEX Optimization Inc. (1992) CPLEX User's guide, Incline Village, NY

Crammer K, Singer Y (2001) On the algorithmic implementation of multiclass kernel-based vector machines. Journal of Machine Learning Research 2: 265–292

Cristianini N, Shawe-Taylor J (2000) An introduction to support vector machines and other kernel-based learning methods. Cambridge University Press, Cambridge, UK

Ferris MC, Munson TS (2000a) "Interior point methods for massive support vector machines." Data Mining Institute Technical Report 00-05, Computer Science Department, University of Wisconsin, Madion, WI

Ferris MC, Munson TS (2000b) Semi-smooth support vector machines. Data Mining Institute Technical Report 00–09, Computer Science Department, University of Wisconsin, Madion, WI

Friedman JH (1994) Flexible metric nearest neighbor classification.Technical Report, Department of Statistics, Stanford University

Friedman JH (1996) Another approach to polychotomous classification. Technical Report, Department of Statistics and Stanford Linear Accelerator Center, Stanford University

Gray RM, Davisson LD (1986) Random processes: a mathematical approach for engineers. Prentice-Hall, Englewood Cliffs, NJ

Hastie TJ, Tibshirani RJ (1996) Discriminant adaptive nearest neighbor classification. IEEE Transactions on Pattern Analysis and Machine Intelligence 18(6): 607–615

Hastie TJ, Tibshirani RJ (1998) Classification by pairwise coupling. In: Jordan MI, Kearns MJ, Solla, SA (eds) Advances in neural information processing systems10, The MIT Press, Cambridge, MA, pp 507–513

Haykin S (1999) Neural networks: a comprehensive foundation. Prentice Hall, Upper Saddle River, NJ

Hsu C–W, Lin CJ (2002) A simple decomposition method for support vector machines. Machine Learning 46: 291–314

Hughes GF (1968) On the mean accuracy of statistical pattern recognizers. IEEE Transactions on Information Theory 14(1): 55–63

Knerr S, Personnaz L, Dreyfus G (1990) Single-layer learning revisited: A stepwise procedure for building and training neural network. In: Neurocomputing: algorithms, architectures and applications, NATO ASI, Springer Verlag, Berlin

Kohavi R, Foster P (1998) Editorial: glossary of terms. Machine Learning 30:271–274
Kreßel U (1999) Pairwise classification and support vector machines. In: Schölkopf B, Burges CJC, Smola AJ (eds), Advances in kernel methods – support vector learning The MIT Press, Cambridge, MA, pp 255–268
Lee Y, Lin Y, Wahba G (2001) Multicategory support vector machines, Technical Report 1043, Department of Statistics, University of Wisconsin, Madison, WI
Leunberger D (1984) Linear and nonlinear programming, 2nd edition. Addison-Wesley, Menlo Park, California
Mangasarian OL, Musicant, DR (1998) Successive overrelaxation for support vector machines. Technical Report, Computer Sciences Department, University of Wisconsin, Madison, Wisconsin.
Mangasarian OL, Musicant DR (2000a) Active support vector machine classification. Technical Report 0004, Data Mining Institute, Computer Sciences Department, University of Wisconsin, Madison, Wisconsin (ftp: //ftp.cs.wisc.edu/pub/dmi/techreports/0004.ps)
Mangasarian OL, Musicant DR (2000b) Lagrangian support vector machines. Technical Report 0006, Data Mining Institute, Computer Sciences Department, University of Wisconsin, Madison, Wisconsin (ftp://ftp.cs.wisc.edu/pub/dmi/techreports/0006.ps)
Mather PM (1999) Computer processing of remotely-sensed images: an introduction 2nd edition. John Wiley and Sons. Chichester, NY
Mercer J (1909) Functions of positive and negative type, and their connection with the theory of integral equations. Transactions of the London Philosophical Society (A) 209: 415–446
Minsky ML, Papert SA (1969) Perceptrons. MIT Press, Cambridge, MA
Murtagh BA, Saunders MA (1987) MINOS 5.1 user's guide. (SOL 83-20R), Stanford University.
Osuna EE, Freund R, Girosi F (1997) Support vector machines: training and applications. A. I. Memo No. 1602, CBCL paper No. 144, Artificial Intelligence laboratory, Massachusetts Institute of Technology, (ftp://publications.ai.mit.edu/ai-publications/pdf/AIM-1602.pdf)
Pal M (2002) Factors influencing the accuracy of remote sensing classifications: a comparative study. Ph. D. Thesis (unpublished), University of Nottingham, UK
Platt JC (1999) Fast training of support vector machines using sequential minimal optimization. In: Schölkopf B, Burges CJC, Smola AJ (eds), Advances in kernel methods – support vector learning The MIT Press, Cambridge, MA, pp 185–208
Platt JC, Cristianini N, Shawe-Taylor J (2000) Large margin DAGs for multiclass classification. In: Solla SA, Leen TK, Müller K-R (eds), Advance in neural information processing systems 12, The MIT Press, Cambridge, MA, pp 547–553
Richards JA, Jia X (1999) Remote sensing digital image analysis: an introduction, 3rd edition. Springer Verlag, Berlin, Heidelberg, New York
Schölkopf B (1997) Support vector learning. PhD thesis, Technische Universität, Berlin.
Schölkopf B, Smola AJ (2002) Learning with kernels – support vector machines, regularization, optimization and beyond.The MIT Press, Cambridge, MA
Takahashi F, Abe S (2002) Decision-tree-based multiclass support vector machines. Proceedings of the 9th International Conference on Neural Information Processing (ICONIP'02), 3, pp 1418–1422
Tso BCK, Mather PM (2001) Classification methods for remotely sensed data. Taylor and Francis, London
Vanderbei RJ (1997) User's manual – version 3.10. (SOR-97-08), Statistics and Operations Research, Princeton University, Princeton, NJ

References

Vapnik VN (1982) Estimation of dependencies based on empirical data. Springer Verlag, Berlin

Vapnik VN (1995) The nature of statistical learning theory. Springer Verlag, New York

Vapnik VN (1998) Statistical learning theory. John Wiley and Sons, New York

Vapnik VN (1999) An overview of statistical learning theory. IEEE Transactions of Neural Networks 10: 988–999

Vapnik VN, Chervonenkis AJ (1971) On the uniform convergence of relative frequencies of events to their probabilities. Theory of Probability and its Applications 17: 264–280

Vapnik VN, Chervonenkis AJ (1974) Theory of pattern recognition (in Russian). Nauka, Moscow

Vapnik VN, Chervonenkis AJ (1979) Theory of pattern recognition (in German). Akademia Verlag, Berlin

Vapnik VN, Chervonenkis AJ (1991) The necessary and sufficient conditions for consistency in the empirical risk minimization method. Pattern Recognition and Image Analysis 1: 283–305

Weston J, Watkins C (1998) Multi-class support vector machines. Technical Report CSD-TR-98-04, Royal Holloway, University of London, UK

Weston J, Watkins C (1999) Multi-class support vector machines. In: Verleysen M (ed), 7th European Symposium on Artificial Neural Networks. Proceedings of ESANN'99, D-Facto, Brussels, Belgium, pp 219–224

CHAPTER 6
Markov Random Field Models

Teerasit Kasetkasem

6.1 Introduction

For decades, Markov random fields (MRF) have been used by statistical physicists to explain various phenomena occurring among neighboring particles because of their ability to describe local interactions between them. In Winkler (1995) and Bremaud (1999), an MRF model is used to explain why neighboring particles are more likely to rotate in the same direction (clockwise or counterclockwise) or why intensity values of adjacent pixels of an image are more likely to be the same than different values. This model is called the Ising model. There are a large number of problems that can be modeled using the Ising model and where an MRF model can be used. Basically, an MRF model is a spatial-domain extension of a temporal Markov chain where an event at the current time instant depends only on events of a few previous time instants. In MRF, the statistical dependence is defined over the neighborhood system, a collection of neighbors, rather than past events as in the Markov chain model. It is obvious that this type of spatial dependence is a common phenomenon in various signal types including images. In general, images are smooth and, therefore, the intensity values of neighboring pixels are highly dependent on each other. Because of its highly theoretical and complex nature, and intensive computational requirements, practical uses of MRF were extremely limited until recently. Due to the dramatic improvements in computer technologies, MRF modeling has become more feasible for numerous applications, for example, image analysis because many image properties, such as texture, seem to fit an MRF model, i. e., intensity values of neighboring pixels of images are known to be highly correlated with each other. The Markovian nature of these textural properties has long been recognized in the image processing community, and has widely been used in a variety of applications (e. g. image compression and image noise removal). However, these applications were limited to empirical studies and were not based on a statistical model such as the MRF model.

The pioneering work by Geman and Geman (1984) introduced a statistical methodology based on an MRF model. Their work inspired a continuous stream of researchers to employ MRF models for a variety of image analysis tasks such as image classification and segmentation. In their paper, a noiseless image was assumed to have MRF properties. The noiseless image was disturbed by an additive noise before being captured by a non-linear photographic device. Then, based on the maximum *a posteriori* (MAP) criterion (Trees 1968; Varshney 1997), an image restoration scheme selected the most likely noiseless image (configuration) given an observed image. Here, a standard iterative search technique such as direct descent cannot be used to search for the solution due to the severe non-convexity of the MAP equation. As a result, this noiseless image may be obtained by using the simulated annealing (SA) algorithm, which iteratively generates a sequence of random images converging to the desired result under the MAP criterion. Here, the SA algorithm visits all the sites (pixels) in the image and updates the corresponding configuration (e.g. intensity value) of a visited site. The new configuration of a pixel is generated through a random number generator whose underlying probability depends on observed information at that site and configurations of its neighboring pixels.

A similar approach to solve the problem under the MAP criterion is implemented in the Metropolis algorithm (Winkler 1995) where a new configuration is proposed randomly and the algorithm can either accept or reject this proposed configuration based on its underlying probability. Hasting (1970) generalized the Metropolis algorithm to the Metropolis-Hasting (MH) algorithm where the proposed configuration can have any well-defined probability density function (PDF). Begas (1986) developed an iterated conditional modes (ICM) algorithm to speed up the image restoration process. The ICM algorithm converges within several iterations whereas the previous algorithms may take hundreds of iterations. However, the ICM algorithm may have lower performance because it converges to the closest saddle point, which is more likely to be a local one. As an alternative to the MAP criterion, Marroguin et al. (1987) chose to maximize the *a posteriori* marginal (MPM) where the goal is to minimize the expectation of a cost function. They claimed that the MPM criterion outperforms the MAP criterion under low signal to noise ratio conditions.

The above methods are based on Monte Carlo procedures. A method not based on the Monte Carlo procedure is the graduated nonconvexity (GNC) algorithm introduced by Nikolova (1999). This algorithm follows a different approach and directly combats the severe nonconvexity of the objective function. The objective function is assumed to be composed of two terms. The first term is convex while the second term is nonconvex. At the beginning, nonconvexity is removed by setting the nonconvex term equal to zero. Then, the corresponding global optimal is determined by using the conventional derivative. Next, nonconvexity is gradually inserted back by multiplying the nonconvex term with a positive constant, $r \in (0, 1)$, and the local optimal closest to the current location is obtained by a gradient search. This process is repeated until the original nonconvex function is reached when $r = 1$. It was shown that under a proper release schedule of nonconvexity, the global optimum point could be reached. However, the GNC algorithm is valid only if an objective function is twice differentiable which might not be true in general.

Before we discuss the implications of MRF models to remote sensing applications in Chaps. 11 and 12, some basic concepts related to the MRF model are presented in this chapter. Section 6.2 is devoted to a detailed discussion of an MRF and its equivalent form, namely a Gibbs field. We also establish the equivalence property between MRFs and Gibbs fields. Next, some possible approaches to employ MRF modeling are examined in Sect. 6.3. Here, we also construct the posterior energy function under the MAP criterion, which is used as an objective function for optimization problems. Then, several widely used optimization methods including simulated annealing are introduced and discussed in Sect. 6.4. Some theoretical results are also presented in this section to make this chapter self-contained.

6.2
MRF and Gibbs Distribution

6.2.1
Random Field and Neighborhood

Consider a stochastic sequence, $\{x_n\}_{n\geq 0}$. It is said to have the Markov property if for $n \geq 1$, x_n is statistically independent of $\{x_k, k = 0, 1, \ldots, n-2\}$ given x_{n-1}. This suggests a local dependency among adjacent time slots. In more general cases, we can extend the notion of dependency to multiple dimensions. But, before defining the MRF in multidimensional systems, we need to first define some terms (Winkler 1995).

6.2.1.1
Random Field

Let $\mathcal{S} = \{s_1, s_2, \cdots, s_M\}$ be a finite set. An element denoted by $s \in \mathcal{S}$ is called a *site*. Let Λ be a finite set called a *phase* or *configuration* space. A random field on $\mathcal{S}$ with phase Λ is defined as a collection $X = \{X(s)\}_{s \in \mathcal{S}}$ of random variables $X(s)$ taking values in the phase space Λ.

A random field can be viewed as a random variable or a random vector taking values in the configuration space $\Lambda^\mathcal{S}$. A configuration $x \in \Lambda^\mathcal{S}$ is of the form $x = (x(s), s \in \mathcal{S})$, where $x(s) \in \Lambda$ for all $s \in \mathcal{S}$. We note that, in image processing applications, $\mathcal{S}$ is normally a subset of Z^2 and Λ represents different intensity values of the image where Z is the set of all integers. Our particular interest in this chapter is on a Markov random field, which is characterized by local interactions. As a result, we need to introduce a neighborhood system on the sites.

6.2.1.2
Neighborhood

A neighborhood system on $\mathcal{S}$ is a family $N = \{N_s\}_{s \in \mathcal{S}}$ of subsets of $\mathcal{S}$ such that for all $s \in \mathcal{S}$,

  i) $s \notin N_s$
  ii) $t \in N_s \Rightarrow s \in N_t$

The subset N_s is called the neighborhood of the site s. The pair $(\mathcal{S}, N)$ is called a graph, or a topology. The boundary of $A \subset \mathcal{S}$ is the set $\partial A = \left(\bigcup_{s \in A} N_s \right) \backslash A$ where $B \backslash A$ denotes the complement of A in B.

6.2.1.3
Markov Random Field

A random field X is called a Markov random field (MRF) with respect to the neighborhood system N if for all sites $s \in \mathcal{S}$, the random variables $X(s)$ and $X(\mathcal{S} \backslash \tilde{N}_s)$ are independent given $X(N_s)$ where $\tilde{N}_s$ is defined as $N_s \cup \{s\}$.

The above definition can also be written in a mathematical form as

$$P\{X(s) = x(s) \,|\, X(\mathcal{S}\backslash s) = x(\mathcal{S}\backslash s)\} = P\{X(s) = x(s) \,|\, X(N_s) = x(N_s)\} \quad (6.1)$$

for all $s \in \mathcal{S}, x \in \Lambda^{\mathcal{S}}$. The above conditional probability is also denoted by

$$\pi^s(x) = P\{X(s) = x(s) \,|\, X(N_s) = x(N_s)\}, \quad (6.2)$$

and is called the *local characteristic* of the MRF at the site s that maps Λ onto an interval $[0, 1]$. The family $\{\pi^s\}_{s \in \mathcal{S}}$ is called the *local specification* of the MRF.

Note that the property in (6.1) is clearly an extension of the general concept of Markov processes.

6.2.1.4
Positivity Condition

The probability distribution π on the finite configuration space $\Lambda^{\mathcal{S}}$ where $\mathcal{S} = \{1, 2, \ldots, K\}$, is said to satisfy the positivity condition if for all $j \in \mathcal{S}, x_j \in \Lambda$

$$(\pi_j(x_j) = 0) \Rightarrow (\pi(x_1, x_2, \ldots, x_{j-1}, x_j, x_{j+1}, \ldots, x_K) = 0) \quad (6.3)$$

for all $x_1, x_2, \ldots, x_{j-1}, x_{j+1}, \ldots, x_K \in \Lambda$, where π_j is the marginal distribution corresponding to the site j.

6.2.2
Cliques, Potential and Gibbs Distributions

The MRF model characterizes the spatial dependence among neighboring sites. However, a direct implementation of (6.1) is not simple because the probabilities can take up any values. As a result, we introduce the Gibbs distribution in this section. The general notation for a Gibbs distribution comes from thermodynamics and statistical physics to explain the phenomenon of spin directions of the particles under a critical temperature. Here, we discuss this distribution

Markov Random Field Models

in the context of image processing. In 1902, Gibbs introduced the following probability distribution,

$$\pi_T(x) = \frac{1}{Z_T} \exp\left\{-\frac{1}{T}E(x)\right\} \qquad (6.4)$$

on the configuration space $\Lambda^\mathcal{S}$, where $T > 0$ is the temperature, $E(x)$ is the energy of the configuration x, and Z_T is the *normalizing constant*, or the *partition function*. Obviously, $E(x)$ can take values in $(-\infty, \infty)$ since $\pi_T(x) \in [0, 1]$. Generally, the energy function $E(x)$ describes local interactions in terms of a potential function which is defined over cliques.

6.2.2.1
Clique

Any singleton $\{s\}$ is a clique. A subset $C \subset \mathcal{S}$ with more than one element is called a clique of the graph $(\mathcal{S}, N)$ if and only if any two distinct sites of C are mutual neighbors. A clique C is called maximal if for any site s, $C \cup \{s\}$ is not a clique.

For the sake of clarification, we provide examples of cliques for two neighborhood systems: 4-neighborhood (Fig. 6.1a) and 8-neighborhood (Fig. 6.1b). The three cliques for the 4-neighborhood system as shown in Fig. 6.1a are: the singleton, the horizontal pair and the vertical pair. Similarly, there are ten cliques for the 8-neighborhood system: one singleton, pixel pairs with different orientations, various combinations of three pixels and the group of four pixels (see Fig. 6.1b). In this example, cliques C_2 and C_3 are the maximal cliques for the 4-neighborhood system, and the clique C_{10} is the maximal clique for the 8-neighborhood system.

6.2.2.2
Gibbs Potential and Gibbs Distribution

A Gibbs potential on $\Lambda^\mathcal{S}$ relative to the neighborhood system N is a collection $\{V_C\}_{C \subset \mathcal{S}}$ of a function $V_C : \Lambda^\mathcal{S} \to \mathbb{R} \cup \{+\infty\}$ such that

i) $V_C \equiv 0$ if C is not a clique,

ii) for all $x, x' \in \Lambda^\mathcal{S}$ and all $C \subset \mathcal{S}$,

$$(x(C) = x'(C)) \Rightarrow (V_C(x) = V_C(x')), \qquad (6.5)$$

where $\mathbb{R}$ is the set of all real numbers.

Hence, the energy function $E : \Lambda^\mathcal{S} \to \mathbb{R} \cup \{+\infty\}$ is said to derive from the potential $\{V_C\}_{C \subset \mathcal{S}}$ if

$$E(x) = \sum_C V_C(x). \qquad (6.6)$$

In this section, we have introduced and defined the Gibbs distribution whose probability density function depends on configurations of neighboring sites. It is clear that both MRF and Gibbs distribution are related.

There are many similarities between a Gibbs field and a Markov random field. A Gibbs field is defined over the neighborhood system N through the potential function and cliques. Likewise, an MRF, by definition, is defined for the same neighborhood system by means of the local characteristics. Hence, in the next theorem, we state that the potential function that leads to a Gibbs field gives rise to a local characteristic. The mathematical proof of the following theorem is beyond the scope of this book and can be found in Winkler (1995). Without loss of generality, we shall assume $T = 1$ throughout this section for notational convenience.

Theorem 6.1 *Gibbs fields are MRFs. Suppose X is a random field with the distribution π where the energy $E(x)$ is derived from a Gibbs potential in $\{V_C\}_{C \in \mathcal{S}}$ relative to the neighborhood system N, then X is Markovian relative to the same neighborhood system N. Moreover, its local specification is given by the formula*

$$\pi^s(x) = \frac{\exp\left\{-\sum_{C \ni s} V_C(x)\right\}}{\sum_{\lambda \in \Lambda} \exp\left\{-\sum_{C \ni s} V_C(\lambda, x(\mathcal{S} \setminus s))\right\}}, \quad (6.7)$$

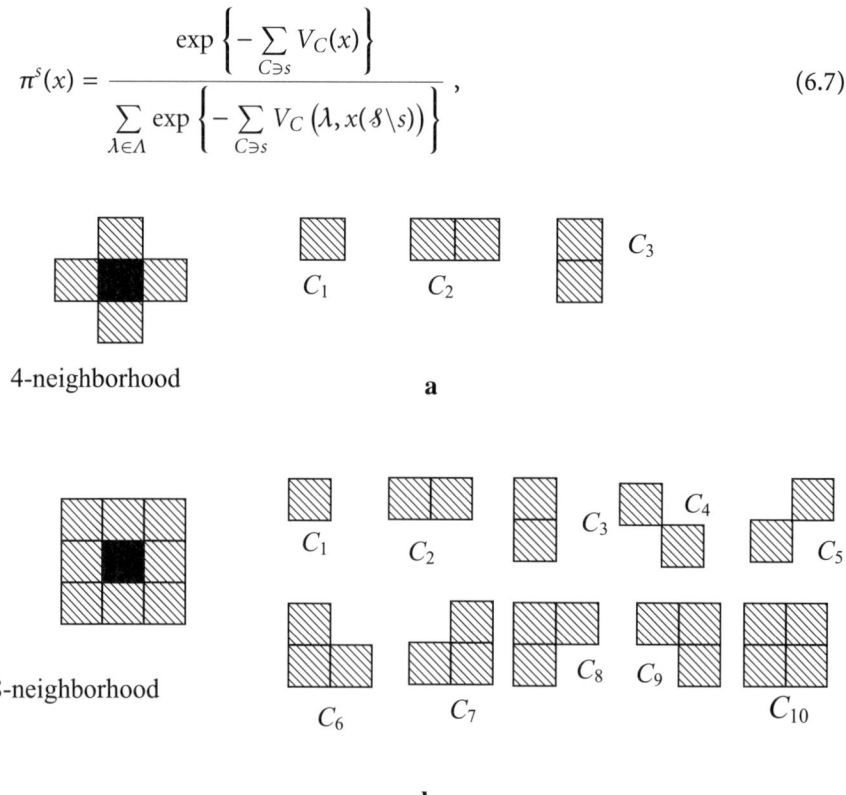

Fig. 6.1a,b. Cliques for **a** 4-neighborhood system **b** 8-neighborhood system

where $\sum_{C \ni s}$ denotes the sum over those sets C that contain the site s. We write $V_C\left(\lambda, x(\mathcal{S}\backslash s)\right)$ to represent the $V_C(x')$ where $x'(s) = \lambda$ and $x'(t) = x(t)$ for all $t \in \mathcal{S}\backslash s$.

Theorem 6.1 has stated the first part of the relationship between MRFs and Gibbs fields. We shall continue with the converse part, which will be formally introduced in Theorem 6.2. Again, the proof of this theorem can be found in Winkler (1995).

Theorem 6.2 (*Hammersley-Clifford Theorem*) *Let $X(\mathcal{S})$ be a set of configurations of an MRF defined on the graph $(\mathcal{S}, N)$, and π be the distribution of a random field $X(\mathcal{S})$ satisfying the positivity condition. Then, $\pi(x)$ for some energy function $E(x)$ derived from a Gibbs potential $\{V_C\}_{C \subset \mathcal{S}}$ associated with the topology $(\mathcal{S}, N)$, is given by*

$$\pi(x) = \frac{1}{Z} \exp\{-E(x)\} .$$

6.3
MRF Modeling in Remote Sensing Applications

So far, we have established the fundamental definitions and relationship of MRFs and Gibbs fields without linking these models to any remote sensing application, which is the main emphasis of this book. Hence, this section examines several approaches for the implementation of MRF modeling in remote sensing applications. There are a number of problems where the MRF models are applicable. However, we limit the scope of this section to a discussion on the application of MRF models for image classification of remotely sensed imagery. In the later chapters, some specific remote sensing problems, namely sub-pixel mapping for hyperspectral imagery (see Chap. 11), and image change detection and fusion (see Chap. 12), where MRF models can be applied, are dealt in great detail. Here, we first refer to Geman and Geman's paper, where MRF models have been applied for image restoration (Geman and Geman 1984). Image classification problem may be formulated as an image restoration problem. In the image restoration problem described in Geman and Geman (1984), the intensity value of an image pixel is disturbed by image noise that results in the mapping of the intensity value to a random variable. Likewise, in the image classification problem such as land cover classification, one land cover class attribute corresponds to a group of independent random vectors that follow an identical probability density function. In this model, we assume that, for a given land cover class, the intensity values of two different pixels (sites) are statistically independent, i.e,

$$\Pr\{Y(\mathcal{S})|X(\mathcal{S})\} = \prod_{s \in \mathcal{S}} \Pr\{y(s)|x(s)\}, \tag{6.8}$$

where $Y(\mathcal{S})$ and $X(\mathcal{S})$ are the observed remote sensing image and the corresponding land cover map, respectively. $y(s)$ and $x(s)$ are the intensity values (vector) of a site s in the observed remote sensing image and the attribute of the land cover class at the corresponding site, respectively. Often, the conditional probability density function (PDF) of the intensity value given a land cover class is assumed to follow a Gaussian distribution. Hence, we have

$$\Pr\{y(s)\,|x(s) = i\} = \frac{\exp\left[-\frac{1}{2}(y(s) - \boldsymbol{\mu}_i)^H \boldsymbol{\Sigma}_i^{-1}(y(s) - \boldsymbol{\mu}_i)\right]}{(2\pi)^{K/2}\sqrt{|\det(\boldsymbol{\Sigma}_i)|}} \tag{6.9}$$

or

$$\Pr\{y(s)\,|x(s)\} = \frac{\exp\left[-E_{MLE}(y(s)\,|x(s))\right]}{(2\pi)^{K/2}}, \tag{6.10}$$

where $\boldsymbol{\mu}_i$ and $\boldsymbol{\Sigma}_i$ are the mean vector and the covariance matrix corresponding to land cover class i, respectively, superscript H represents the Hermitian operation, and

$$E_{MLE}(y(s)\,|x(s) = i) = \frac{1}{2}(y(s) - \boldsymbol{\mu}_i)^H \boldsymbol{\Sigma}_i^{-1}(y(s) - \boldsymbol{\mu}_i) + \frac{1}{2}\log|\det(\boldsymbol{\Sigma}_i)|. \tag{6.11}$$

Here, we observe that a higher value of E_{MLE} results in a lower probability while a lower value of E_{MLE} corresponds to a higher probability. The conventional maximum likelihood classifier (MLC) labels a pixel s with a class attribute that minimizes E_{MLE}. Unfortunately, this approach is often inadequate since it does not consider the fact that the same land cover class is more likely to occur in neighboring pixels. In other words, the spatial structure in a remotely sensed land cover classification is neglected in the MLC.

The MRF models allow this spatial dependence to be integrated into the classification process. Here, the prior PDF of land cover maps is assumed to have the Gibbs distribution, i.e.,

$$\Pr(X(\mathcal{S})) = \frac{1}{Z}\exp\left[-\sum_{C \subset \mathcal{S}} V_C(X)\right]. \tag{6.12}$$

Based on (6.11) and (6.12), the posterior probability is given by

$$\Pr(X(\mathcal{S})\,|Y(\mathcal{S})) = \frac{1}{Z}\exp\left[-\sum_{C \subset \mathcal{S}} V_C(X) - \sum_{s \in \mathcal{S}} E_{MLE}(y(s)\,|x(s))\right]$$

$$= \frac{1}{Z}\exp\left[-E_{\text{post}}\right], \tag{6.13}$$

where

$$E_{\text{post}} = \sum_{C \subset \mathcal{S}} V_C(X) + \sum_{s \in \mathcal{S}} E_{MLE}(y(s)\,|x(s)). \tag{6.14}$$

The term in (6.14) is called the posterior energy function. Like in the case of E_{MLE}, higher and lower values of E_{post} correspond to lower and higher values of posterior probabilities. By choosing the configurations $X(\delta)$ that minimize E_{post}, we actually maximize the posterior probability. In other words, the maximum *a posteriori* (MAP) criterion is used. Under this criterion, the probability of error, i.e., that of choosing incorrect classified maps is minimized. We note that there are other optimization criteria such as the MPM (maximization of the posterior marginals) developed in Marroguin et al. (1987) that can be employed. However, the MAP criterion is designed to optimize a statistical quantity, namely the probability of error. Hence, the MAP criterion is chosen as the optimization criterion for all the MRF-based image analysis problems in this book.

Although the MAP criterion is meaningful and powerful in solving various image analysis problems, it does not provide any systematic method to find its solutions. If E_{post} is differentiable and convex (i.e. it has only one saddle point), any gradient-based optimization algorithm can be used to search for the MAP solutions. Unfortunately, this is not the case in general. Hence, other optimization techniques that are capable of handling non-convex functions must be employed. In the next section, we introduce several optimization algorithms for finding the MAP solutions.

6.4
Optimization Algorithms

In the previous two sections, we have established the statistical models of the MRF and its equivalent form, Gibbs field. This model is used to describe spatial properties of an image. Depending upon the problem at hand, these images can be classified images (Solberg et al. 1996), noiseless images (Geman and Geman 1984), or change images (Bruzzone and Preito 2000; Kasetkasem and Varshney 2002). Then, based on MAP criterion, the best (optimum) image is chosen. If the image model is simple, the optimum solution can be determined by using simple exhaustive search or gradient search approaches. However, since the MRF model is complex, (i.e. its marginal distribution is non-concave), the optimum solution under the MAP criterion can no longer be obtained by just a gradient search approach. Furthermore, for most cases, the image space is too large for exhaustive search methods to handle in an efficient manner. For example, there are more than 2^{4000} possible binary images (i.e. the intensity value of a pixel can either be 0 or 1) of size 64×64. Hence, the need for more efficient optimization algorithms is obvious. In this section, we discuss several optimization algorithms that are widely used for solving MRF model based image analysis optimization problems. These include the simulated annealing (SA), Metropolis, and iterated conditional modes (ICM) algorithms.

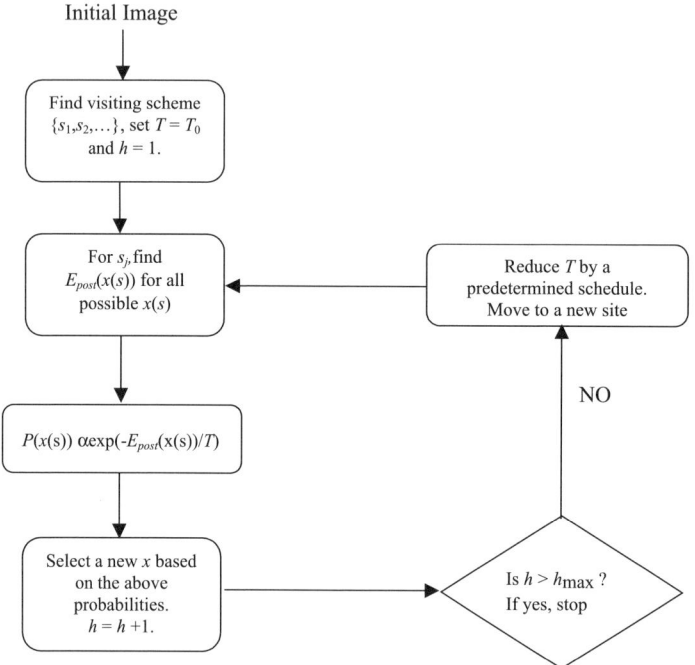

Fig. 6.2. Block diagram of the simulated annealing algorithm

6.4.1
Simulated Annealing

The SA algorithm, shown in Fig. 6.2, is a computationally efficient mathematical tool for the implementation of the MAP detector/estimator because a direct search in the configuration space is computationally expensive for large images. The idea of the SA algorithm is to generate a random sequence of configurations (images) that eventually converges to solutions of the MAP problem. This method involves an inhomogeneous Markov chain and its time-dependent Markov kernel. The inhomogeneity is introduced via the temperature parameter. Here, the temperature decreases as the number of updates increases. In the limit, it eventually reaches zero when the number of times a site is visited approaches infinity. Before going into the details of the SA algorithm, we shall investigate the statistical behavior of the Gibbs field for arbitrarily low temperatures.

Proposition 6.1 *Let $\pi_T(x) = \frac{1}{Z_T} \exp\left(-\frac{1}{T}E(x)\right)$ be a Gibbs distribution. Then*

i) $$\lim_{T \to 0} \pi_T(x) = \begin{cases} \frac{1}{\|x_m\|} & \text{if } x \in x_m \\ 0 & \text{otherwise} \end{cases}, \quad (6.15)$$

where x_m denotes the set of all minima of E, and $\|a\|$ denotes the cardinality of a set a.

ii) For all $T' < T'' < \varepsilon$ and some $\varepsilon > 0$, if $x \in x_m$, then $\pi_{T'} > \pi_{T''}$ and if $x \notin x_m$, then $\pi_{T'} < \pi_{T''}$.

Proof: Let E_m denote the minimum value of E. Then

$$\pi_T(x) = \frac{\exp\left(-\frac{1}{T}E(x)\right)}{\sum_{\lambda \in \Lambda^s} \exp\left(-\frac{1}{T}E(\lambda)\right)} \tag{6.16}$$

$$= \frac{\exp\left(-\frac{1}{T}\left(E(x) - E_m\right)\right)}{\sum_{\lambda \in x_m} \exp\left(-\frac{1}{T}\left(E(\lambda) - E_m\right)\right) + \sum_{\lambda \notin x_m} \exp\left(-\frac{1}{T}\left(E(\lambda) - E_m\right)\right)}$$

$$= \frac{\exp\left(-\frac{1}{T}\left(E(x) - E_m\right)\right)}{\|x_m\| + \sum_{\lambda \notin x_m} \exp\left(-\frac{1}{T}\left(E(\lambda) - E_m\right)\right)}. \tag{6.17}$$

When x or $\lambda \in x_m$ then the respective exponent vanishes as $T \to 0$. As a result, the corresponding exponential term becomes one and we obtain the left term in the denominator of (6.17). However, for all x or $\lambda \notin x_m$, the respective exponent is always negative. Therefore, the corresponding exponential decreases to zero. This completes the proof of Part 1 of the proposition.

Now, we will prove the second part. First, let $x \notin x_m$, and $a(y) = E(y) - E(x)$. Then, we rewrite (6.16) in the following form:

$$\pi_T(x) = \frac{1}{\left[\|\{y : E(y) = E(x)\}\| + \sum_{y : a(y) < 0} \exp\left(-\frac{1}{T}a(y)\right) + \sum_{y : a(y) > 0} \exp\left(-\frac{1}{T}a(y)\right)\right]}.$$

Furthermore, if Part 2 is true, then $T' < T'' < \varepsilon$ implies $\pi_{T'} < \pi_{T''}$ or $\frac{d\pi_T}{dT} > 0$ for $\forall T \leq \varepsilon$. Hence,

$$\frac{d\pi_T}{dT} = -\frac{\sum_{y:a(y)<0} \frac{a(y)}{T^2} \exp\left(-\frac{1}{T}a(y)\right) + \sum_{y:a(y)>0} \frac{a(y)}{T^2} \exp\left(-\frac{1}{T}a(y)\right)}{\left[\|\{y : E(y) = E(x)\}\| + \sum_{y:a(y)<0} \exp\left(-\frac{1}{T}a(y)\right) + \sum_{y:a(y)>0} \exp\left(-\frac{1}{T}a(y)\right)\right]^2}.$$

Since the denominator of the above equation is always positive, we only have to consider the numerator:

$$\frac{1}{T^2}\left\{-\sum_{y:a(y)<0} a(y)\exp\left(-\frac{1}{T}a(y)\right) - \sum_{y:a(y)>0} a(y)\exp\left(-\frac{1}{T}a(y)\right)\right\}.$$

The second term tends to zero, and the first term tends to infinity as $T \to 0$. Therefore, there must exist some ε such that $\frac{d\pi_T}{dT} > 0$ for $\forall T \leq \varepsilon$.

For $x \in x_m$, the distribution is

$$\pi_T(x) = \frac{\exp\left(-\frac{1}{T}\left(E(x) - E_m\right)\right)}{\|x_m\| + \sum_{\lambda \notin x_m} \exp\left(-\frac{1}{T}\left(E(\lambda) - E_m\right)\right)}$$

$$= \frac{1}{\|x_m\| + \sum_{\lambda \notin x_m} \exp\left(-\frac{1}{T}\left(E(\lambda) - E_m\right)\right)}.$$

Because for $\lambda \notin x_m$, $E(\lambda) - E_m$ is always positive, we have

$$\exp\left(-\frac{1}{T'}\left(E(\lambda) - E_m\right)\right) < \exp\left(-\frac{1}{T''}\left(E(\lambda) - E_m\right)\right) \tag{6.18}$$

for $0 < T' < T'' < \varepsilon$. Therefore,

$$\sum_{\lambda \notin x_m} \exp\left(-\frac{1}{T'}\left(E(\lambda) - E_m\right)\right) < \sum_{\lambda \notin x_m} \exp\left(-\frac{1}{T''}\left(E(\lambda) - E_m\right)\right)$$

for $0 < T' < T'' < \varepsilon$. Hence, there exists an ε such that

$$\pi_{T'}(x) > \pi_{T''}(x)$$

for all $x \in x_m$. Q.E.D.

From the above proposition, we observe that the Gibbs distribution at zero temperature is uniformly distributed among the global optima. If the zero temperature stage can be reached from arbitrary starting points (configurations), the optimum solution under the MAP criterion can also be obtained regardless of the initial configurations. The procedure described in Fig. 6.2 attempts to produce an inhomogenous Markov chain of images (X) that eventually converges to the limiting distribution in (6.15).

To successfully approach the result given by (6.15), we first need to determine the *visiting scheme* defined as $\mathcal{S} = \{1, 2, \ldots, M\}$. Different visiting schemes may result in different rates of convergence of the SA algorithm. From practical considerations, the row wise visiting scheme may be employed due to its simplicity. However, in the literature (e.g. Bremaud 1999), more random schemes that may result in faster convergence rates than the row wise visiting scheme, have also been used. In addition, the "*cooling schedule*" which is a decreasing sequence of the positive value $T(n)$ that eventually becomes zero needs to be determined. To guarantee convergence, the cooling sequence must decrease to zero at the rate at least

$$T(n) \geq \frac{M\Delta}{\ln(n)}, \tag{6.19}$$

where $\Delta = \max_{s,x,y} \left\{ \left| E(y) - E(x) \right| : x_{\mathcal{S}\setminus s} = y_{\mathcal{S}\setminus s} \right\}$ and $M = |\mathcal{S}|$.

Markov Random Field Models

After determining a visiting scheme and a cooling schedule, the algorithm sets the temperature to the initial value, say T_0, and either randomly or deterministically picks an initial configuration (image X). Then, a sweep of an entire image is performed. During a sweep, the algorithm visits a site at a time until all sites are visited. Without loss of generality, we assume that a site s_1 is being visited. Here, the algorithm computes the energy E_{post} for all possible configurations at s_1 while keeping the configurations at all the other sites fixed. As a result, a set of posterior probabilities associated with all the possible configurations at s_1 is obtained. Next, a new configuration at s_1 is generated based on the computed posterior probabilities (e.g. if $P_{\text{post}}(x(s_1) = 0) = 0.7$ and $P_{\text{post}}(x(s_1) = 1) = 0.3$, then there are 70% and 30% chances that the new configuration at s_1 is 0 or 1, respectively). After the update at s_1, the algorithm moves to the next site in the visiting scheme, increases the counter, and decreases the temperature parameter. The above procedure is repeated until the counter exceeds a predetermined value at which point the algorithm is terminated.

In general, the cooling schedule given in (6.19) is computationally not feasible. For instance, when $\Delta = 1$ and $M = 64 \times 64$, it will take more than e^{4000} sweeps before the temperature reduces to 1 from the lower bound given by (6.19). Therefore, in practice, we usually start with a much lower temperature (e.g. between 1 and 5) so that the steady state is reached sooner. Of course, this violates the condition given in (6.19) and, therefore, convergence is not guaranteed. But in most cases, we are able to start from initial images that are fairly close to one of the minima of E such that the SA procedure converges in a reasonable amount of time.

Example

We present a simple example to illustrate the implementation of the SA algorithm. In this example, we consider a Gibbs field that consists of two sites $\{1, 2\}$ whose configurations can be either -1 or 1. The Gibbs distribution associated with this system is given by

$$\pi_T = \frac{1}{Z} \exp\left(-I(x_1, x_2)\right), \tag{6.20}$$

where

$$I(a, b) = \begin{cases} -1 & \text{if } a = b \\ 1 & \text{if } a \neq b \end{cases} \tag{6.21}$$

and $Z = \sum_{x_2=\{-1,1\}} \sum_{x_1=\{-1,1\}} \exp\left(-I(x_1, x_2)\right) = 6.17$. The prior probabilities associated with each configuration can be computed from (6.20) and are given in Table 6.1.

Furthermore, let us assume that the realization of the configuration is $\{1, 1\}$, but because of identical, independent Gaussian noise with zero mean and unit variance, we actually obtain the real number $\{y_1, y_2\} = \{0.5, -0.15\}$. The

Table 6.1. Prior probabilities of $\{x_1, x_2\}$

$\{x_1, x_2\}$	$\{-1,-1\}$	$\{-1,1\}$	$\{1,-1\}$	$\{1,1\}$
$\pi(x_1, x_2)$	0.44	0.06	0.06	0.44

MAP detector is chosen here to estimate the configurations from the observed data $\{y_1, y_2\}$. The goal of the MAP detector is to select $\{\hat{x}_1, \hat{x}_2\}$ such that the *a posteriori* probability is maximum, i.e.,

$$\{\hat{x}_1, \hat{x}_2\} = \arg\left[\max_{\{x_1, x_2\}} \left(\Pr(x_1, x_2 | y_1, y_2)\right)\right]. \tag{6.22}$$

The MAP detector is optimum under the minimum probability of error criterion (Trees 1968; Varshney 1997). The posterior probability in this example is given by

$$\Pr(y_1|\hat{x}_1)\Pr(y_2|\hat{x}_2)\pi(\hat{x}_1, \hat{x}_2)$$
$$= \frac{1}{Z(2\pi)} \exp\left[-\frac{1}{2}(y_1 - \hat{x}_1)^2 - \frac{1}{2}(y_2 - \hat{x}_2)^2 - I(\hat{x}_1, \hat{x}_2)\right].$$

Under the MAP criterion, it is equivalent to choose $\{\hat{x}_1, \hat{x}_2\}$ such that

$$E(\hat{x}_1, \hat{x}_2) = \frac{1}{2}(y_1 - \hat{x}_1)^2 + \frac{1}{2}(y_2 - \hat{x}_2)^2 + I(\hat{x}_1, \hat{x}_2)$$

is minimum. The energies associated with different $\{\hat{x}_1, \hat{x}_2\}$ can be computed using the above equation and are shown in Table 6.2. The minimum energy occurs correctly at $\{\hat{x}_1, \hat{x}_2\} = \{1, 1\}$.

The SA algorithm is used to solve this problem with initial configuration $\{1, -1\}$ and cooling schedule

$$T(n) = \frac{2.2}{\log(n+1)}.$$

The results are plotted in Fig. 6.3a,b for x_1 and x_2, respectively. We observe that during the early stages of the algorithm, configurations frequently switch between -1 and 1. This implies the extreme randomness of the SA algorithm at high temperature values. After around 700 sweeps, the configurations essentially converge to the desired value, $\{1, 1\}$.

Table 6.2. The Gibbs energy associated with all possible configurations

$E(x_1, x_2)$	$x_1 = -1$	$x_1 = 1$
$x_2 = -1$	0.4864	1.4863
$x_2 = 1$	2.7862	-0.2138

Markov Random Field Models

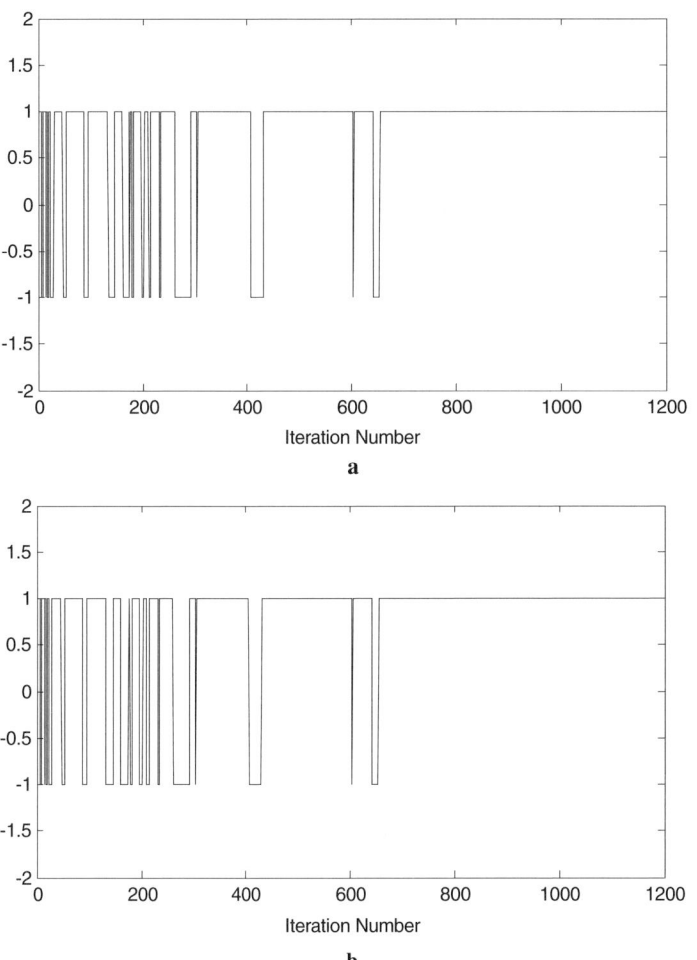

Fig. 6.3a,b. The result of implementing the SA algorithm for Example **a** x_1, **b** x_2

6.4.2
Metropolis Algorithm

A popular alternative to the SA algorithm is the Metropolis algorithm. Unlike the SA algorithm that uses a Gibbs sampler to generate a new configuration, the Metropolis algorithm randomly proposes a new configuration and evaluates the energy function corresponding to this configuration. If the new configuration provides a lower energy function than the current configuration, it will be accepted with probability one otherwise with a certain probability less than one. Let E denote the energy function of interest and let x be the current configuration, the two steps of Metropolis algorithm, shown in Fig. 6.4, can be written as,

1. The propose step: A new configuration x_1 is proposed by sampling a PDF $G(x_2, x_1)$ on Λ.

2. The acceptance step: If $E_{\text{post}}(x_1) \leq E_{\text{post}}(x_2)$ then x_1 is accepted as a new configuration with probability one. However, if $E_{\text{post}}(x_1) > E_{\text{post}}(x_2)$ then x_1 is accepted with probability

$$\alpha(x_2, x_1) = \exp\left[-\frac{1}{T}\left(E_{\text{post}}(x_1) - E_{\text{post}}(x_2)\right)\right]. \quad (6.23)$$

The matrix G is called the proposal or exploration matrix. A new configuration x_1 is not automatically rejected if it is less favorable than the current configuration x_2. Instead, a new configuration x_1 is accepted with probability that decreases with the increment of difference in energies $E_{\text{post}}(x_2) - E_{\text{post}}(x_1)$. This procedure allows the algorithm to escape from the local minimum points.

The obvious difference between the SA and the Metropolis algorithms is that the SA algorithm needs to compute the probabilities associated with all possible configurations at a given site while the Metropolis algorithm only requires the difference between energies of the current and proposed configurations in order to update the configuration of the site of interest. This makes the Metropolis algorithm more efficient when a configuration

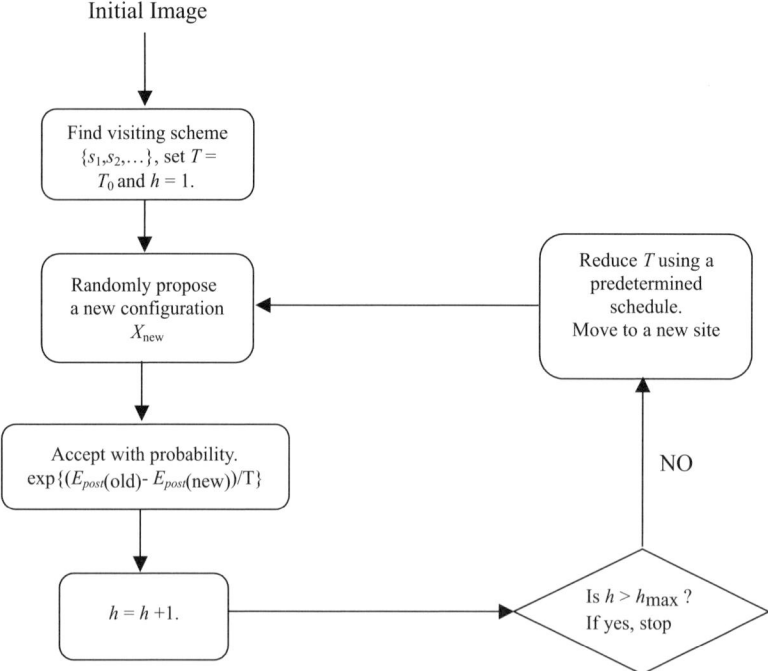

Fig. 6.4. Block diagram of the Metropolis algorithm

Markov Random Field Models

can take many different values. However, since we used the proposal matrix G to propose a new configuration at a given site, there is a possibility that some configurations may not be reachable after completion of a sweep (complete update of all pixels.). This causes the transition matrix to have zero elements which may jeopardize the positivity conditions required for convergence of the induced Markov chain. As a result, the sufficient conditions for the Metropolis algorithm must be modified to make the positivity condition valid. The following definition provides this sufficient condition for a proposal matrix that will guarantee the positivity conditions for the transition matrix.

Definition: A Markov kernel G on Λ^s is called irreducible if for all $x, y \in \Lambda^s$ there exists a chain $x = u_0, u_1, \ldots, u_{\sigma(x,y)} = y$ in Λ^s such that $G(u_{j-1}, u_j) > 0$ for $1 \leq j \leq \sigma(x, y) < \infty$. The corresponding homogeneous Markov chain is also called irreducible as well. Next, we shall call $y \in \Lambda^s$ a neighbor of $x \in \Lambda^s$ if $G(x, y) > 0$, i.e.,

$$N(x) = \left\{ y \in \Lambda^s : x \neq y, G(x, y) > 0 \right\}. \tag{6.24}$$

The cooling schedule for the Metropolis algorithm is different from the SA algorithm due to the fact that, in the SA algorithm, any possible configurations can be reached after one sweep of an entire image while several sweeps may be required in the Metropolis algorithm. The optimum cooling schedule is given by

$$\frac{\tau \Delta}{\ln(n)},$$

where $\tau = \max \left\{ \sigma(x, y) : x, y \in \Lambda^s \right\}$.

6.4.3 Iterated Conditional Modes Algorithm

For many remote sensing applications, we often deal with images of very large size (e.g. data from hyperspectral sensors or high resolution images). These data are too large for a global optimization algorithm to handle in an efficient manner, even with highly efficient algorithms such as the SA and Metropolis algorithms. In such cases, it is preferable to deal with a simpler and less computationally intensive optimization algorithm that may only guarantee local optima rather than a more accurate and complex global optimization algorithm (such as the SA and Metropolis algorithms). Among suboptimum algorithms, the iterated conditional modes (ICM) algorithm, proposed by Begas (1986), has received a great deal of attention because of its simplicity and fast convergence rate. In this algorithm, the MAP equation is still the objective function to be optimized. However, unlike the previous two algorithms where an image model can take any valid forms (having a finite PDF), the ICM algorithm only permits the prior probability of an image to be the multiplication of local characteristics of all sites, i.e.,

$$\Pr(X) = \prod_{s \in \mathcal{S}} \Pr\left(X(s) \,|\, X\left(\mathcal{S} \backslash s\right)\right). \tag{6.25}$$

The above assumption does not fit the conventional MRF models, described in Sects. 6.2 and 6.3. Nevertheless, it still considers the local interactions among adjacent pixels through a statistical model while allowing the optimization process to be completed at a faster rate because it deals with individual local characteristics rather than the Gibbs distribution of an entire image. In addition, Begas pointed out that an image analysis algorithm based on the conventional MRF model may not always produce an accurate result because the conventional MRF model tends to favor single color cases (e.g. an entire image composed of only one intensity value) of X over the more realistic multiple color cases (multiple land cover class attributes) of X. Single or multiple color cases of X do not have significant influence on the prior probability under the Begas' model as long as a spatial dependence is characterized correctly. As a result, the model in (6.25) may be more suitable for the multiple color cases.

There are several approaches to find the optimum solutions of the MAP problem under the assumption given in (6.25). The simplest one is to visit one pixel (e.g. s) at a time, and try to replace the configuration (e.g. $x(s)$) of this pixel with the configuration that maximizes the posterior probability. As mentioned above, computation of the posterior probability can be obtained easily under (6.25) than the conventional MRF model because, for one update, we need to determine only the summation over all the possible configurations of a site rather than the summation of all the possible configurations of the entire image. Furthermore, the convergence of this method is guaranteed as long as the posterior probability is well defined, since the posterior probability always increases as the number of iterations increases, and is bounded by one. The above procedure is summarized in Fig. 6.5. Another approach is to update the configuration of a site by only considering its local characteristic, i.e.,

$$x_{\text{new}}(s) = \arg \left[\max_{l} \left(\Pr\left(y(s) \,|\, l\right) \Pr\left(l \,|\, N_{s,\text{old}}\right) \right) \right], \tag{6.26}$$

where the subscripts new and old indicate configurations of the current and previous iterations, respectively. For this second approach, the convergence may not always be achieved since a higher posterior probability may not be attained after one complete update. Moreover, based on the same reason, the configurations may oscillate between two or more values. However, this approach allows the entire image to be updated at the same time because the update at a site has no effect on other sites.

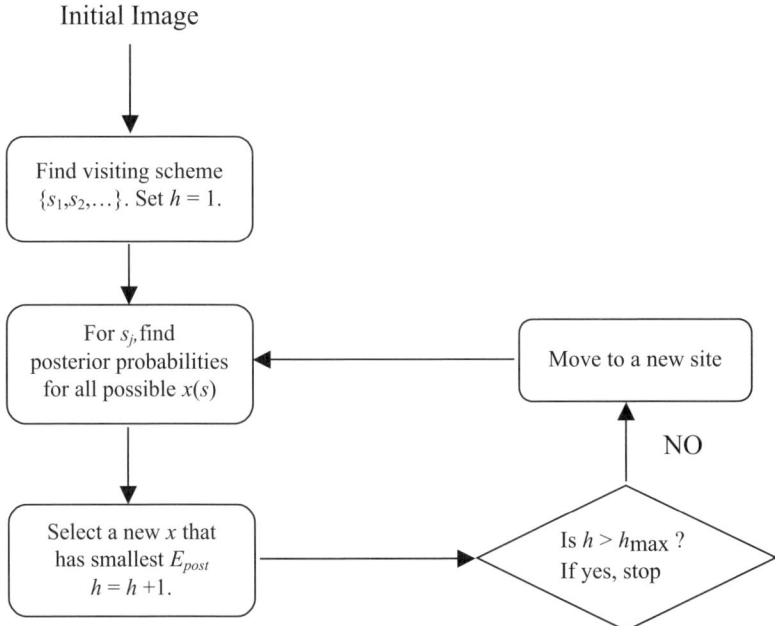

Fig. 6.5. Block diagram of the ICM algorithm

6.5
Summary

In this chapter, we have introduced the concept of Markov random field models. Here, statistical correlations among neighboring pixels or sites are quantified through the Gibbs energy functions (e.g. Ising model). The low and high energy values are associated with similarity and dissimilarity between the configurations (intensity values) of neighboring sites. The exponent of negative energy functions was defined as the Gibbs distribution, which is equivalent to the MRF model. Based on the MRF models, the maximum *a posteriori* (MAP) criterion was selected to solve image analysis problems. Here, the most likely image given the observed image was chosen as the optimum solution. The simulated annealing and Metropolis algorithms were proposed as the suitable choices to find the optimum solution under the MAP criterion because image spaces are generally very large and the *a posteriori* probability is also extremely concave for exhaustive search algorithms or gradient-based approaches to handle efficiently. Both algorithms generate a sequence of random images that converge to the global optima. Convergence usually occurs after hundreds of iterations. To speed up the convergence rate, a suboptimum approach, namely the ICM algorithm, was also introduced in this chapter to reduce computational time. The ICM algorithm looks for the closest saddle point of the *a posteriori* probability.

References

Begas J (1986) On the statistical analysis of dirty pictures. Journal of Royal Statistical Society B 48(3): 259–302

Bremaud P (1999) Markov chains Gibbs field, Monte Carlo simulation and queues. Springer Verlag, New York

Bruzzone L, Prieto DF (2000) Automatic analysis of the difference image for unsupervised change detection. IEEE Transactions on Geoscience and Remote Sensing 38(3):1171–1182

Geman S, Geman D (1984) Stochastic relaxation, Gibbs distributions and the Bayesian restoration of images. IEEE Transactions on Pattern Analysis and Machine Intelligence PAMI-6(6): 721–741

Hasting WK (1970) Monte Carlo sampling method using Markov chains and their applications. Biometrika 57(1): 97–109

Kasetkasem T, Varshney PK (2002) An image change detection algorithm based on Markov random field models. IEEE Transactions on Geoscience and Remote Sensing 40(8):1815–1823

Marroguin J, Mitter S, Poggio T (1987) Probabilistic solution of ill-posed problems in computer vision. Journal of the American Statistical Association 82: 76–89

Nikolova M (1999) Markovian reconstruction using a GNC approach. IEEE Transactions on Image Processing 8(9): 1204–1220

Solberg AHS, Taxt T, Jain AK (1996) A Markov random field model for classification of multisource satellite imagery. IEEE Transactions on Geoscience and Remote Sensing 34:100–113

Van Trees HL (1968) Detection, estimation and modulation theory. Wiley, New York

Varshney PK (1997) Distributed detection and data fusion. Springer Verlag, New York

Winkler G (1995) Image analysis random fields and dynamic Monte Carlo methods. Springer Verlag, New York

Part III
Applications

CHAPTER 7

MI Based Registration of Multi-Sensor and Multi-Temporal Images

Hua-mei Chen, Pramod K. Varshney

7.1
Introduction

The necessity of accurate registration and geometric rectification arises due to the presence of a number of distortions (errors) in remote sensing images that occur as a result of variations in platform positions, rotation of earth and relief displacements etc. (see Chap. 2). The behavior of most of these distortions is systematic and, thus, can be easily removed at data acquisition centers. It is also generally expedient to procure systematic-corrected images as these are already geometrically rectified to a map projection system such as the Universal Transverse Mercator (UTM). However, systematic corrections are normally performed on the basis of platform ephemeris data obtained from the header information, which may be relatively inaccurate. Therefore, some distortions may still be present in the systematic-corrected data. This can be illustrated with the help of agency supplied Landsat TM images taken at two different times as shown in Fig. 7.1. The images were systematically corrected and geometrically rectified to UTM at the data acquisition center. It can, however, be seen that the two points (marked by "+") having the same UTM coordinates in both images are located at different positions indicating the presence of non-systematic registration errors.

Traditionally, image registration for remote sensing applications is performed by selecting a few features, also known as ground control points (GCP), in both images. The GCP are matched by pair and transformation parameters to register the images are computed (see Chap. 2 for details). A resampling procedure is then employed to estimate the intensity values at the new sampled positions of the reference image. The feature based registration technique via manual selection of GCP is, however, laborious, time intensive and a complex task. Some automatic algorithms have been developed to automate the selection of GCP to improve efficiency (Toth and Schenk 1992; Dai and Khorram 1999). However, the extraction of GCP may still suffer from the fact that sometimes too few a points will be selected, and the extracted points may be inaccurate and unevenly distributed over the images. This may lead to large registration errors. On the other hand, intensity-based registration techniques do not involve feature identification and extraction. Hence, automatic intensity based registration techniques may be more appropriate than the feature

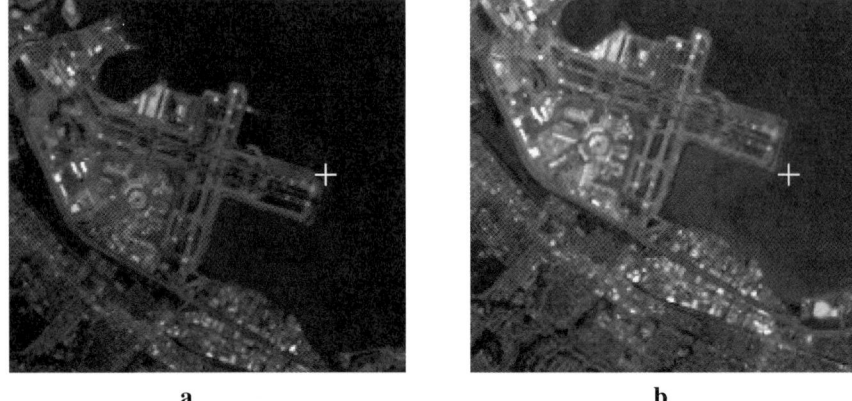

Fig. 7.1a,b. Multi-temporal remote sensing images. **a** Landsat TM band 1 images taken in 1997. **b** Landsat TM band 1 images taken in 1998. The two crosses denote the points of the same UTM coordinates in each image. The noticeable shift demonstrates the presence of registration errors in systematic-corrected data

based techniques. In this chapter, the use of MI as a similarity measure for intensity based registration of a variety of remote sensing images is presented. This chapter is built on the theoretical concepts discussed in Chap. 3. Three registration scenarios are considered here,

1. Multi-sensor registration (Chen et al. 2003a)
   (a) Registration of a pair of multi-sensor images, having a large difference in spatial resolution

   (b) Registration of a pair of multi-sensor images, having a small difference in spatial resolution

2. Multi-temporal registration (Chen et al. 2003b)

For multi-sensor registration, two cases are considered. In the first case, images with a large difference in spatial resolution are used to demonstrate a commonly used optimization procedure for image registration, namely multi-scale optimization (Pluim et al. 1998; Chen and Varshney 2000; Thevenaz and Unser 2000). In the second example, we use a pair of multi-sensor images with similar spatial resolutions to compare the performance of several MI based image registration algorithms. Each algorithm has its own unique joint histogram estimation method. As for multi-temporal image registration, we use the GPVE algorithm introduced in Chap. 3 to overcome the artifact problem, which is frequently encountered in many multi-temporal registration applications. The advantage of the MI based registration technique over the traditional intensity based registration techniques using normalized cross-correlation and mean squared error as similarity measures for multi-temporal registration problems is also demonstrated. Since no reference data is available for the images used in the experiments, registration consistency described in the next section is employed to evaluate the performances of different algorithms.

7.2
Registration Consistency

In the absence of reference data, registration consistency (Holden et al. 2000) may be used as a measure to evaluate the performance of different intensity-based image registration algorithms. Let $T_{A,B}$ ($T_{B,A}$) be the transformation obtained using image A (B) as the floating image and image B (A) as the reference image, the registration consistency (dp) of $T_{A,B}$ and $T_{B,A}$ can be defined as

$$\langle dp \rangle_{AB} = (1/N_A) \sum_{(x,y) \in (I_A \cap I_{A,B})} \| (x,y) - T_{B,A} \circ T_{A,B}(x,y) \|, \qquad (7.1a)$$

$$\langle dp \rangle_{BA} = (1/N_B) \sum_{(x,y) \in (I_B \cap I_{A,B})} \| (x,y) - T_{A,B} \circ T_{B,A}(x,y) \|, \qquad (7.1b)$$

where the composition $T_{A,B} \circ T_{B,A}$ represents the transformation that applies $T_{B,A}$ first and then $T_{A,B}$. The overlap region of images A and B is defined as $I_{A,B}$. The discrete domains of images A and B are I_A and I_B respectively, N_A and N_B are the number of pixels of image A and image B within the overlap region. The registration consistency defined in (7.1a) specifies the mean distance of the two mapped points $T_{A,B}(p)$ and $T_{B,A}^{-1}(p)$, where p is a pixel in image A. Similarly, (7.1b) represents the mean distance of the two mapped points $T_{B,A}(p)$ and $T_{A,B}^{-1}(p)$, where p is a pixel in image B. In general, it is expected that the two values from (7.1a) and (7.1b) will practically be the same. Similarly, a three-date registration consistency can be defined as

$$\langle dp \rangle_3 = \frac{1}{2N_A} \sum_{(x,y) \in (I_A \cap I_{A,B,C})} \| (x,y) - T_{C,A} \circ T_{B,C} \circ T_{A,B}(x,y) \|$$
$$+ \| (x,y) - T_{B,A} \circ T_{C,B} \circ T_{A,C}(x,y) \|, \qquad (7.2)$$

where $I_{A,B,C}$ is the overlap region of images A, B and C.

In cases where the transformation required to register the two images is just a displacement, as we will encounter in multi-temporal registration (see Sect. 7.4), the transformation $T_{A,B}$ can be represented as a 2D vector $v_{A,B} = (\Delta x, \Delta y)$, where Δx is the vertical displacement and Δy is the horizontal displacement. For this case, a simplified two-date registration consistency can be defined as

$$\langle dp \rangle_2 = \| v_{A,B} + v_{B,A} \|. \qquad (7.3)$$

Similarly, the three-date registration consistency can be defined as

$$\langle dp \rangle_3 = \left(\left\| v_{A,B} + v_{B,C} + v_{C,A} \right\| + \left\| v_{A,C} + v_{C,B} + v_{B,A} \right\| \right) / 2 . \tag{7.4}$$

Though the registration consistency defined above may not be treated as a measure of registration accuracy, it is one of the means to quantitatively assess the quality and reliability of the registration method employed because a reliable registration algorithm should result in a consistent result, no matter which image is served as the floating image and which is served as the reference image.

7.3
Multi-Sensor Registration

The primary advantage of using MI as a similarity measure lies in its ability to measure the similarity of two images taken from different types of sensors. Multi-sensor registration refers to the alignment of two images of the same scene acquired from different types of sensors. Accurate multi-sensor registration is essential for image fusion and image classification to improve the quality of extracted information for multi-sensor remote sensing data. For example, integrating images from different spectral bands may improve the ability to extract objects such as buildings, roads, vehicles, and type of vegetation from the data (Brown 1992). In this section, two multi-sensor registration cases are considered. They are; registration of multi-sensor images having a large difference in spatial resolution and registration of multi-sensor images having similar spatial resolution.

7.3.1
Registration of Images Having a Large Difference in Spatial Resolution

We consider a multi-sensor registration problem where we register a pair of images having a large difference in spatial resolution. We utilize a commonly used optimization procedure called multi-scale optimization to maximize the mutual information measure. Multi-scale optimization has been shown to be quite robust as the chances of getting trapped in a local optimum are lower, and is efficient (Pluim et al. 1998; Chen and Varshney 2000; Thevenaz and Unser 2000).

The data consists of two remote sensing images: one from a digital aerial photograph with spatial resolution 15 cm obtained from Eastman Kodak (Fig. 7.2a) and the other from the airborne HyMap sensor with spatial resolution 6.8 m (Fig. 7.2b). The spatial resolution of these images covering about the same area is so distinct that the sizes of these images are quite different (i.e. 100 × 100 pixels for the HyMap image and 4000 × 4000 pixels for the digital aerial photograph). Intuitively, one may choose either image as the floating image and the other as the reference image. However, our experiment shows that it is not

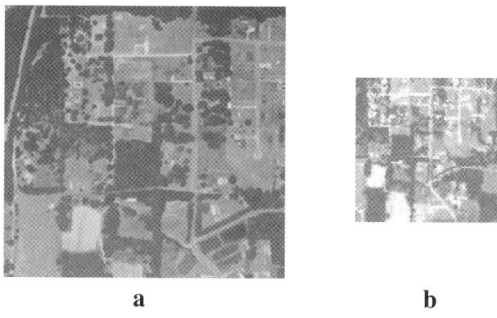

Fig. 7.2a,b. Images used in the experiment. **a** Digital aerial photograph (*red band*) from Eastman Kodak and **b** airborne HyMap image (band 9)

feasible to use the low resolution image, the HyMap image, as the floating image because the resulting MI registration function is very rough, which makes the optimization barely possible.

Figure 7.3 shows the 2D MI registration function using i) HyMap image (low resolution) as the floating image and ii) digital aerial photograph (high resolution) as the floating image. Clearly, the registration function shown in Fig. 7.3a is very rough and it is extremely difficult to find the global optimum. On the other hand, the registration function shown in Fig. 7.3b is very smooth and it is much easier to find the global optimum. For this reason, we use aerial photograph as the floating image in our experiment. The multi-scale optimization procedure developed for the registration of these two images is described next.

First, an image pyramid is constructed for the aerial photograph using Haar wavelet decomposition. This is done by repeating the following procedure at each resolution level of the image pyramid: convolving the image with an averaging filter of size 2×2 pixels followed by subsampling (taking every other pixel) along both vertical and horizontal directions to generate the image of the next (lower) resolution level. Many algorithms exist to generate an

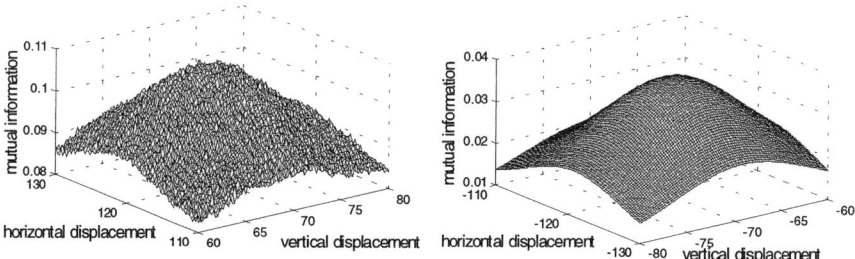

Fig. 7.3a,b. 2D registration function using **a** HyMap image and **b** Digital aerial photograph as the floating image. The displacements are shown in meters

Table 7.1. Image pyramid constructed from digital aerial photograph

Level	Image size [pixel × pixel]	Resolution (m)	Scaling factor
0	4000×4000	0.1508	0.0222
1	2000×2000	0.3016	0.0444
2	1000×1000	0.6032	0.0887
3	500×500	1.2064	0.1774
4	250×250	2.4128	0.3548
5	125×125	4.8256	0.7096

image pyramid (Burt and Adelson 1983; Burt 1984; Toet 1992). Here we used Haar wavelet decomposition to accomplish this task because of its simplicity. For the experiment, an image pyramid of 6 levels (level 0 $\sim$ level 5, level 0 is the original resolution) was constructed. Table 7.1 lists the image size and corresponding spatial resolution at each level. To apply the multi-scale optimization strategy, we start with the registration of the HyMap image and the aerial photograph at level 5. The aerial photograph at level 5 is served as the floating image because of the reason mentioned previously. We denote the transformation parameter set obtained at this level (i.e. scale, rotation, x-displacement (in pixel), y-displacement (in pixel)) as $[s_5, r_5, dx_5, dy_5]$, then the parameter set $[s_5/2, r_5, dx_5, dy_5]$ is used as the initial search point to optimize the MI similarity measure between the HyMap image (reference image) and the aerial photograph at level 4 (floating image). The reason for doubling the scaling factor is that the resolution of the digital aerial photograph is doubled when we move from level 5 to level 4, as a result, the scaling factor between the photograph at level 4 and the HyMap image is decreased by the same factor 2. The above procedure is repeated until the digital aerial photograph at level 0 (original resolution) is reached. Figure 7.4 summarizes the multi-scale optimization procedure stated above.

In Fig. 7.4, $\{F_i\}_{i=0}^{L-1}$ forms an image pyramid of L levels and the image at each level is used in consecutive iterations, from $i = L - 1$ to $i = 0$, as the

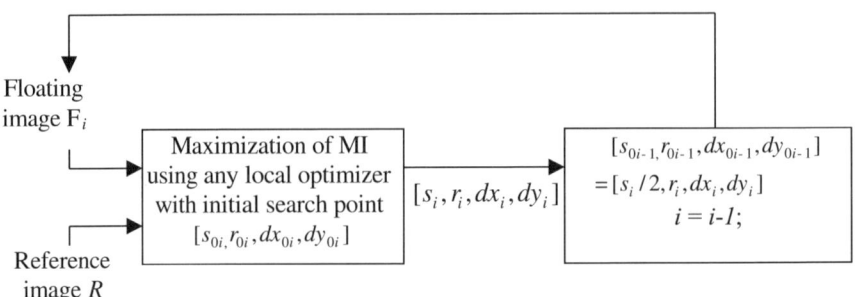

Fig. 7.4. The multi-scale registration procedure used in the experiment

floating image to be registered with the reference image R. Parameter set $[s_{0i}, r_{0i}, dx_{0i}, dy_{0i}]$ is the initial search point used at each level i. When $i = L - 1$, the initial search point is arbitrarily chosen. Parameter set $[s_i, r_i, dx_i, dy_i]$ represents the position of the optimum found by the local optimizer at level i. The initial search point at level $i - 1$ is then determined by the parameter set obtained at the previous level i by dividing the scaling factor s by 2.

In our experiments, the spatial resolutions of the two original images were given in the associated header files. Therefore, the scaling factors at each level are known. They are listed in the rightmost column of Table 7.1. The simplex search algorithm was employed as the local optimizer at each resolution level (Nelder and Mead 1965). In addition, the header files showed no rotational difference between the two images because they were both rectified to UTM map projection system. Therefore, only the two displacements are to be determined to register the two images. The success of this multi-scale optimization procedure strongly relies on the success of the optimization performed at the lowest resolution level $i = L - 1$. Unfortunately, in our experiment, by choosing zero displacements along both directions as the initial search point did not converge to the desired optimum and resulted in large registration error as observed from visual inspection. To overcome this problem, we adopted the hybrid global optimization procedure introduced in Sect. 3.6 at the lowest resolution level. First, the floating image is partitioned into four equal-sized subimages. After that, each of them is used as the floating image to register with the reference image. The initial search point generator shown in Fig. 3.15 (see Chap. 3), is designed in the following manner:

1. Randomly generate 30 points within the search space. In the experiment, the search space is defined as $[dx = \pm 30$ pixel, $dy = \pm 30$ pixel$]$

2. Choose the point that results in the maximum mutual information measure as the initial search point.

Using this hybrid global optimization procedure, the images can be registered successfully at the lowest resolution level, and therefore, at the consecutive higher levels of the image pyramid. Table 7.2 shows the registration results at each resolution level and Fig. 7.5 shows the registered image pair and the superimposed image. From Table 7.2 we observe that the registration results at

Table 7.2. Numerical results for HyMap and digital aerial photograph registration

Level (i)	dx (m)	dy (m)
5	−71.1938	−119.5488
4	−71.0765	−119.6574
3	−71.1267	−119.4965
2	−71.1059	−119.5197
1	−71.1327	−119.5900
0	−71.1482	−119.4564

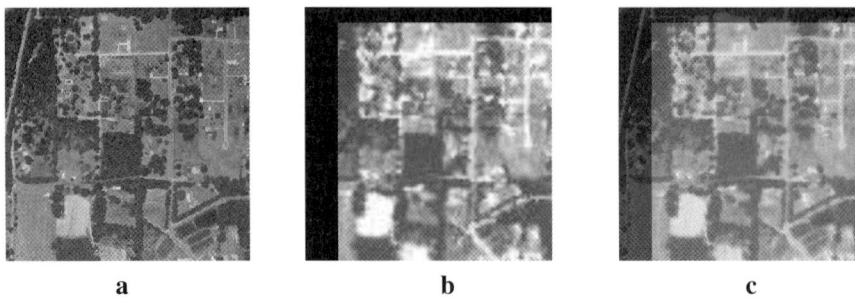

Fig. 7.5a–c. Registration result. **a** Floating image. **b** Resampled (registered) reference image. **c** Superimposed image

different levels are almost the same. The possible reason for this is that, in this experiment, only the floating image is used to construct the image pyramid for multi-scale optimization. In this manner, the resolution of the reference image is fixed throughout the entire optimization process. However, if both images have approximately the same resolution and both are used to construct the image pyramids for multi-scale optimization, we expect improved registration accuracy as we move to higher resolution levels as shown in (Chen and Varshney 2000). Nevertheless, the major advantage of this multi-scale optimization approach is its computational efficiency. Significant speedups by using the multi-scale optimization strategy were reported in Maes et al. (1999) and Pluim et al. (2001). This is due to the fact that the initial search point at each resolution level (except for at the lowest resolution level) is in the vicinity of the optimum at that resolution level. This accelerates the convergence rate of the local optimizer in use. In this experiment, the MI measure was computed through the PVI algorithm. In the next section, performances of different joint histogram estimation methods are compared using a pair of multi-sensor images having similar spatial resolution.

7.3.2
Registration of Images Having Similar Spatial Resolutions

In this section, we present registration results of a multi-sensor image pair using four different joint histogram estimation methods and compare their performances based on registration consistency. Though the registration problem considered in Sect. 7.3.1 is also multi-sensor registration, the performances of different MI registration algorithms were not compared. This is due to the fact that the use of the HyMap image as the floating image is problematic as demonstrated in Fig. 7.3, and the registration consistency measure can not be computed.

The image pair used in the experiment consists of an IRS PAN image and a Radarsat SAR image at approximately the same resolution (i.e. 5.8 m of PAN and 6.25 m of SAR). The image dimensions were [360 × 360] pixels for both PAN and SAR images (Fig. 7.6a,b). Since joint histogram is the only

Fig. 7.6a,b. Remote sensing images used in the experiment. **a** IRS PAN image. **b** Radarsat SAR image

quantity required to compute the mutual information similarity measure (see Chap. 3), different joint histogram estimation algorithms constitute different MI based registration methods. Here, four joint histogram estimation methods are considered,

1. Nearest neighbor interpolation (NN)
2. Linear interpolation
3. Cubic convolution interpolation (CC), and
4. 2D implementation of partial volume interpolation (PVI).

The first three methods belong to the two-step joint histogram estimation category (see Sect. 3.3.1), in which, an intermediate resampled image is produced and is employed to compute mutual information. The fourth method falls in the one-step joint histogram estimation category. Since for multi-sensor registration, images are acquired from different types of sensors placed on different platforms, spatial resolutions are seldom exactly the same. Therefore, the phenomenon of interpolation induced artifacts (see Sect. 3.4) is not an issue. Hence, the GPVE algorithm is not compared here in this experiment. For each method, the simplex search procedure (Nelder and Mead 1965) is used to find the global optimum. Since the simplex search procedure is just a local optimizer, there are instances when it may fail to find the position of the global optimum. In this situation (observed by visual inspection), we may change the initial search points until the correct optimum is obtained or adopt the hybrid global optimization procedure as illustrated in the previous section.

Table 7.3 lists the registration results for the IRS PAN and Radarsat SAR images shown in Fig. 7.6 using different joint histogram estimation methods. The transformation parameters ($T_{A,B}$) representing rotation (in degrees), vertical and horizontal displacements (in meters), and the registration consistencies obtained using various interpolation algorithms are shown in this table as well.

From Table 7.3, by comparing the transformation parameters, it can be observed that the registration results from different interpolation algorithms are very close to each other. However, partial volume interpolation method

Table 7.3. Results from the registration of IRS PAN and Radarsat SAR images. T denotes the transformation parameters (rotation (degree), vertical displacement (m) and horizontal displacement (m))

Method	$T_{PAN,SAR}$ [degree,m,m]	$T_{SAR,PAN}$ [degree,m,m]	dp (m)
NN	[20.60,69.14,35.38]	[−20.50,−52.63,−56.45]	1.65
Linear	[20.68,67.68,33.76]	[−20.62,−49.78,−53.26]	2.79
CC	[20.63,65.04,33.63]	[−20.75,−49.47,−56.24]	2.28
PVI	[20.62,66.85,35.14]	[−20.60,−50.44,−56.49]	0.36

obtained the most consistent results, as observed from the registration consistency shown in the last column of the table. Notice that the nearest neighbor interpolation algorithm yields performance comparable to the other algorithms. Hence, nearest neighbor interpolation may be suitable for joint histogram estimation for MI based multi-sensor registration because of its computational efficiency.

The experimental results presented in this section have demonstrated the ability of an intensity based image registration technique using MI as a similarity measure for the multi-sensor registration problem. Next, we consider the use of the MI based registration technique for multi-temporal registration problems.

7.4
Multi-Temporal Registration

Accurate registration of multi-temporal remote sensing images is quite essential for various change detection applications based on the processing of data collected at different times (i. e. see Chapter 2). A variety of change detection algorithms based on techniques such as image differencing (Castelli et al. 1998), principal component analysis (Mas 1999), change vector analysis (Lambin and Strahler 1994), Markov Random Fields (Kasetkasem and Varshney 2002) and neural networks (Liu and Lathrop 2002) may be used. A fundamental requirement for the success of all these algorithms is, however, accurate registration of images taken at different times. For example, registration accuracy of less than one-fifth of a pixel is required to achieve a change detection error of less than 10% (Dai and Khorram 1998).

Conventionally, mean square difference (MSD) and normalized cross-correlation (NCC) are used as similarity measures for multi-temporal image registration problems (Brown 1992). However, while using these similarity measures, images are assumed radiometrically corrected. In some cases, even though accurate radiometric correction has been applied, registration results using MSD or NCC as similarity measures may still not be reliable. This occurs, for example, when the scene undergoes a significant amount of change between two times. This can be illustrated by registering two small portions

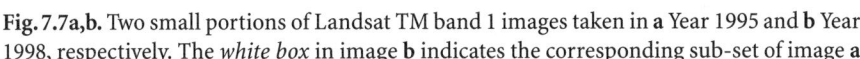

Fig. 7.7a,b. Two small portions of Landsat TM band 1 images taken in **a** Year 1995 and **b** Year 1998, respectively. The *white box* in image **b** indicates the corresponding sub-set of image **a**

(128 × 128 pixels and 228 × 228 pixels) of Landsat TM band 1 images taken in years 1995 and 1998 (see Fig. 7.7). The images are systematically corrected, and thus contain no radiometric errors. From this figure, a significant change between the intensity values of the two images can be observed which is probably due to the change in crop types in the region. The 2D registration functions obtained by using MSD and NCC as similarity measures, to register images in Fig. 7.7, are shown in Fig. 7.8a,b. If both MSD and NCC were suitable similarity measures in this case, we would expect a global minimum in Fig. 7.8a and

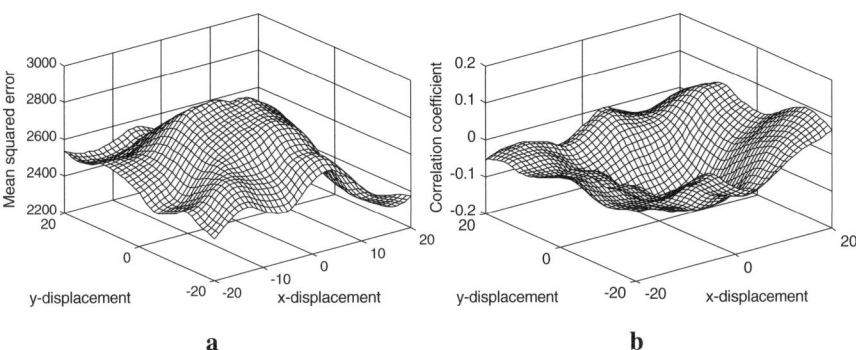

Fig. 7.8a,b. Two dimensional registration functions while registering images shown in Fig. 7.7 using **a** MSD and **b** NCC as similarity measures

a global maximum in Fig. 7.8b both at the position $(0,0)$. Instead, it occurs at $(20,-14)$ for MSD and at $(20,-12)$ for NCC. Thus, these two similarity measures fail to register the two images accurately. In order to overcome this potential difficulty pertaining to MSD and NCC similarity measures, a more robust similarity measure is required. Since MI has been successful as a similarity measure for many multi-sensor registration problems, it is natural to examine its efficacy for registration of images acquired at different times.

Figure 7.9 illustrates the 2D registration function using MI as the similarity measure for the registration of images shown in Fig. 7.7. It is clear from Fig. 7.9 that the maximum occurs at the position $(0,0)$, which demonstrates that MI is able to successfully register the two images. This was not the case either with the use of NCC or MSD as similarity measures. Now, we evaluate the performance of mutual information (MI) as the similarity measure for multi-temporal remote sensing image registration.

Three multi-temporal image data sets are used in our experiments. These are Landsat TM band 1 images of three different dates (1995, 1997 and 1998), IRS PAN images of two different dates (1997 and 1998), and Radarsat SAR images of two different dates (1997 and 1998). Figure 7.10 shows the images used in the experiments. The images obtained from the source were already systematically corrected. Since the images from the same sensor have the same spatial resolution, interpolation artifacts are expected to be present in these experiments. The header files associated with each image indicate that there is no rotational difference in Landsat TM images, a slight rotational difference of $0.02°$ in IRS PAN images and a significant amount of rotational difference (i.e. $18.57°$) in the SAR images. Therefore, it is anticipated that interpolation-induced artifacts may be more pronounced while registering multi-temporal Landsat TM and IRS PAN images than while registering multi-temporal Radarsat SAR images. This is because in the case of SAR image registration, when the two images are nearly registered, the significant amount of rotational difference makes the agreement between the grid points of the two images rare upon registration.

We evaluate the performance of four joint histogram estimation algorithms to compute mutual information. These are linear interpolation, the first order

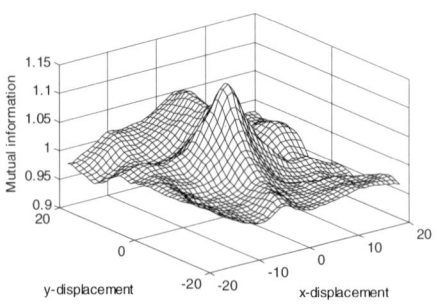

Fig. 7.9. 2D registration function while registering images shown in Fig. 7.7 using MI as the similarity measure

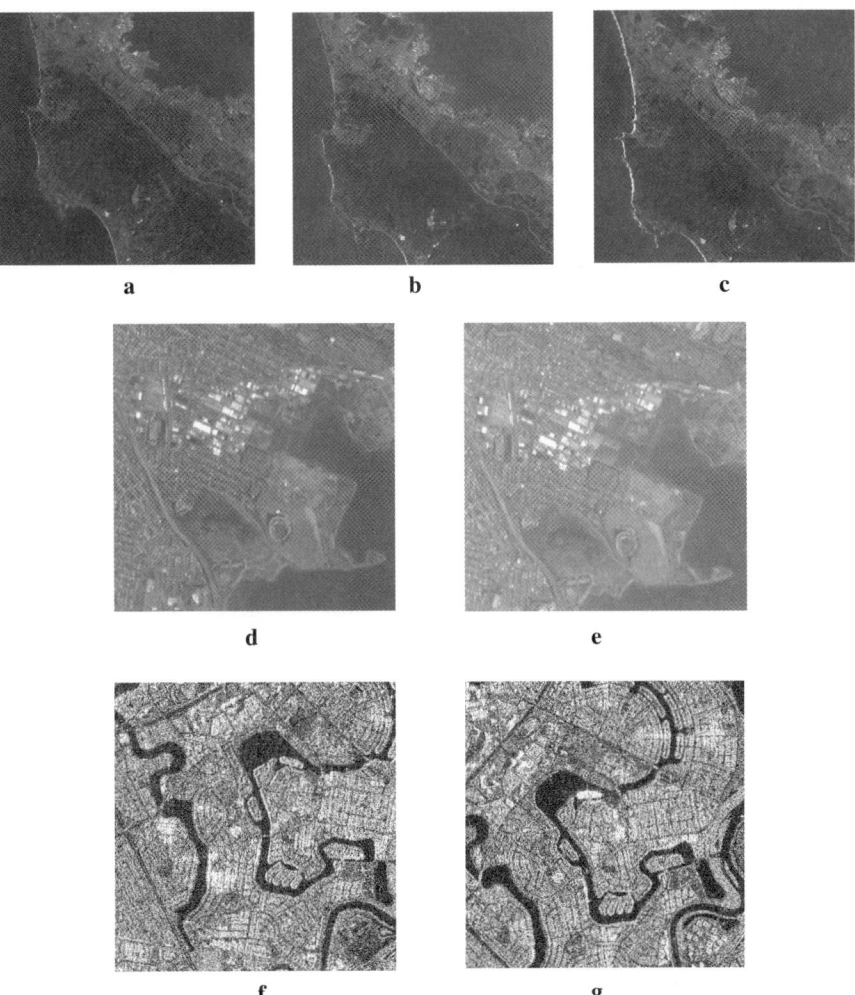

Fig. 7.10a–g. Landsat TM band 1 images of **a** Year 1995, **b** Year 1997 and **c** Year 1998. IRS PAN images of **d** Year 1997, **e** Year 1998 and Radarsat SAR images of **f** Year 1997 and **g** Year 1998

GPVE algorithm, which is actually the PVI algorithm, and the second and third order GPVE algorithms. Again, the simplex search procedure is used to find the global optimum in all cases. When the global optimum position is not found, we change the initial search points until the correct optimum is obtained. Table 7.4 to 7.9 show all the experimental results obtained while registering different pairs of multi-temporal remote sensing images. The first date in each pair serves as the floating image and the second date as the reference image. Since, Landsat TM images are obtained for three dates, three sets of registration are performed. Table 7.4 to 7.6 show the registration results in the form of

Table 7.4. Registration results for Landsat TM 1995 and 1997 images. T denotes the transformation parameters (vertical and horizontal displacements)

Algorithm	$95 \rightarrow 97$ $T_{95,97}$ [pixel,pixel]	$97 \rightarrow 95$ $T_{97,95}$ [pixel,pixel]	2-date consistency
Linear	[2.9109, 62.4256]	[−3.0757, −62.4665]	0.1698
1st order	[3.0000, 62.0000]	[−3.0000, −62.0000]	0.0000
2nd order	[2.9812, 62.3717]	[−2.9819, −62.3739]	0.0024
3rd order	[2.9837, 62.3907]	[−2.9838, −62.3930]	0.0023
MSD	[3.2801, 62.4177]	[−3.0000, −62.3163]	0.2979
NCC	[3.1412, 62.3719]	[−3.1039, −62.3604]	0.0390

Table 7.5. Registration results for Landsat TM 1995 and 1998 images. T denotes the transformation parameters (vertical and horizontal displacements)

Algorithm	$95 \rightarrow 98$ $T_{95,98}$ [pixel,pixel]	$98 \rightarrow 95$ $T_{98,95}$ [pixel,pixel]	2-date consistency
Linear	[33.3183, −13.5464]	[−33.3238, 13.5576]	0.0125
1st order	[33.0001, −13.9990]	[−33.0000, 14.0000]	0.0010
2nd order	[33.2304, −13.6165]	[−33.2307, 13.6169]	0.0005
3rd order	[33.2504, −13.5969]	[−33.2522, 13.5975]	0.0020
MSD	[33.3777, −13.6033]	[−33.3438, 13.6517]	0.0592
NCC	[33.3507, −13.6290]	[−33.3489, 13.6486]	0.0197

Table 7.6. Registration results for Landsat TM 1997 and 1998 images. T denotes the transformation parameters (vertical and horizontal displacements)

Algorithm	$97 \rightarrow 98$ $T_{97,98}$ [pixel,pixel]	$98 \rightarrow 97$ $T_{98,97}$ [pixel,pixel]	2-date consistency
Linear	[30.3784, −76.0752]	[−30.3745, 76.0577]	0.0179
1st order	[30.0000, −76.0000]	[−30.0010, 76.0003]	0.0011
2nd order	[30.2734, −75.9989]	[−30.2723, 76.0115]	0.0127
3rd order	[30.2953, −75.9996]	[−30.2938, 76.0000]	0.0015
MSD	[30.3637, −75.9647]	[−30.4245, 75.7822]	0.1923
NCC	[30.3818, −75.8891]	[−30.3867, 76.1127]	0.2237

transformation parameters (i.e. displacements in x and y directions) along with the corresponding two-date registration consistency. Table 7.7 shows the corresponding three-date registration consistencies of each algorithm to register these images.

A glance at the first four rows of Table 7.4 to 7.6 shows no significant difference between the performances of each algorithm as all of them perform fairly well with the linear interpolation resulting in the worst registration consistency. Further, on close inspection of transformation parameters for the

Table 7.7. Three-date registration consistency for registration of Landsat TM 1995, 1997 and 1998 images

Algorithm	3-date Registration Consistency
Linear	0.1187
1st order	0.0008
2nd order	0.0289
3rd order	0.0291
MSD	0.2314
NCC	0.2023

Table 7.8. Registration results for IRS PAN 1997 and 1998 images. T denotes the transformation parameters (rotation angle, vertical and horizontal displacements)

Algorithm	$T_{97,98}$ [degree,pixel,pixel]	$T_{98,97}$ [degree,pixel,pixel]	2-date consistency
Linear	[−0.1158, 27.9863, −8.8709]	[0.1290, −27.9837, 8.9155]	0.0470
1st order	[−0.0005, 28.0001, −8.9993]	[0.0008, −28.0023, 8.9990]	0.0024
2nd order	[−0.1017, 28.0313, −8.8341]	[0.0997, −28.0241, 8.8878]	0.0104
3rd order	[−0.1025, 28.0575, −8.8460]	[0.1015, −28.0360, 8.9002]	0.0071
MSD	[−0.1206, 28.0089, −8.8800]	[0.1281, −27.9905, 8.9639]	0.0309
NCC	[−0.1222, 27.9752, −8.9126]	[0.1270, −27.9441, 8.9919]	0.0241

Table 7.9. Registration results for Radarsat SAR 1997 and 1998 images. T denotes the transformation parameters (vertical and horizontal displacements)

Algorithm	$T_{97,98}$ [degree,pixel,pixel]	$T_{98,97}$ [degree,pixel,pixel]	2-date consistency
Linear	[−18.5656, 76.4484, −56.0337]	[18.5284, −54.6218, 77.4446]	0.1150
1st order	[−18.5345, 76.3701, −56.1286]	[18.5186, −54.5811, 77.5116]	0.0533
2nd order	[−18.5163, 76.3780, −56.0975]	[18.5172, −54.5888, 77.4628]	0.0237
3rd order	[−18.5208, 76.3863, −56.0444]	[18.5210, −54.5977, 77.4538]	0.0565
MSD	[−18.5228, 76.2855, −55.6907]	[18.5264, −54.6726, 77.1890]	0.1504
NCC	[−18.5257, 76.2932, −55.7384]	[18.5200, −54.6581, 77.1451]	0.0641

four joint histogram estimation algorithms in Table 7.4 to 7.7, we notice that the PVI algorithm results in almost perfect registration consistency. However, if we plot the registration function corresponding to the PVI algorithm for a pair of images to be registered, we can clearly observe interpolation-induced artifacts. One such illustration is provided in Fig. 7.11a and 7.11b for the registration of Landsat TM images of 1995 and 1997. This explains the perfect registration consistency of the PVI algorithm: it is a consequence of the artifact pattern. Therefore, we have a good reason to question the registration accuracy achieved by PVI in this case. In contrast, the second and higher order GPVE do

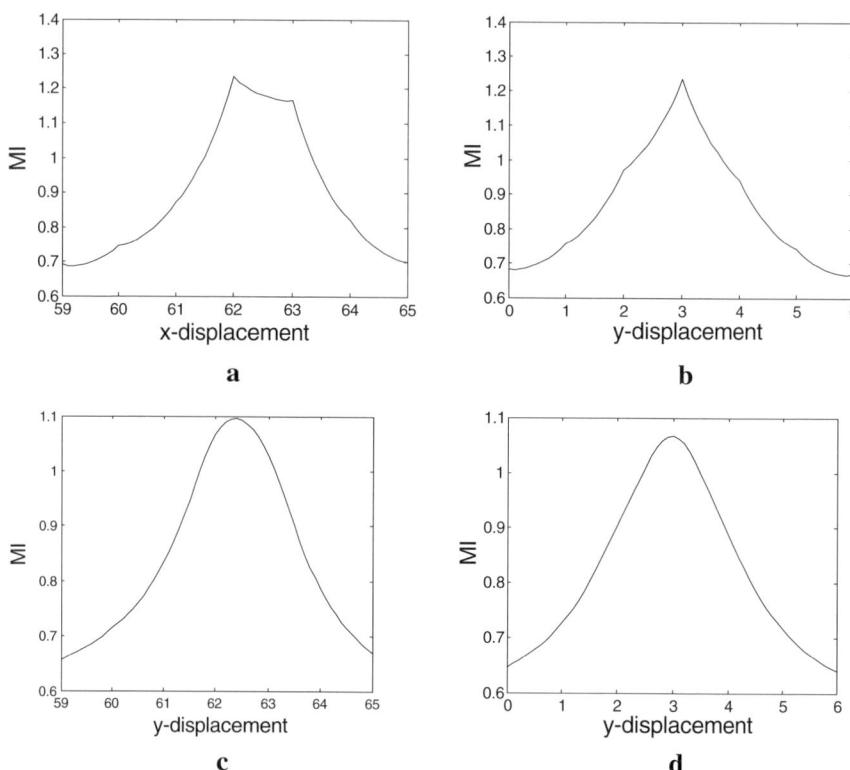

Fig. 7.11a–d. 1D registration functions resulting from PVI (**a** and **b**) and 2nd order GPVE (**c** and **d**) in registering Landsat TM images of 1995 and 1997

not result in any artifact patterns (see Fig. 7.11c and 7.11d, for second order as an example). Thus, higher order GPVE algorithms clearly have an advantage over linear and PVI algorithms since the resulting registration function is very smooth.

Similar conclusions can be drawn from the registration results of IRS PAN images, which are reported in Table 7.8. In case of the registration of Radarsat SAR images (Table 7.9), we did not observe any artifact patterns when either linear interpolation or PVI algorithm is used. This is because of the rotational difference between the two images. In this case, linear interpolation again results in relatively poor registration consistency while PVI and higher order GPVEs have similar performance. Thus, joint histogram estimation using higher order GPVE algorithms produces more reliable registration results than using linear or PVI algorithms.

Further, in order to evaluate the performance of our MI based registration algorithm, the registration results of the 3rd order GPVE algorithm are compared with those obtained from image registration using MSD and NCC as similarity measures. Linear interpolation is used when implementing the algorithms based upon these two similarity measures. The registration results obtained via MSD and NCC similarity measures are shown in the last two rows of Table 7.4 to 7.9. From our experiments, it is hard to determine which similarity measure results in better registration accuracy due to a lack of ground data. However, on the basis of registration consistency it can be clearly seen that MI based registration implemented through higher order GPVE algorithms outperforms the registration obtained with MSD and NCC as similarity measures.

7.5
Summary

In this chapter, we applied the MI based registration technique introduced in Chap. 3 for multi-sensor and multi-temporal registrations. Multi-sensor registration was performed for two sets of remote sensing data: the images from two sensors having large difference in spatial resolutions and the images from two sensors having similar spatial resolution. For the former case, multi-scale optimization strategy was introduced to speedup the whole process. For the latter case, four joint histogram estimation algorithms were used to compute the MI measure. They were nearest neighbor interpolation, linear interpolation, cubic convolution interpolation and PVI interpolation. Registration consistency was used to evaluate registration performance. Our experiments show that PVI produced the most consistent result and surprisingly, nearest neighbor interpolation outperformed linear interpolation and cubic interpolation in most cases. Since the sizes of remote sensing images are often very large and nearest neighbor interpolation is computationally most efficient, it seems reasonable to adopt this algorithm when registering images of large sizes using an MI based approach. It should be noted that this is appropriate only when the two images involved have different spatial resolutions, which is true in most multi-sensor registration applications; otherwise interpolation-induced artifacts have to be taken into account.

For multi-temporal registration, images to be registered often have the same spatial resolution and artifacts are likely to be present. To overcome this problem, higher order GPVE was used to implement the MI based registration technique. Although, a precise evaluation of registration accuracy is not possible without the availability of accurate ground data, we have shown that MI based registration implemented through the higher order GPVE algorithm results in better registration consistency than the registration performed using MSD and NCC as the similarity measures.

References

Brown LG (1992) A survey of image registration techniques. ACM Computing Surveys 24: 325–376

Burt PJ, Adelson E (1983) The Laplacian pyramid as a compact image code. IEEE Transactions on Communications, Com-31(4): 532–540

Burt PJ (1984) The pyramid as structure for efficient computation. In: Multiresolution Image Processing and Analysis, Springer-Verlag, pp 6–35 edited by A. Rosenfeld

Castelli V, Elvidge CD, Li CS, Turek JJ (1998) Classification-based change detection: Theory and applications to the NALC data set. In: Lunetta RS, Elvidge CD (eds), Remote sensing change detection: environmental monitoring methods and applications, Ann Arbor Press, Michigan, pp 53–74

Chen H, Varshney PK (2000) A pyramid approach for multimodality image registration based on mutual information. Proceedings of 3rd International Conference on Information Fusion, 1, pp MoD3 9–15

Chen H, Varshney PK, Arora MK (2003a) Mutual information based image registration for remote sensing data. International Journal of Remote Sensing 24(18): 3701–3706

Chen H, Varshney PK, Arora MK (2003b) Automated registration of multi-temporal remote sensing images using mutual information. IEEE Transactions on Geoscience and Remote Sensing 41(11): 2445–2454

Dai X, Khorram S (1999) A feature-based image registration algorithm using improved chain-code representation combined with invariant moments. IEEE Transactions on Geoscience and Remote Sensing 37(9): 2351–2362

Dai X, Khorram S (1998) The effects of image misregistration on the accuracy of remotely sensed change detection. IEEE Transactions on Geoscience and Remote Sensing 36: 1566–77

Holden M, Hill DLG, Denton ERE, Jarosz JM, Cox TCS, Rohlfing T, Goodey J, Hawkes DJ (2000) Voxel similarity measures for 3-D serial MR brain image registration. IEEE Transactions on Medical Imaging 19: 94–102

Kasetkasem T, Varshney PK (2002) An image change detection algorithm based on Markov random field models. IEEE Transactions on Geoscience and Remote Sensing 40: 1815–1823

Lambin EF, Strahler AH (1994) Indicators of land-cover change for change-vector analysis in multitemporal space at coarse spatial scales. International Journal of Remote Sensing 15: 2099–2119

Liu X, Lathrop RG (2002) Urban change detection based on an artificial neural network. International Journal of Remote Sensing 23: 2513–2518

Maes F, Vandermeulen D, Suetens P (1999) Comparative evaluation of multiresolution optimization strategies for multimodality image registration of mutual information. Medical Image Analysis 3(4): 373–386

Mas JF (1999) Monitoring land-cover changes: a comparison of change detection techniques. International Journal of Remote Sensing 20: 139–152

Nelder JA, Mead R (1965) A simplex method for function minimization. The Computer Journal 7: 308–313

Pluim JPW, Antoine Maintz JB, Max AV (1998) A multiscale approach to mutual information matching. Proceedings of SPIE Conference on Image Processing, San Diego, California, 3338, pp 1334–1344

Pluim JPW, Maintz JBA, Viergever MA (2001) Mutual information matching in multiresolution contexts. Image Vision and Computing 19(1–2): 45–52

Thevenaz P, Unser M (2000) Optimization of mutual information for multiresolution image registration. IEEE Transactions on Image Processing 9(12): 2083–2089

Toet A (1992) Multiscale contrast enhancement with application to image fusion. Optical Engineering 31(5): 1026–1031

Toth CK, Schenk T (1992) Feature-based matching for automatic image registration. ITC Journal 1: 40–46

CHAPTER 8

Feature Extraction from Hyperspectral Data Using ICA

Stefan A. Robila, Pramod K. Varshney

8.1
Introduction

Most of the image processing techniques for multispectral or hyperspectral data have complexity that depends directly on the number of spectral bands in the acquired data (Swain and Davis 1978). Due to the large number of bands involved in the hyperspectral images, it is of interest to find methods that transform the image cube into one with reduced dimensionality while, at the same time, maintaining as much information content as possible. These techniques are known under the general name of *feature extraction* (Richards and Jia 1999). The term *feature* is used to refer to the spectral bands or other transforms derived from combinations of bands.

In Chap. 4, we have seen that independent component analysis (ICA) (Hyvärinen et al. 2001) has the promise of being an efficient feature extraction tool. There were, however, several drawbacks. Application of ICA to the full data set provides a way of separating the class information in different bands. Unfortunately, the number of bands in the hyperspectral imagery makes the ICA method extremely time consuming and decreases its attractiveness for use with these datasets. In addition, automated identification of the bands associated with the relevant independent components (i.e. the ones containing useful class information) is difficult. Previously known ranking methods (based on signal to noise ratio, kurtosis, entropy, distances between the transform vectors, etc.) are not efficient for such identification (Robila et al. 2000). Moreover, a very large number of observations may also decrease the efficiency of the algorithm since ICA will try to produce more independent components than the actual number of components in the dataset.

To overcome the above deficiencies, we suggest two ICA based approaches for unsupervised feature extraction in this chapter. The first one is a hybrid PCA/ICA algorithm, where PCA is used to decorrelate and reduce the data. ICA is then applied to the reduced data cube. This hybrid ICA based feature extraction algorithm (ICA-FE) is a direct improvement over the PCA based feature extraction algorithm, the resulting bands being as independent as possible.

In the second approach, we use PCA only for decorrelation of data without performing any band reduction. To reduce the data, we design a new ICA based algorithm that, unlike the original one, allows the extraction of fewer sources than the number of observations. This algorithm is known as the under complete ICA-FE algorithm (UICA-FE). There is a clear need for such an algorithm in the context of hyperspectral imagery, where the number of spectral bands available far exceeds the number of independent classes contained in the image. Designing a method that produces only these useful components will constitute a major improvement.

In both approaches, the number of independent components to be generated is computed based on the eigenvalue information produced in the PCA step. It corresponds to the number of highest eigenvalues that make up approximately 99% of the variance. This criterion is frequently used in PCA for reducing data dimensionality.

This chapter is organized as follows. In Sect. 8.2, we provide a brief assessment of the advantages of ICA over PCA when used for feature extraction. In Sect. 8.3, we present the two ICA based feature extraction algorithms. Section 8.4 contains an assessment of the efficiency of the two methods based on a practical experiment that uses hyperspectral data from AVIRIS sensor. The performance metrics used are: comparative execution times, mutual information and visual inspection of the extracted features (bands) as well as the accuracy of unsupervised classification. The results obtained by the use of PCA are also presented. We note that the usual quantitative measures (distance between the means, distance between the distributions, etc.) have not been used for assessing feature extraction algorithms as they employ information from training data sets. In the case of unsupervised processing, this information is not available.

Given the fact that in the context of hyperspectral data, the term *feature* is similar to the term *band*, and that in ICA, the *components* are associated with features, all these three terms will be used interchangeably throughout this chapter.

8.2
PCA vs ICA for Feature Extraction

Traditionally, feature extraction algorithms focus on the increase of the separation between classes within each feature. The separation can be measured using class information such as distance between means, distance between probabilities, etc. Analogous to classification, depending on whether or not prior class information is available (in the form of training pixel vectors), feature extraction can be either *supervised* or *unsupervised* (Richards and Jia 1999) (see Chap. 2). In the case of supervised feature extraction, the availability of training data allows the computation of the statistics and distance measures for the classes. Feature extraction is performed, in fact, to increase the separability of the classes based on the training data. Since further processing (such as supervised classification) is based on the same training data, the increased separability leads to increased accuracy of the results (Richards and Jia 1999). Unfortunately, in many cases, reference data may either not be available or may not be error-free, so the quality of the training data may be poor in that it may not accurately characterize the classes. This would decrease the quality of both feature extraction and classification (Swain and King 1973).

Alternatively, when no prior information about the classes is available, one can perform unsupervised feature extraction. In this case, the statistics and distance measures between classes cannot be computed or estimated and the main goal of feature extraction shifts to reduction of data redundancy. The narrow bandwidths associated with the spectral bands of hyperspectral data lead to correlation between the adjacent bands resulting in a relatively high level of redundancy in the data (Richards and Jia 1999). Based on this observation, one can simply proceed to perform *feature selection* by analyzing the correlation matrix and selecting only a few bands from each group of highly correlated bands. A better approach is to transform the data such that the resulting features are decorrelated. The variance of the individual components is considered to be an indicator of information content; large values suggest high levels of information and low values indicate the presence of mostly noise. Based on this, only the features with high variance are selected for further processing (Richards and Jia 1999).

Both PCA and ICA are multivariate methods that, given a random vector, proceed in such a way that the resulting components have increased class separability. In case of PCA, the separability is achieved through decorrelation whereas in ICA, it is via independence (Lee 1998).

When considering hyperspectral data, each band corresponds to a component of random pixel vectors and constitutes a feature of the data. In this context, the image cube resulting after PCA processing has the bands decorrelated and sorted according to their variance. This indicates that most of the information can be retrieved from the first few features and that most of the last features have close to zero variance (indicating a lack of information). Based on this observation, PCA is frequently used for feature extraction by dropping the lowest variance components (Richards and Jia 1999).

Decorrelation based feature extraction has been observed to be less efficient when dealing with small classes (i.e. classes having small spatial extents) in the image. In other words, small classes are sparsely represented in the data (i.e. small classes contain very few pixels). Due to their size, these classes tend to have little influence on the band variance leading to the possibility of being discarded in the lower variance bands. In the context of target detection, loss of information regarding small targets (that correspond to small classes in the image) affects the accuracy of the feature extraction algorithms (Achalakul and Taylor 2000; Tu et al. 2001). Therefore, there is no guarantee that band reduction using PCA will correctly preserve the entire information content of the image cube (Richards and Jia 1999; Tu et al. 2001).

The PCA model assumes that each original band is a linear mixture of unknown *uncorrelated* bands, and proceeds to recover them such that their variance is maximized. In ICA, the assumption is that each original band is, a linear mixture of unknown *independent* bands (Robila et al. 2000). The goal is to find the unmixing matrix and to perform the inversion, in order to recover the independent bands (Robila and Varshney 2002). In the case of ICA based feature extraction, it is assumed that the unmixing will lead to identification of independent features or components.

We have developed two feature extraction algorithms based on independent component analysis that are presented next.

8.3
Independent Component Analysis Based Feature Extraction Algorithm (ICA-FE)

When applied to hyperspectral imagery, ICA produces only a few independent components that contain important information. Most of the other components are mainly associated with either noise or artifacts introduced by the sensor or the data acquisition conditions. In the ICA-FE algorithm, first the data dimensionality is reduced using PCA and then ICA is applied. This is due to the fact that only a reduced number of components are relevant for further processing by ICA, which can be recovered from the first few components of the PCA processed data. The underlying assumption is that these components contribute significantly to the variance, and are thus pushed in the highest eigenvalue PCA components (Swain and Davis 1978). In this case, reduction of the number of bands after PCA should not significantly affect the recovery of the classes. We can, therefore, proceed to apply the ICA algorithm to the first few principal components. The result will have the bands (components) as independent as possible.

The steps involved in the independent component analysis feature extraction algorithm (ICA-FE) are shown in Fig. 8.1. First, PCA is applied to the n-band data for dimensionality reduction on the basis of eigenvalues. Following the computation of the eigenvalues and eigenvectors for the covariance matrix of x, the number of bands to be retained is determined by taking the highest eigenvalues that make up a pre-specified percentage of the sum of all the eigenvalues.

The corresponding principal components can be directly obtained by transforming the data through the eigenvectors associated with the selected eigenvalues. The ICA step described in Chap. 4 (see Sect. 4.3.2) is then applied. The algorithm results in m-independent features (bands).

The ICA-FE algorithm is very fast, compared to the direct application of ICA to the full data. In both cases, PCA is used as the preprocessing step. When PCA is employed in conjunction with band reduction, the complexity of an ICA iteration is reduced from $O(n^2 p)$ to $O(m^2 p)$ where n is the number of original bands, m is the number of resulting PCA components, and p is the number of pixel vectors.

Another difference between the ICA-FE algorithm and ICA applied on the full data set is the fact that we no longer need to perform band selection on the results obtained by ICA. This may, however, be a major drawback. Since we are using only the high variance bands, we assume that all the independent components contribute to them. In the case of a low variance independent component, it is possible that, its contribution may be relegated to a lower

Feature Extraction from Hyperspectral Data Using ICA

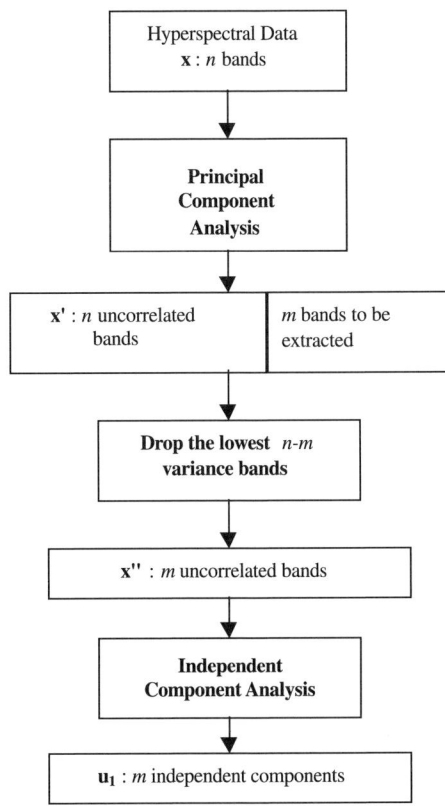

Fig. 8.1. Independent component analysis feature extraction (ICA-FE) algorithm

variance PCA band and the algorithm may fail to recover it. In that case, the PCA step may need to be modified such that all the principal components with nonzero variance are selected for processing by ICA. However, since the hyperspectral data contain noise, zero variance bands are seldom found. The ICA-FE algorithm will then reduce to the application of ICA on the original data and will not provide any improvement in computational speed. These issues are further discussed when an example is considered in Sect. 8.5.

8.4
Undercomplete Independent Component Analysis Based Feature Extraction Algorithm (UICA-FE)

Sometimes the hyperspectral images may contain independent components with low variance. For example, in automatic target detection problem (see Sect. 2.9 of Chap. 2), a small target in the image is generally characterized by a limited number of pixels that will not contribute significantly to the overall band variance. Therefore, if PCA is applied on this dataset, the component containing information on the target will have low variance and it is highly likely that this component may be discarded thereby missing important information in the dataset. Hence, the ICA-FE algorithm may not work accurately for the datasets containing small targets. In this case, we would like to use all the PCA derived components to produce only those independent components that will correspond to the classes present in the image. Therefore, a modification in the design of the ICA algorithm is desired.

Consider the minimum mutual information (MMI) framework for the solution of ICA (see Sect. 5.3.2):

$$u = Wx, \tag{8.1}$$

where x is an n-dimensional random vector, u is an m-dimensional random vector and W is an $m \times n$ matrix. Intuitively, when the number of observations is larger than the number of sources ($n > m$), the data observed can be represented as a linear combination of a smaller set of components (the data is *undercompletely* represented) (Shriki et al. 2000). Thus, for $m < n$ the problem is called *undercomplete MMI (undercomplete ICA)*.

We derive a gradient-based algorithm to solve the undercomplete MMI problem. Unfortunately, we cannot directly apply the information maximization algorithm (refer to Sect. 4.3.2), since we had assumed $m = n$. In that case, W was a full rank matrix, and the probability density function (pdf) of u was directly expressed in terms of the pdf of x as:

$$p(u) = \frac{p(x)}{|\det W|}. \tag{8.2}$$

In the case of $m < n$, expressing the pdf of the random vector u using the conditional pdfs, we have,

$$p(u) = \int p(u|x) p(x) \, dx. \tag{8.3}$$

Assume now that the output random vector u is perturbed by additive independent Gaussian noise and consider the limit when the variance of the noise goes to zero. In this case, we can use the multivariate Gaussian distribution to express the conditional pdf as (Shriki et al. 2000):

$$p(u|x) = \lim_{\sigma^2 \to 0} \frac{1}{\left(\sqrt{2\pi\sigma^2}\right)^m} e^{-\frac{1}{2\sigma^2} \|u - Wx\|^2}. \tag{8.4}$$

When the realization of the random vector u is in the image of Wx there exists a vector x_0 such that:

$$u = Wx_0. \tag{8.5}$$

In this case, the norm involved in (8.4) becomes:

$$\|u - Wx\|^2 = \|Wx_0 - Wx\|^2 = \|W(x_0 - x)\|^2$$
$$= (W(x_0 - x))^T (W(x_0 - x))$$
$$= (x_0 - x)^T (W^T W)(x_0 - x). \tag{8.6}$$

The conditional pdf can then be rewritten as:

$$p(u|x) = \lim_{\sigma^2 \to 0} \frac{1}{\left(\sqrt{2\pi\sigma^2}\right)^m} e^{-\frac{1}{2\sigma^2}(x_0-x)^T(W^TW)(x_0-x)}. \tag{8.7}$$

We multiply and divide the entire expression by $\sqrt{\det(WW^T)}$:

$$p(u|x) = \lim_{\sigma^2 \to 0} \left(\frac{1}{\left(\sqrt{2\pi\sigma^2}\right)^m} e^{-\frac{1}{2\sigma^2}(x_0-x)^T(W^TW)(x_0-x)} \frac{\sqrt{\det(WW^T)}}{\sqrt{\det(WW^T)}} \right)$$

$$= \frac{1}{\sqrt{\det(WW^T)}} \lim_{\sigma^2 \to 0} \left(\frac{\sqrt{\det(WW^T)}}{\left(\sqrt{2\pi\sigma^2}\right)^m} e^{-\frac{1}{2\sigma^2}(x_0-x)^T(W^TW)(x_0-x)} \right)$$

$$= \frac{1}{\sqrt{\det(WW^T)}}$$

$$\times \lim_{\sigma^2 \to 0} \left(\frac{1}{\left(\sqrt{2\pi}\right)^m} \frac{1}{\sqrt{\det(\sigma^2(WW^T)^{-1})}} e^{-\frac{1}{2}(x_0-x)^T\left(\frac{W^TW}{\sigma^2}\right)(x_0-x)} \right). \tag{8.8}$$

For $x = x_0$, we have:

$$\lim_{\sigma^2 \to 0} \left(\frac{1}{\sqrt{2\pi}^m} \frac{1}{\sqrt{\det\left(\sigma^2(WW^T)^{-1}\right)}} e^0 \right) = \infty. \tag{8.9}$$

The limit corresponds to a dirac delta function in x around x_0. Putting this result into (8.3), for u in the image of Wx, we obtain:

$$p(u) = \int \frac{1}{\sqrt{\det(WW^T)}} \delta(x - x_0) p(x) dx = \frac{p(x_0)}{\sqrt{\det(WW^T)}}. \tag{8.10}$$

Note that in the case when W is a square matrix, i.e., when $m = n$, we get back the relationship in (8.2) since $\det(W) = \det(W^T)$.

Let us define V as:

$$V = \begin{pmatrix} v_{11} & \cdots & v_{1m} \\ \vdots & & \vdots \\ v_{m1} & \cdots & v_{mm} \end{pmatrix} = WW^T$$

$$= \begin{pmatrix} \sum_{k=1}^{n} w_{1k}w_{1k} & \cdots & \sum_{k=1}^{n} w_{1k}w_{mk} \\ \vdots & & \vdots \\ \sum_{k=1}^{n} w_{mk}w_{1k} & \cdots & \sum_{k=1}^{n} w_{mk}w_{mk} \end{pmatrix}. \qquad (8.11)$$

We proceed with the computation of the gradient:

$$\begin{aligned}
\frac{\partial E\{\log(p(\mathbf{u}))\}}{\partial W} &= \frac{\partial E\left\{\log\left(\frac{p(\mathbf{x})}{\sqrt{\det(WW^T)}}\right)\right\}}{\partial W} \\
&= \frac{\partial E\{\log(p(\mathbf{x}))\}}{\partial W} - \frac{\partial E\left\{\log\left(\sqrt{\det(WW^T)}\right)\right\}}{\partial W} \\
&= -\frac{\partial E\left\{\log\left(\sqrt{\det(V)}\right)\right\}}{\partial W} \\
&= -\frac{\partial \log\left(\sqrt{\det(V)}\right)}{\partial W} \\
&= \frac{1}{2}\frac{\partial \log(\det(V))}{\partial W} = \frac{1}{2}\frac{1}{\det(V)}\frac{\partial \det(V)}{\partial W}. \qquad (8.12)
\end{aligned}$$

We compute the gradient for each of the elements w_{ij}. Let V_{it} denote the determinant of the $(m-1) \times (m-1)$ matrix obtained by eliminating the line i and the column t from the matrix $V = WW^T$. Then, we can express the determinant of V using the cofactors associated with the elements in the line i:

$$\begin{aligned}
\frac{\partial \det(V)}{\partial w_{ij}} &= \sum_{t=1}^{m} \frac{\partial \left(v_{it}(-1)^{i+t}V_{it}\right)}{\partial w_{ij}} \\
&= \sum_{t=1}^{m} \left(\frac{\partial v_{it}}{\partial w_{ij}}(-1)^{i+t}V_{it} + (-1)^{i+t}v_{it}\frac{\partial V_{it}}{\partial w_{ij}}\right) \\
&= \sum_{t=1}^{m} \left(\frac{\partial v_{it}}{\partial w_{ij}}(-1)^{i+t}V_{it}\right) + \sum_{t=1}^{m}\left((-1)^{i+t}v_{it}\frac{\partial V_{it}}{\partial w_{ij}}\right). \qquad (8.13)
\end{aligned}$$

Following several derivation steps (detailed in Robila 2002) we obtain the gradient with respect to w_{ij} as:

$$\frac{\partial \det(V)}{\partial w_{ij}} = \sum_{t=1}^{m}\left(w_{tj}(-1)^{i+t}V_{it}\right) + w_{ij}(-1)^{i+i}V_{ii}$$

$$+ \sum_{p=1}^{i-1} w_{pj}(-1)^{i+p}V_{pi} + \sum_{p=i+1}^{m} w_{pj}(-1)^{i+p}V_{pi}$$

$$= 2\sum_{p=1}^{m}\left(w_{pj}(-1)^{i+p}V_{pi}\right). \tag{8.14}$$

Finally, we express the gradient in W in a more convenient matrix form:

$$\frac{\partial E\{\log(p(u))\}}{\partial W} = \frac{1}{2}\frac{1}{\det(WW^T)}\frac{\partial \det(WW^T)}{\partial W}$$

$$= \frac{1}{\det(WW^T)}adj\left(WW^T\right)W = (WW^T)^{-1}W. \tag{8.15}$$

If W is a square matrix, then the above expression will reduce to $(W^T)^{-1}$. Using an approach similar to the one in Chap. 4 (see Sect. 4.3.2), we get the update formula:

$$\Delta W = \left(WW^T\right)^{-1}W + E\left\{g(u)x^T\right\}, \tag{8.16}$$

where $g(u)$ is the vector formed as $(g(u_1),\ldots,g(u_n))$, and $g(\cdot)$ is a nonlinear function that can be closely approximated by:

$$g(u_i) \approx \frac{1}{p(u_i)}\frac{\partial p(u_i)}{\partial u_i}. \tag{8.17}$$

Note that we can no longer use the natural gradient to eliminate the inverse of the matrix. The steps of complete algorithm are presented in Fig. 8.2. It starts by randomly initializing W and then proceeds to compute the update step described in (8.16). In the algorithm described in the figure, k is an update coefficient (smaller than one) and controls the convergence speed. W_{old} and W_{new} correspond to the matrix W, prior to and after the update step. The algorithm stops when the mutual information does not change significantly (based on the value of a presepecified α). The result consists of m independent components.

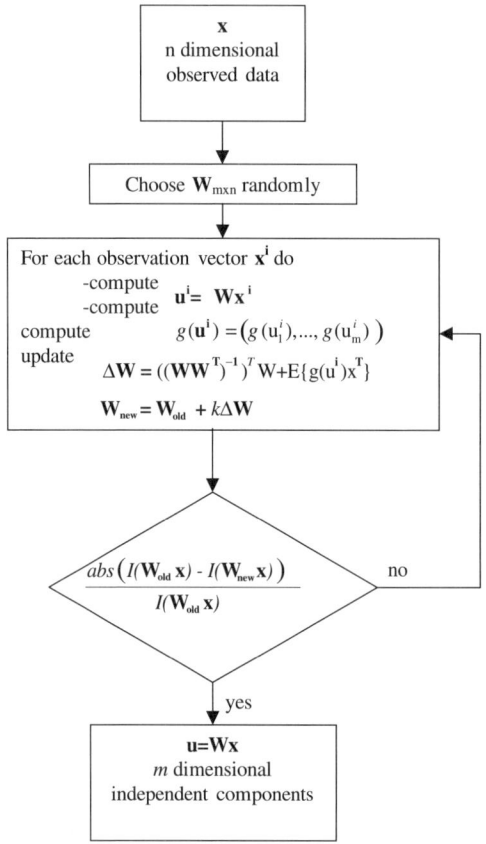

Fig. 8.2. Undercomplete independent component analysis algorithm. The n dimensional random vector x is transformed into a m dimensional random vector u with components that are independent, and $m < n$

The complexity of the algorithm depends on the number of iterations needed to converge. For a single iteration, the complexity can be approximated by $O(pnm)$ where n is the number of original bands, m the number of produced components (bands), and p the number of pixels in each band.

We use this ICA algorithm to design a feature extraction algorithm for hyperspectral imagery (UICA-FE, Fig. 8.3). Prior to applying ICA, we preprocess the data using PCA, which provides us an estimate of the number of components to be computed. In our experiments, this corresponds to the number of PCA produced components with eigenvalues summing upto approximately 99% of all the eigenvalues.

In Fig. 8.4, we pictorially summarize the two algorithms discussed here. For comparison purposes, feature extraction via PCA is also presented. It can be seen from this figure that in comparison to PCA, ICA-FE (Fig. 8.4b) contains an additional step required to generate independent components. In the case

Feature Extraction from Hyperspectral Data Using ICA

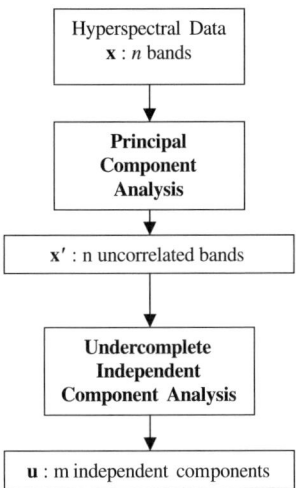

Fig. 8.3. Undercomplete independent component analysis based feature extraction (UICA-FE) algorithm

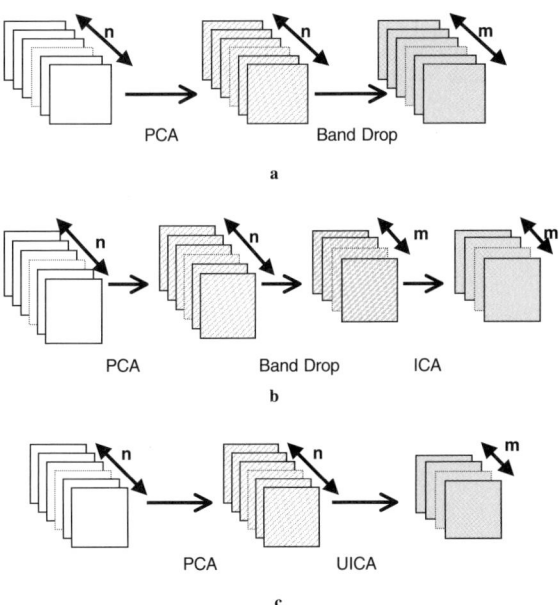

Fig. 8.4a–c. Schematic description of the two ICA based feature extraction algorithms along with the original PCA algorithm. **a** PCA, **b** ICA-FE, **c** UICA-FE

of UICA-FE (Fig. 8.4c), the band elimination step is replaced by the direct ICA computation. Even though this step is time consuming, it has the potential of extracting more informative features than PCA for further processing.

8.5
Experimental Results

To assess the effectiveness of the proposed algorithms we have conducted several experiments using AVIRIS data (Sect. 4.4.3 of Chap. 4 for details on this dataset) (Fig. 8.5).

Several of the original bands were eliminated (due to sensor malfunctioning, water absorption, and artifacts not related to the scene) reducing the data to 186 bands. Then, we applied the ICA-FE and UICA-FE algorithms and compared the results with those produced by PCA.

After applying PCA, it was found that the variance of the first ten components contributes over 99% to the cumulative variance of all the components. Therefore, only ten components were further used in the ICA-FE algorithm. Figure 8.6a–j displays the 10 components obtained from the PCA algorithm. A concentration of the class information can be noticed in the first four components (Fig. 8.6a–d). However, a number of classes can also be discerned in lower ranked components. For example, the classes trees and grass (showing dark in left and lower part in Fig. 8.6e) and grass pasture (showing bright in the middle left part of Fig. 8.6g) are clearly identifiable in fifth through seventh components.

Figure 8.7 displays the results of the ICA-FE algorithm when run on the 10 PCA derived components. A clear separation of the classes present in the image is noticeable, with several of the classes being projected in different components. The class soybean is projected in first component (bright in the top area of Fig. 8.7a), the roads and the stone–steel towers are separated in the third component (dark in Fig. 8.7c) and the corn has been projected in fourth component (bright, right hand side in Fig. 8.7d). It is also interesting to point out that both the wheat and grass pasture show up together (dark in lower left side in Fig. 8.7b) indicating that not enough information is available to separate them. It may be mentioned that both the ICA and PCA generated components are orthogonal. Therefore, when the dataset contains several

Fig. 8.5. Single band of AVIRIS hyperspectral scene

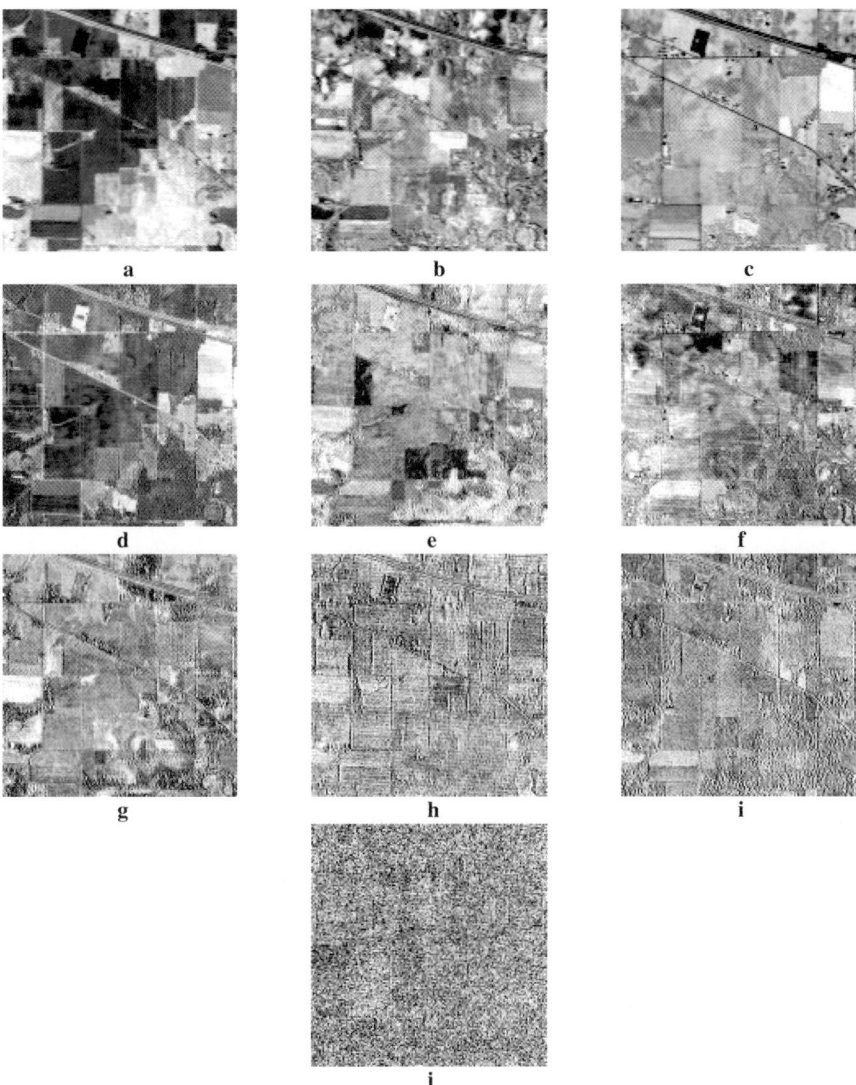

Fig. 8.6a–j. First 10 components with the highest variance data processed with PCA

classes, it is unlikely that each of them will be represented by the orthogonal data.

Next, we applied the UICA-FE algorithm. The resulting components are presented in Fig. 8.8. There is a considerable difference in class separation between the ICA-FE and the UICA-FE results, with the latter producing components that are almost identical to the result obtained when ICA was applied to the full image cube (see Fig. 4.9). It is also interesting to note that UICA-FE is able

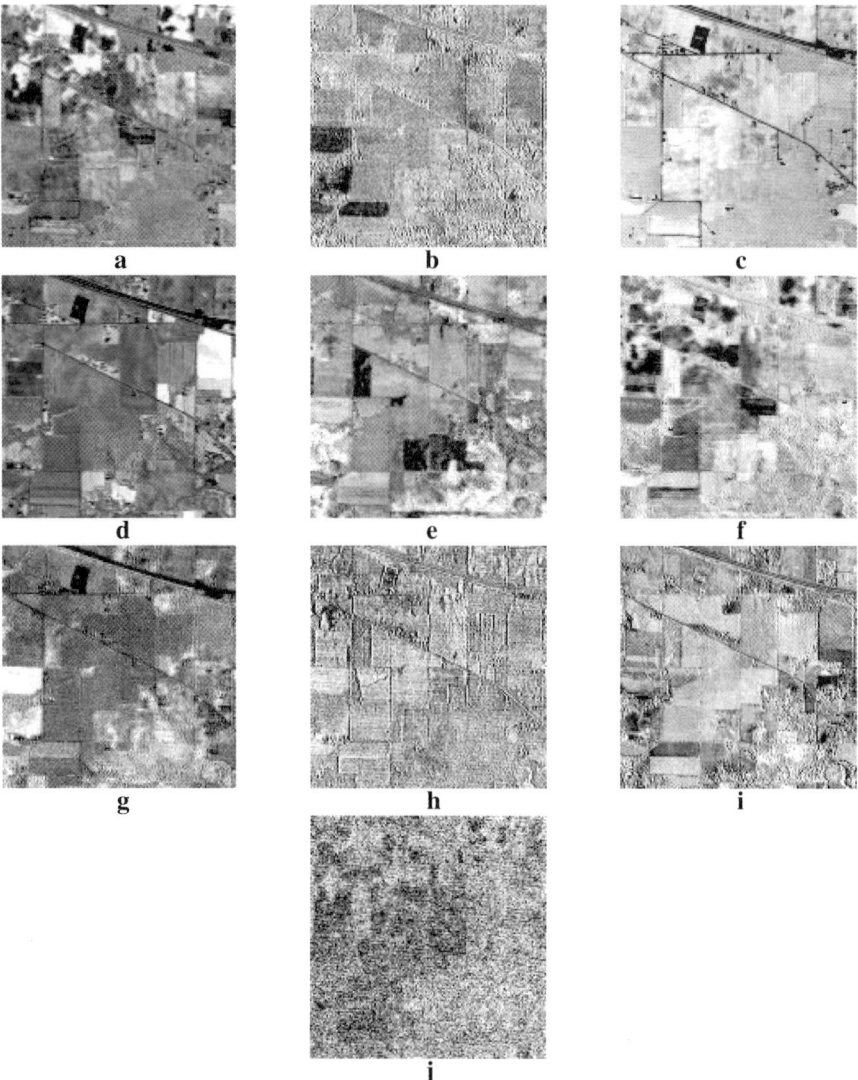

Fig. 8.7a–j. Components produced by the ICA-FE algorithm. **a** soybean; **b** wheat and grass pasture; **c** roads and towers; **d** corn; **e** trees; **g** pasture

to differentiate between the classes wheat and the grass pasture, which was not the case with ICA-FE. The wheat was projected in dark, in the lower side of the Fig. 8.8f, while the grass pasture has been projected in bright in the left side of the Fig. 8.8c.

We employ mutual information as a quantitative measure of how well the classes have been separated among various components. For an m dimensional

Feature Extraction from Hyperspectral Data Using ICA

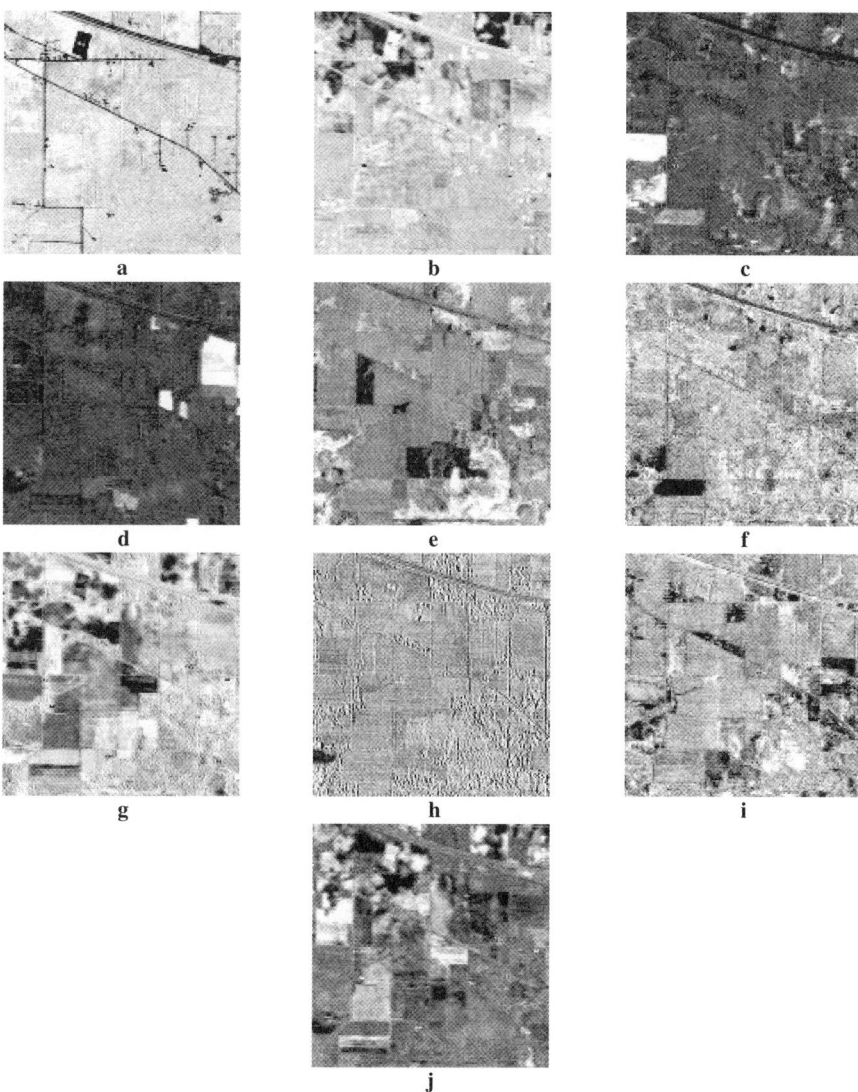

Fig. 8.8a–j. Components produced by the UICA-FE algorithm. **a** roads/towers; **b** soybean clean/mowed grass; **c** pasture; **d** hay, alfalfa; **e** trees; **f** wheat

random vector $\boldsymbol{u}$, the mutual information of its components is defined as:

$$I\left(u_1, \ldots, u_m\right) = E\left\{\log \frac{p(\boldsymbol{u})}{\prod_{i=1}^{m} p\left(u_i\right)}\right\}. \tag{8.18}$$

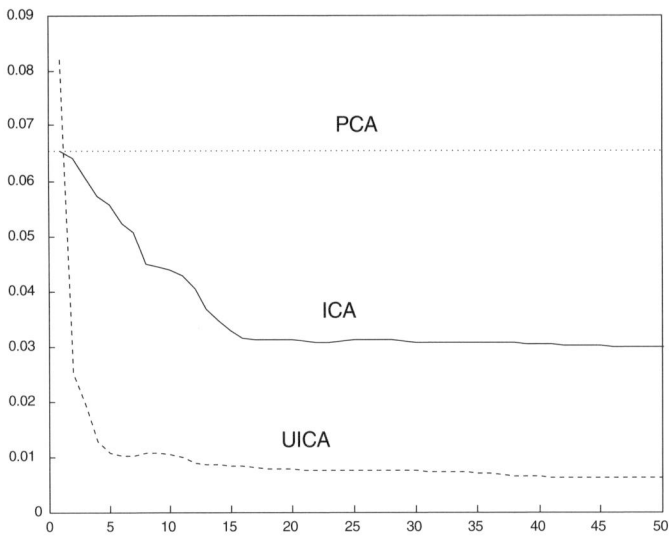

Fig. 8.9. Graph of mutual information for PCA, ICA and UICA when applied on the AVIRIS hyperspectral data

A higher value of the mutual information means a lower class separation. The mutual information is the highest for PCA and the lowest for UICA-FE (see Fig. 8.9). Our results based on mutual information confirm the visual inspection of the resulting components. UICA-FE performs better than ICA-FE that, in turn, outperforms PCA.

Both UICA-FE and ICA-FE converge rather fast, after less than 20 iterations. At the same time, UICA-FE is the slower of the two algorithms. When run on an Intel Pentium Xeon, 2.2 GHz, 1 GB machine, each ICA-FE iteration took 0.85 seconds whereas each UICA-FE iteration took 14 seconds. The times for ICA-FE and UICA-FE to iterate can be considered overhead compared with PCA since they use PCA as the preprocessing step. In our complexity estimates, we found that the ICA-FE iteration was $O(m^2 p)$ where p is the number of pixels in the image and m is the number of components. At the same time, the UICA-FE complexity for one iteration was estimated as $O(mnp)$ where n is the number of original components. Overall, the ratio between the UICA-FE and ICA-FE algorithms per iteration would be close to $O(n/m)$. In our case, the reduction is from $n = 186$ components to $m = 10$ components leading to a slowdown of 18.6. In our experiments, this rate is $14\,\text{s}/0.85\,\text{s} = 16.67$, which is close to the original estimate. This shows, that, indeed, the complexity of computing the ICA transform is linear in the number of bands initially available, confirming our complexity estimates.

Finally, we performed k-means clustering (see Sect. 2.6.2 in Chap. 2) on features extracted by each of the three algorithms (PCA, ICA-FE, UICA-FE). For this we used the Research Systems' ENVI 3.6 unsupervised classification tool (k-means) with 16 classes, and 1% change margin (the minimum percentage of change needed to continue the clustering). In this context, and given the available reference data, the PCA derived components provided a classification accuracy of 43%, ICA-FE an accuracy of 53% and UICA-FE an accuracy of 58%, indicating superior performance of our ICA based algorithms. There are several potential reasons for the lower values of classification accuracy. First, the reference data contained large areas that were not classified, including the road and railroad segments. Additionally, several of the crops were further detailed in various sub-classes (such as soybeans – notill/min/clean). This finer classification is useful in other experiments but these classes could not always be clearly distinguished when only unsupervised classification is employed.

Another set of experiments using the same algorithms was performed on HYDICE data (Robila and Varshney 2003). In that case, due to a lack of reference data, the evaluation was based only on visual inspection or was coupled with target detection (Robila and Varshney 2003; Robila 2002). The results were consistent with those provided by the above experiment and, therefore, are not described here.

8.6
Summary

In this chapter, we have presented two ICA-based algorithms for feature extraction. The first (ICA-FE) algorithm uses PCA to reduce the number of bands and then proceeds to transform the data such that the features are as independent as possible. The second (UICA-FE) algorithm produces the same number of components, relying on PCA only for determination of the number and decorrelation of components. In the UICA-FE algorithm, the data are projected into lower dimensionality directly through the ICA transform. Basically, both algorithms have the same goal: independence of the features to be extracted. However, by using the information from all the components and not only the ones selected by PCA reduction, UICA-FE is able to increase the separability of the classes in the derived independent features. This was validated through experiments on hyperspectral data. The UICA-FE derived features display an increased class separation as compared to ICA-FE. Thus, both ICA-FE and UICA-FE provide attractive approaches for unsupervised feature extraction from hyperspectral data.

References

Achalakul T, Taylor S (2000) A concurrent spectral-screening PCT algorithm for remote sensing applications. Information Fusion 1(2): 89–97

Hyvärinen A, Karhunen J, Oja E (2001) Independent component analysis. John Wiley and Sons, New York

Lee TW (1998) Independent component analysis: theory and applications. Kluwer Academic Publishers, Boston

Rencher AC (1995) Methods of multivariate analysis. John Wiley and Sons, New York

Richards JA, Jia X (1999) Remote sensing digital image analysis: an introduction. Springer-Verlag, Berlin

Robila SA, (2002) Independent component analysis feature extraction for hyperspectral images, PhD Thesis, Syracuse University, Syracuse, NY

Robila SA, Varshney PK (2002) Target detection in hyperspectral images based on independent component analysis. Proceedings of SPIE Automatic Target Recognition XII, 4726, pp 173–182

Robila SA, Varshney PK (2003) Further results in the use of independent component analysis for target detection in hyperspectral Images. Proceedings of SPIE Automatic Target Recognition XIII, 5094, pp 186–195

Robila SA, Haaland P, Achalakul T, Taylor S (2000) Exploring independent component analysis for remote sensing. Proceedings of the Workshop on Multi/Hyperspectral Sensors, Measurements, Modeling and Simulation, Redstone Arsenal, Alabama, U.S. Army Aviation and Missile Command, CD

Shriki O, Sompolinski H, Lee DD (2001) An information maximization approach to overcomplete and recurrent representations. In Advances in Neural Information Processing Systems 13, Leen TK, Dietterich TG, and Tresp V, Eds., 13, pp 612–618

Swain PH, King RC (1973) Two effective feature selection criteria for multispectral remote sensing. Proceedings of First International Conference on Pattern Recognition, November 1973, pp 536–540

Swain PH, Davis SM (eds) (1978) Remote sensing: the quantitative approach, McGraw Hill, New York

Tu TM, Huang PS, Chen PY (2001) Blind separation of spectral signatures in hyperspectral imagery. IEEE Proceedings on Vision, Image and Signal Processing 148(4): 217–226

CHAPTER 9
Hyperspectral Classification Using ICA Based Mixture Model

Chintan A. Shah

9.1 Introduction

Unsupervised classification of remote sensing images is typically based on a mixture model, where the distribution of the entire data is modeled as a weighted sum of the class-component densities (Duda et al. 2000). When the class-component densities are assumed to be multivariate Gaussian, the mixture model is known as the Gaussian mixture model. The K-means and the ISODATA algorithms that are widely used in remote sensing are based on the Gaussian mixture model. These Gaussian mixture model based classification algorithms often perform unsatisfactorily. This stems from the Gaussian distribution assumption for the class-component densities. Gaussianity is only an assumption, rather than a demonstrable property of natural spectral classes, and has been widely accepted due to its analytical tractability and mathematical simplicity. However, if a class happens to be multimodal, it is no longer appropriate to model the class with a multivariate Gaussian distribution. Therefore, the use of the Gaussian mixture model in such cases may lead to unsatisfactory performance.

This underlying Gaussian mixture model assumption is limited as it exploits only the second order statistics of the observed data to estimate the posterior densities. Limitations of using second order statistics (mean and covariance), in characterizing multivariate data, have been discussed in Shah (2003). From this discussion, one can identify the importance of employing higher order statistics, to represent the data more completely (Shah et al. 2004). In fact, it is well established that in theory, a complete description of an arbitrary distribution can be made by the use of statistics of all orders, as in an infinite series (Stark and Woods 1994). Independent component analysis (ICA) exploits higher order statistics in multivariate data (e. g. Common 1994; Hyvärinen and Oja) and has been applied to hyperspectral image analysis for feature extraction and target detection (e. g. Robila and Varshney 2002a,b; Robila 2002; Chang et al. 2002).

In Chap. 8, Robila and Varshney have investigated the feasibility of ICA for unsupervised feature extraction from hyperspectral imagery. Their results have demonstrated that ICA achieves significantly better performance in extracting features as compared to traditional PCA that is based only on second order statistics. Chang et. al. (2002) have considered an application of ICA to linear spectral mixture analysis, referred to as ICA-based linear spectral random mixture analysis (LSRMA). They model an image pixel as a random source resulting from a random composition of multiple spectral signatures of distinct materials (or classes) in the image. Their experimental results have demonstrated that the proposed LSRMA method is an effective unsupervised approach for feature extraction and hence for classification and target detection problems. However, in these research efforts, the application of ICA has been limited to being a feature extraction algorithm.

It is important to note that it may be inappropriate to apply a classification or a target detection algorithm based on second order statistics on the features extracted by ICA. This is due to the fact that these algorithms do not possess the statistical properties that they can complement the enhancement of the information content of the features (extracted based on higher order statistics as is done in ICA).

The work presented in this chapter is the application of a relatively new approach, the ICA mixture model (ICAMM) algorithm (Lee et al. 2000), derived from ICA, for an unsupervised classification of non-Gaussian classes from remote sensing data. So far, to the best of our knowledge, the ICAMM algorithm has been employed for unsupervised classification problems in other applications such as speech signals (Lee et al. 2000), learning efficient codes of images (Lee and Lewicki 2002) and blind signal separation in teleconferencing (Bae et al. 2000) but not for the classification of remote sensing data.

In the proposed approach, we model each pixel of a hyperspectral image as a random vector of intensity values that can be described by the ICA mixture model. Unlike the approach adopted by Robila and Varshney (2002a,b), Chang et. al. (2002) and many others who have employed the ICA model for feature extraction, we employ K ICA model to explain the data generation properties and propose an ICAMM algorithm for unsupervised classification. Here, K is the prespecified number of spectral classes.

The ICAMM algorithm views the observed hyperspectral data as a mixture of several mutually exclusive classes. Each of these classes is described by a linear combination of independent components with non-Gaussian (leptokurtic or platykurtic) probability density functions (see Sect. 4.2 in Chap. 4). The ICAMM algorithm finds independent components and the mixing matrix for each class using an extended information-maximization learning algorithm and computes the class membership probabilities for each pixel. The pixel is allocated to the class with the highest posterior class probability to produce a classification map.

This chapter is organized in the following manner. In Sect. 9.2, we formulate the ICAMM algorithm for unsupervised classification of non-Gaussian classes. The steps involved in the ICAMM algorithm are also provided in this section. In Sect. 9.3, we describe the proposed experimental methodology. The performance of the algorithm is evaluated by conducting experiments on a hyperspectral dataset in Sect. 9.4. Finally, we provide summary of the experimental results.

9.2
Independent Component Analysis Mixture Model (ICAMM) – Theory

Given only sensor observations that are assumed to be linear mixtures of the unobserved, statistically independent source signals, the problem of blind source separation is to recover these independent source signals. The term "blind" indicates that both the source signals and the way the signals were mixed are unknown. ICA provides a solution to the blind source separation problem (see Sect. 4.1 in Chap. 4). The goal of ICA is to perform a linear transformation of the observed sensor signals, such that the resulting transformed signals are as statistically independent from each other as possible. When compared to correlation-based transformations such as PCA, ICA not only decorrelates the sensor observations composed of mixed signals (in terms of second order statistics), but also reduces the higher order statistical dependencies between them. Chapter 4 provided a detailed description of ICA, where various approaches for the implementation of ICA were discussed. In this chapter, we will employ an extended version of the information maximization (infomax) algorithm known as the extended infomax algorithm. A detailed description of this algorithm can be found in Lee et al. (1999).

Consider a random vector x_t, whose elements are the intensity values $x_{i,t}$ of the pixel t in the spectral band i of a hyperspectral image such that $x_t = [x_{1,t}, \ldots, x_{N,t}]'$, where $[\cdot]'$ denotes the transpose. Assume that the pixel vectors (for $t = 1$ to T) are obtained probabilistically from the set of classes $\{\omega_1, \ldots, \omega_K\}$. Class ω_j is selected with prior class probability $P(\omega_j)$, so that the probability density function for x_t may be expressed by the mixture density model as (Duda et al. 2000),

$$p(x_t|\Theta) = \sum_{j=1}^{K} p(x_t|\omega_j, \theta_j) P(\omega_j), \qquad (9.1)$$

where, $\Theta = (\theta_1, \ldots, \theta_K)$ are the K class parameter vectors. The class-conditional densities $p(x_t|\omega_j, \theta_j)$ are referred to as class-component densities, and the prior class probabilities $P(\omega_j)$ are the mixing parameters. Let, $X = \{x_1, \ldots, x_T\}$ be a set of T unlabeled pixel vectors, drawn independently from the mixture model in (9.1). The likelihood of these observed samples, by definition, is the joint density, given by,

$$p(X|\Theta) = \prod_{t=1}^{T} p(x_t|\Theta). \qquad (9.2)$$

The maximum likelihood estimate of Θ is that value of Θ which maximizes $p(X|\Theta)$. When the class-component densities in (9.1) are modeled as multivariate Gaussian, the mixture density model is known as a Gaussian mixture model. However, in the ICAMM algorithm, the class-component densities are assumed to be non-Gaussian. The data within each class can be described by the ICA model (Lee et al. 1999) with the addition of an N-dimensional bias vector b_j as,

$$x_t = A_j s_{j,t} + b_j . \tag{9.3}$$

In (9.3), vector x_t, corresponds to the N-dimensional random vector with its dimensionality reduced by an appropriate feature extraction technique such that $N = M$, where M is the number of unobserved independent sources. A_j is a $N \times N$, full rank, real-valued mixing matrix. Additionally, the vector $s_{j,t} = [s_{j,1,T}, \ldots, s_{j,N,T}]'$ for class j in (9.3) corresponds to the random vector of N independent, unobserved, real-valued, source signals $s = [s_1, \ldots, s_N]'$ as defined for the ICA model (Lee et al. 1999) (see Chap. 4 for more details). Since the hyperspectral data X is modeled as a mixture of K classes, and each of these K classes are described by a linear combination of N independent, non-Gaussian sources, the generative model in (9.3) depicts one of the K ways for generating the random pixel vector x_t. Thus, depending on the values of the class parameters $\theta_j = \{A_j, b_j\}$, and the unobserved independent sources $s_{j,t}$ corresponding to each class j, there are K ways for viewing x_t (i.e. $x_t = A_1 s_{1,t} + b_1, \ldots, x_t = A_j s_{j,t} + b_j, \ldots, x_t = A_K s_{K,t} + b_K$). Additionally, it is significant to note that all the assumptions and restrictions (see Sect. 4.2) necessary for the identifiability of the ICA solution are implicitly inherited by the ICAMM algorithm as well.

Expressing (9.3) as

$$s_{j,t} = A_j^{-1} (x_t - b_j) , \tag{9.4}$$

we get further insight into the generative model. Here we subtract the bias for class j from the random pixel vector x_t, which is then linearly transformed by the unmixing matrix $W_j = A_j^{-1}$, to obtain the unobserved independent sources $s_{j,t}$, corresponding to class j. As mentioned before, we assume that the observed hyperspectral sensor observations X are a mixture of several mutually exclusive classes, and that the maximum likelihood estimation results in the model that best fits the data.

The steps involved in the ICAMM algorithm for an unsupervised classification of remote sensing images are provided in the following discussion.

9.2.1
ICAMM Classification Algorithm

A detailed mathematical derivation of the ICAMM algorithm can be found in (Shah 2003; Lee et al. 2000; Lee and Lewicki 2002). In the present discussion, we only provide the steps involved in the ICAMM classification algorithm.

Step 1:

1. Initialize the total number of classes K.

2. Randomly initialize the mixing matrix A_j and the bias vector b_j, for each class j, where $j = 1$ to K.

For $t = 1$ to T do

Step 2: Estimate the vector of independent components $s_{j,t}$ as given in (9.4).

Step 3: Compute the log of the class-component density. From (9.3), the class-component density can be expressed as (Papoulis 1991),

$$p(x_t|\omega_j, \theta_j) = \frac{p(s_{j,t})}{|\det(A_j)|} . \quad (9.5)$$

Thus,

$$\log[p(x_t|\omega_j, \theta_j)] = \log[p(s_{j,t})] - \log[|\det(A_j)|] . \quad (9.6)$$

Log of the prior probability $p(s_{j,t})$, in (9.6) can be approximated as (Lee et al. 2000),

$$\log[p(s_{j,t})] \propto -\sum_{i=1}^{M}\left\{\phi_{j,i}\log[\cosh(s_{j,i,t})] - \frac{s_{j,i,t}^2}{2}\right\}, \quad (9.7)$$

where $\phi_{j,i}$ is defined as the sign of the kurtosis of the ith independent component corresponding to class j,

$$\phi_{j,i} = \text{sign}[kurt(s_{j,i})] . \quad (9.8)$$

Step 4: Compute the posterior class probability, for each pixel vector x_t, using the Bayes theorem,

$$P(\omega_j|x_t, \Theta) = \frac{p(x_t|\omega_j, \theta_j) P(\omega_j)}{\sum_{j=1}^{K} p(x_t|\omega_j, \theta_j) P(\omega_j)} . \quad (9.9)$$

In (9.9), the prior class probabilities are assumed to be equal.

Step 5: Adapt A_j for each class j, using gradient ascent, where the gradient is approximated using an extended information-maximization learning rule (Lee et al. 1999),

$$\Delta A_j \propto P(\omega_j|x_t, \Theta) A_j \left[I - \Phi_j \tanh(s_{j,t}) s'_{j,t} - s_{j,t}s'_{j,t}\right], \quad (9.10)$$

where I is an identity matrix, and Φ_j is an N-dimensional diagonal matrix corresponding to class j and is composed of $\phi_{j,i}$'s, as defined in (9.8).

Step 6: Update the bias b_j by using the pixel vectors from $\tau = 1$ to t, where t is the index of the current pixel vector being processed

$$b_j = \frac{\sum_{\tau=1}^{t} x_\tau P(\omega_j | x_\tau, \Theta)}{\sum_{\tau=1}^{t} P(\omega_j | x_\tau, \Theta)} . \qquad (9.11)$$

End For

Step 7: If the log-likelihood of the observed hyperspectral data, given as

$$\sum_{t=1}^{T} \log [p(x_t|\Theta)] = \sum_{t=1}^{T} \log \left[\sum_{j=1}^{K} p(x_t|\omega_j, \theta_j) P(\omega_j) \right], \qquad (9.12)$$

remains unchanged, go to **Step 9**, else continue.

Step 8: Repeat **Step 2** to **Step 7**.

Step 9: Use Bayes decision rule to determine class allocation for pixel vector as,

$$x_t \in \omega_i, \quad \text{if } P(\omega_i|x_t, \Theta) > P(\omega_j|x_t, \Theta), \quad \text{for all } j \neq i. \qquad (9.13)$$

The ICAMM classification algorithm as described above is an iterative algorithm, where the adaptation loop (**Step 2** to **Step 7**) is repeated until the convergence is achieved. The convergence is measured in terms of the log-likelihood (i. e. $\log[p(X|\Theta)]$) of the observed data. The adaptation is ceased once the log-likelihood of the observed data stabilizes with increasing number of iterations. This adaptation procedure, however, becomes computationally intensive with an increase in data dimensionality (Shah 2003; Shah et al. 2004).

In the next section, we will outline the proposed experimental methodology in classifying the hyperspectral datasets. We will also discuss the feature extraction techniques employed for preprocessing the hyperspectral data in order to satisfy the assumption $N = M$.

9.3
Experimental Methodology

The steps involved in our experimental methodology for classification of hyperspectral data using the ICAMM algorithm are depicted in Fig. 9.1. First, we remove the water absorption bands and some noisy bands as observed from visual inspection of the dataset. Next, we perform preprocessing of data, also known as feature extraction to reduce the dimensionality further in order to satisfy $N = M$ assumption for implementing the ICAMM algorithm.

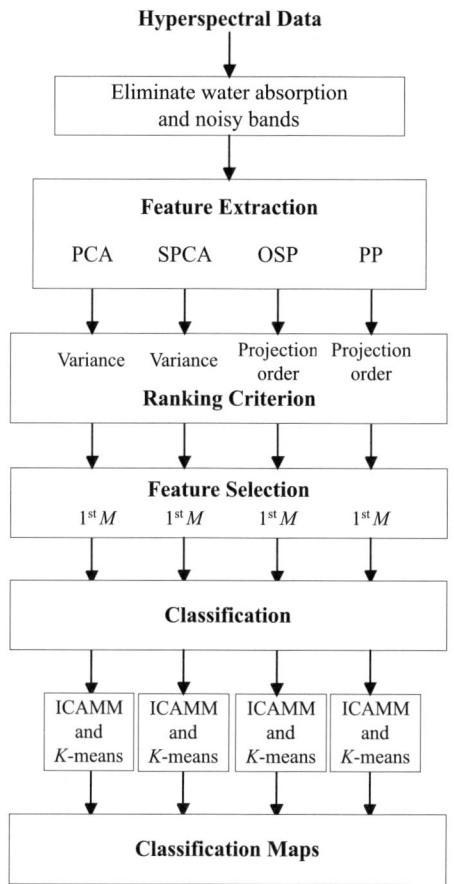

Fig. 9.1. Proposed experimental methodology for unsupervised classification of hyperspectral data

9.3.1
Feature Extraction Techniques

Due to a large number of narrow spectral bands and their contiguous nature, there is a significant degree of redundancy in the data acquired by hyperspectral sensors (Landgrebe 2002). In classification studies involving data of high spectral dimension, it is desirable to select an optimum subset of bands, in order to avoid the Hughes phenomenon (Hughes 1968) and parameter estimation problems due to interband correlation, as well as to reduce computational requirements (Landgrebe 2002; Shaw and Manolakis 2002; Tadjudin and Landgrebe 2001; Richards and Jia 1999). This gives rise to the need to develop algorithms for reducing the data volume significantly. From the perspective of statistical pattern recognition, feature extraction refers to a process, whereby a data space is transformed into a feature space, in which the original data is represented by a reduced number of effective features, retaining most of the intrinsic information content (Duda et al. 2000). Feature extraction from remote sensing data relies on the ability to separate classes based on their spectral characteristics (Richards and Jia 1999a).

In our experiments, we investigate four unsupervised feature extraction techniques: principal component analysis (PCA) (Richards and Jia 1999a), segmented principal component analysis (SPCA) (Richards and Jia 1999b), orthogonal subspace projection (OSP) (Harsayani and Chang 1994), and projection pursuit (PP) (Ifarraguerri and Chang 2000), to reduce the data volume so as to increase the efficiency of the ICAMM algorithm.

PCA, a correlation based linear transformation, seeks a projection that leads to a representation with high data variance. It projects N-dimensional data onto a new subspace, spanned by the principal axes, leading to principal components. SPCA, makes use of the block structure of the correlation matrix so that PCA is conducted on data of smaller dimensionality (Richards and Jia 1999b). For OSP, a two-step iterative process is followed, i.e., first a pixel vector with the maximum length as a class signature S_i is selected and then the orthogonal projections of all image pixels onto an orthogonal complement space of S_i are found. PP, also a two-step iterative process, first searches for a projection that maximizes a projection index, based on the information divergence of the projection's estimated probability distribution from the Gaussian distribution. Next, it reduces the rank by projecting the data onto the subspace orthogonal to the previous projection.

9.3.2
Feature Ranking

Feature extraction is followed by feature ranking, where a ranking criterion is used to rank the features extracted by a particular feature extraction technique. The ranking criterion is established based on its ability to rank the features in the order of significance of information content of the extracted features (Landgrebe 2002). Features obtained by PCA and SPCA, are ranked based on variance as a measure of information content. For OSP and PP, features are ranked based on the order of projections.

9.3.3
Feature Selection

Once the features are ranked in the order of significance, we perform the task of feature selection, i.e., we determine the optimal number of features to be retained, which lead to a high compression ratio and at the same time minimize the reconstruction error. This is indicated by the feature selection stage in Fig. 9.1, where we select the first M features from the features extracted by each of the feature extraction techniques. Recall our discussion in Sect. 9.2, pertaining to the assumption about the restriction on the number of sources (M) and the number of sensor observations (N) in the ICAMM algorithm. We have made an assumption that the number of sources is equal to the number of sensor observations ($N = M$). This simplification is justified due to the fact that if $N > M$, the dimensions of the sensor observation vector x_t can always be reduced so that $N = M$. Such a reduction in dimensionality can be achieved by all of the preprocessing or feature extraction techniques discussed above. However, a detailed review of these feature extraction techniques reveals that, other than PCA, none of the above mentioned feature extraction techniques have a well-established theory on estimating the optimal number of ranked features to be retained. In the case of PCA, the criterion for feature selection can be stated as – sum of the variance of the retained principal components should exceed 99% of the total data variance. Hence, we consider PCA as a technique for estimating the intrinsic dimensionality of the hyperspectral data. PCA, though optimal in the mean square sense, suffers from various limitations (Duda et al. 2000), and therefore, there is a need to investigate the performance of other feature extraction techniques for hyperspectral data. Thus, in addition to PCA, we consider SPCA, OSP and PP as feature extraction techniques.

9.3.4
Unsupervised Classification

The features obtained by a particular feature extraction technique are employed by the ICAMM algorithm as well as by the most widely used K-means algorithm for classification. The accuracy of the ICAMM algorithm derived classification is then compared with that obtained from the K-means classification algorithm.

In the next section, we apply and evaluate the performance of the ICAMM algorithm for classification of data acquired from AVIRIS hyperspectral sensor.

9.4
Experimental Results and Analysis

To comparatively evaluate the classification performance we utilize the measure, overall accuracy that is computed as the percentage of correctly classified pixels among all the pixels. The overall accuracy, obtained by the ICAMM algorithm, varies for each run. This is due to the random initialization of the mixing matrix and the bias vectors for the classes. Hence, we present the results averaged over 100 runs of the ICAMM algorithm. We also provide further insight into the ICAMM derived classification obtained from one run by critically analyzing the classification map and its corresponding error matrix. Individual class accuracies have been assessed through producer's and user's accuracy measures (see Sect. 2.6.5 of Chap. 2 for details on accuracy measures).

Data from the AVIRIS sensor, has been used for this experiment (see Sect. 4.4.3 of Chap. 4 for details on this dataset). For computational efficiency, a sub-image of size 80 x 40 pixels has been selected here. Some of the bands (centered at 0.48 μm, 1.59 μm, and 2.16 μ m) are depicted in Fig. 9.2a–c. The corresponding reference image consisting of 4 classes is shown in Fig. 9.2d. The total number of pixels belonging to each class in this reference image is given in Table 9.1. Here, similar to earlier studies on classification of this dataset (e. g. Tadjudin and Landgrebe 2001), twenty-one water absorption bands numbered 104 – 108 (1.36 – 1.40 μm), 150 – 163 (1.82 – 1.93 μm), and 220 (2.50 μm) have been discarded. Besides this, fourteen bands containing noise, as observed from visual inspection, have also been removed. Thus, a total of 185 AVIRIS bands have been used in this investigation.

The mean intensity values of the classes in each of the spectral bands are plotted in Fig. 9.2e. It can be inferred from the plot that the two classes – corn-notill and soybeans-min have almost equal mean spectral responses in many of the bands, thus indicating a non-linear separability of the two classes from each other. An explanation for this may be the fact that the data has been acquired in the early growing season of the year (Tadjudin and Landgrebe 2001). The

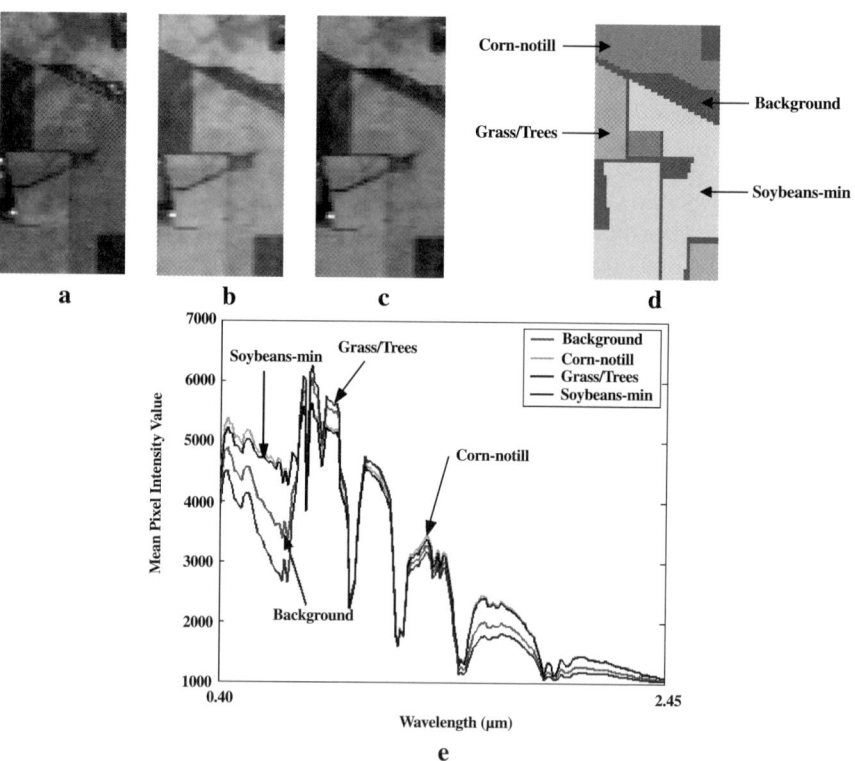

Fig. 9.2a–e. Three images (80 × 40 pixels) centered at **a** 0.48 μm, **b** 1.59 μm, and **c** 2.16 μm. These images are a subset of the data acquired by the AVIRIS sensor. **d** Corresponding reference map for this dataset consisting of four classes: background, corn-notill, grass/trees and soybeans-min. **e** Mean intensity value of each class plotted as a function of the wavelength in μm. For a colored version of this figure, see the end of the book

Table 9.1. Classes and the number of pixels in each class

Class	Number of pixels
Background	594
Corn-notill	640
Grass/Trees	383
Soybeans-min	1583
Total	3200

crop canopies had approximately 5% cover, the rest being soil, covered with the residue of previous year's crop. Given this low canopy cover, the variation in spectral response due to – i) the soil type variations and ii) the varying amount of residue from last season's crop, may have a much greater influence upon the net pixel spectral response than the variation in the type of class. The other two classes, which are background and grass/trees, are distinguishable from each other as well as from the other two classes in most wavelength regions.

Figure 9.3a–d show the results of the first six features produced by PCA, SPCA, OSP and PP respectively. The first 10 principal components obtained from the PCA have a total data variance of 99.67%, and thus have been used for classification by the ICAMM algorithm, with $N = M = 10$. Similarly, the features extracted by the remaining three techniques have been ranked based on their corresponding ranking criterion. We select the first 10 features extracted by each of these feature extraction techniques.

The first six features obtained by SPCA, correspond to the first two principal components obtained by performing PCA transformation on each block of correlated bands. It can be seen that the first feature obtained by both PCA and SPCA extract almost the same information. We investigate the cause of this similarity. Since PCA has a significant property that it concentrates the data variability to the maximum extent possible into the first few components and the variability of the data is scale-dependant, PCA is sensitive to the scaling of the data to which it is applied.

For example, if the intensity values of one or some of the hyperspectral bands are arbitrarily doubled, their contribution to the variance of the dataset will be increased fourfold, and they will therefore be found to contribute more to the earlier eigenvalues and eigenvectors. With this remark, it is of interest to carefully observe the first feature in Fig. 9.3a,b as well as the image of a spectral band centered at 0.48 μm (Fig. 9.2a). These three images appear almost identical and hence we conclude that it is one or some of the bands in the first block of highly correlated bands employed for SPCA, which have contributed the most to the first feature obtained by PCA performed on the entire data.

The improvement provided by SPCA over PCA, in enhancing interclass separability as determined from visual inspection, can be comprehended through the fact that the third, fourth, fifth and the sixth features generated by PCA do not exhibit high data variability, while all the six features (shown in Fig. 9.3b)

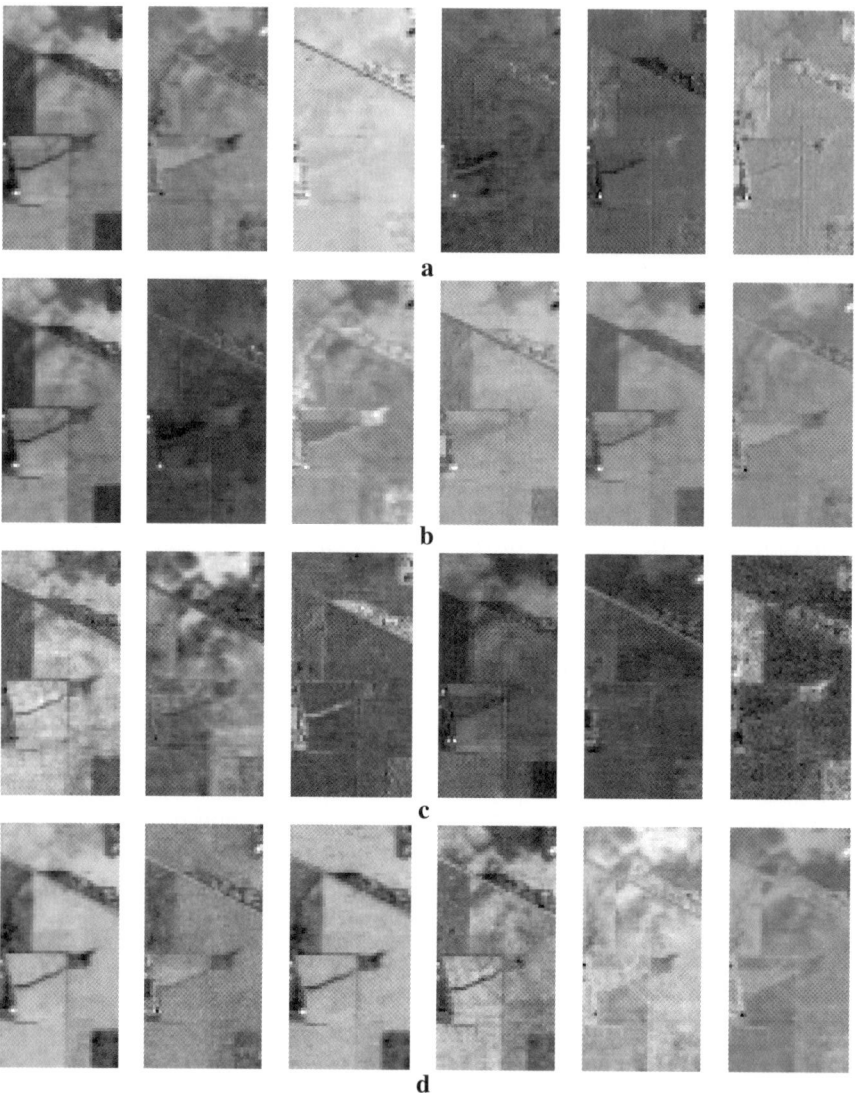

Fig. 9.3a–d. First six features extracted by **a** PCA, **b** SPCA, **c** OSP, and **d** PP applied on the AVIRIS data. The features extracted by PCA and SPCA, are ranked in decreasing order of variance. For OSP and PP, the features are ranked in the order of projections

obtained through SPCA capture significant class information. A visual inspection of the features generated by OSP reveals that its performance is comparable to that of PCA and SPCA in its ability to discriminate between classes. Of particular interest is the third feature generated by OSP (Fig. 9.3c), in which it detects the signature corresponding to class background. For PCA and SPCA, information pertaining to this class is split into several features. The first three features from PP (Fig. 9.3d) reveal that though the information pertaining to classes background and grass/trees has been significantly enhanced, it does not capture enough information needed to distinguish the remaining two classes (i.e. corn-notill and soybeans-min). However, it is interesting to note that the fourth feature generated by PP exhibits a significant amount of information when compared with the fourth feature generated by each of PCA, SPCA and OSP. A remark applicable to each of these feature extraction techniques is that they show excellent performance in extracting the information corresponding to classes background and grass/trees, but provide not much variability between the remaining two classes. This observation confirms the earlier discussion, where we had noted that corn-notill and soybeans-min have almost equal mean spectral responses in many of the bands, leading to a non-linear separability between these two classes.

The accuracy of classification produced from the ICAMM and the K-means algorithm are presented in Table 9.2. It can be seen from this table that the SPCA technique leads to the highest mean overall accuracy (i.e. 61.4%) as compared to those achieved by the other feature extraction techniques (i.e. PCA, OSP and PP). The minimum overall accuracy obtained by SPCA is also higher than the others. The maximum overall accuracy (i.e. 76.4%) has been achieved by OSP but is insignificantly higher in comparison to the maximum overall accuracy achieved by SPCA.

On comparing the classification accuracy of the ICAMM algorithm with that obtained by the K-means algorithm, it can be seen from Table 9.2 that the mean overall classification accuracies obtained by the ICAMM algorithm for all the feature extraction techniques are significantly higher than those obtained from the K-means classification algorithm. In particular, for the dataset preprocessed by SPCA, the minimum (53.4%), maximum (75.9%) and mean (61.4%) overall accuracies attained by the ICAMM algorithm, far exceed the overall accuracy produced by the K-means algorithm, (i.e. 48.3%).

We further analyze the performances of the ICAMM and the K-means classification algorithm through their classification maps and the corresponding error matrices. In Fig. 9.4a–d and Fig. 9.4e–h, we present the classification

Table 9.2. Comparison of overall classification accuracy (%) achieved by the ICAMM and K-means algorithms applied on the AVIRIS dataset

Feature extraction technique	ICAMM classification Overall Accuracy (%) based on 100 runs			K-means classification Overall Accuracy (%)
	Minimum	Mean	Maximum	
PCA	45.7	58.9	70.7	50.5
SPCA	53.4	61.4	75.9	48.3
OSP	51.5	60.7	76.4	51.7
PP	46.9	56.7	63.5	52.5

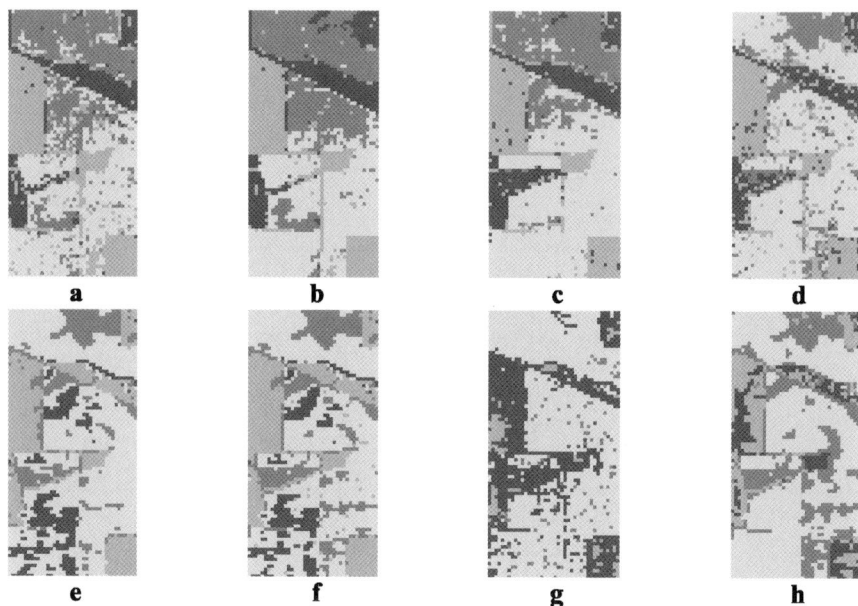

Fig. 9.4a–h. Classification maps obtained by the ICAMM classification algorithm applied on data preprocessed by **a** PCA (Overall Accuracy = 68.6%), **b** SPCA (Overall Accuracy = 71.7%), **c** OSP (Overall Accuracy = 73.1%), and **d** PP (Overall Accuracy = 61.6%). Classification maps obtained by the K-means classifier applied on data preprocessed by **e** PCA (Overall Accuracy = 50.5%), **f** SPCA (Overall Accuracy = 48.3%), **g** OSP (Overall Accuracy = 51.7%), and **h** PP (Overall Accuracy = 52.5%)

maps obtained by the ICAMM and the K-means algorithm applied on the AVIRIS data preprocessed by each of the feature extraction techniques. In Table 9.3 to Table 9.6, we present the error matrices from classifications obtained by performing ICAMM and K-means algorithms on the features extracted by each of PCA, SPCA, OSP and PP respectively.

Though, there is no optimal measure for selecting a subset of features from those obtained by OSP, it performs considerably better than PCA and PP in terms of their minimum, mean, and maximum overall classification accuracies. PCA detects all the four classes and its mean overall accuracy is comparable to OSP. However, its classification map as shown in Fig. 9.4a reveals a 'noisy' classification, leading to lower overall accuracies. In the case of PP, once again there is no optimal measure for determining the number of feature to be selected, and a mean overall accuracy of 56.7% is obtained. Fig. 9.4d shows its poor performance in classifying corn-notill, a major portion of which is classified as soybeans-min. This was expected from the earlier analysis that the features produced by PP failed to capture information pertinent to these two classes.

On comparing the classification maps obtained by the ICAMM algorithm, via all the four feature extraction techniques with those obtained by the K-

Table 9.3. Error matrix for the classification of data, preprocessed by PCA. **a** ICAMM classification algorithm and **b** *K*-means classification algorithm

a

ICAMM (PCA)		Reference map				Total
		C 1	C 2	C 3	C 4	
	C 1	321	23	3	63	410
Classification	C 2	29	489	0	291	809
Map	C 3	201	53	380	223	857
	C 4	43	75	0	1006	1124
Total		594	640	383	1583	3200
Producer's Accuracy (%)		54.0	76.4	99.2	63.6	
User's Accuracy (%)		78.3	60.5	44.3	89.5	
		Overall Accuracy (%) = 68.6				

C 1 – Background, C 2 – Corn-notill, C 3 – Grass/Trees, C 4 – Soybeans-min

b

K-means (PCA)		Reference map				Total
		C 1	C 2	C 3	C 4	
	C 1	43	0	0	331	374
Classification	C 2	129	154	21	176	480
Map	C 3	293	1	362	19	675
	C 4	129	485	0	1057	1671
Total		594	640	383	1583	3200
Producer's Accuracy (%)		7.2	24.1	94.5	66.8	
User's Accuracy (%)		11.5	32.1	53.6	63.3	
		Overall Accuracy (%) = 50.5				

C 1 – Background, C 2 – Corn-notill, C 3 – Grass/Trees, C 4 – Soybeans-min

means algorithm, it can be seen that our algorithm performs significantly better. In most cases, the *K*-means algorithm fails to differentiate between the two classes, background and tree/grass, which have been correctly classified by the ICAMM algorithm used with all the four feature extraction techniques. Table 9.3a,b reveal that the ICAMM and the *K*-means algorithm classify grass/trees with significantly high producer's accuracy of 99.2% and 94.5% respectively. As noted before, class corn-notill and soybeans-min exhibit similar spectral properties and this has a profound impact on the producer's and the user's accuracy of class corn-notill for the *K*-means classification map. ICAMM, however, exhibits comparatively higher producer's and user's accuracy of 76.4% and 60.5% respectively for this class.

As pointed out earlier, the *K*-means algorithm classifies majority of the pixels belonging to class corn-notill as soybeans-min. This can be verified from the

Table 9.4. Error matrix for the classification of data, preprocessed by SPCA. **a** ICAMM classification algorithm and **b** K-means classification algorithm

a

ICAMM (SPCA)		Reference map				Total
		C 1	C 2	C 3	C 4	
	C 1	350	60	1	52	463
Classification	C 2	35	521	0	390	946
Map	C 3	154	18	382	99	653
	C 4	55	41	0	1042	1138
Total		594	640	383	1583	3200
Producer's Accuracy (%)		58.9	81.4	99.7	65.8	
User's Accuracy (%)		75.6	55.1	58.5	91.6	
		Overall Accuracy (%) = 71.7				

C 1 – Background, C 2 – Corn-notill, C 3 – Grass/Trees, C 4 – Soybeans-min

b

K-means (SPCA)		Reference map				Total
		C 1	C 2	C 3	C 4	
	C 1	42	0	0	288	330
Classification	C 2	129	162	10	271	572
Map	C 3	311	1	373	55	740
	C 4	112	477	0	969	1558
Total		594	640	383	1583	3200
Producer's Accuracy (%)		7.1	25.3	97.4	61.2	
User's Accuracy (%)		12.7	28.3	50.4	62.2	
		Overall Accuracy (%) = 48.3				

C 1 – Background, C 2 – Corn-notill, C 3 – Grass/Trees, C 4 – Soybeans-min

error matrix (Table 9.3b), where 485 pixels out of the total 640 pixels belonging to class corn-notill have been classified as soybeans-min. This misclassification caused by the K-means algorithm is compensated by significantly higher accuracies (both producer's and user's) for class soybeans-min. The lowest producer's and user's accuracies for the K-means algorithm are obtained for the class background (7.2% and 11.5% respectively). In case of the K-means algorithm applied on the data preprocessed by the SPCA (Table 9.4b), similar conclusions can be drawn about the accuracies of the class background. These accuracies are as low as 7.1% (producer's) and 12.7% (user's). However, these accuracies obtained by the K-means algorithm for the class background are relatively higher for the OSP and PP preprocessed data. For the SPCA preprocessed data, ICAMM exhibits very high accuracy (99.7%) in classifying grass/trees.

Thus, based on the classification accuracies (Table 9.2), the classification maps (Fig. 9.4) and the corresponding error matrices (Table 9.3 to Table 9.6),

Table 9.5. Error matrix for the classification of data, preprocessed by OSP. **a** ICAMM classification algorithm and **b** K-means classification algorithm

a

ICAMM (OSP)		Reference map				Total
		C 1	C 2	C 3	C 4	
Classification Map	C 1	430	48	59	184	721
	C 2	18	431	0	160	609
	C 3	89	8	324	84	505
	C 4	57	153	0	1155	1365
Total		594	640	383	1583	3200
Producer's Accuracy (%)		72.4	67.3	84.6	73	
User's Accuracy (%)		59.6	70.8	64.2	84.6	
		Overall Accuracy (%) = 73.1				

C 1 – Background, C 2 – Corn-notill, C 3 – Grass/Trees, C 4 – Soybeans-min

b

K-means (OSP)		Reference map				Total
		C 1	C 2	C 3	C 4	
Classification Map	C 1	369	18	306	295	988
	C 2	3	14	0	89	106
	C 3	51	0	76	4	131
	C 4	171	608	1	1195	1975
Total		594	640	383	1583	3200
Producer's Accuracy (%)		62.1	2.2	19.8	75.5	
User's Accuracy (%)		37.4	13.2	58.0	60.5	
		Overall Accuracy (%) = 51.7				

C 1 – Background, C 2 – Corn-notill, C 3 – Grass/Trees, C 4 – Soybeans-min

we conclude that the AVIRIS data preprocessed by SPCA and classified by the ICAMM algorithm, exhibits an improved overall classification performance. Additionally, these results clearly show that the ICAMM algorithm outperforms the conventional K-means algorithm for the classification of AVIRIS data considered in this experiment.

9.5 Summary

The primary issue that motivated this research is the limitation of Gaussian mixture model based classification algorithms. The underlying Gaussian distribution assumption is limited in the sense that the Gaussian mixture model exploits only second order statistics of the observed data to estimate the poste-

Table 9.6. Error matrix for the classification of data, preprocessed by PP **a** ICAMM classification algorithm and **b** K-means classification algorithm

a

ICAMM (PP)		Reference map				Total
		C 1	C 2	C 3	C 4	
	C 1	275	8	26	71	380
Classification	C 2	27	191	0	192	410
Map	C 3	211	47	357	173	788
	C 4	81	394	0	1147	1622
Total		594	640	383	1583	3200
Producer's Accuracy (%)		46.3	29.8	93.2	72.5	
User's Accuracy (%)		72.4	46.6	45.3	70.7	
		Overall Accuracy (%) = 61.6				

C 1 – Background, C 2 – Corn-notill, C 3 – Grass/Trees, C 4 – Soybeans-min

b

K-means (PP)		Reference map				Total
		C 1	C 2	C 3	C 4	
	C 1	127	0	161	1	289
Classification	C 2	182	171	3	401	757
Map	C 3	166	9	219	18	412
	C 4	119	460	0	1163	1742
Total		594	640	383	1583	3200
Producer's Accuracy (%)		21.4	26.7	57.2	73.5	
User's Accuracy (%)		43.9	22.6	53.2	66.8	
		Overall Accuracy (%) = 52.5				

C 1 – Background, C 2 – Corn-notill, C 3 – Grass/Trees, C 4 – Soybeans-min

rior densities. Quantifying the higher order statistics of each class, rather than just estimating the mean and covariance to better fit the data into a parametric class probability density function seems desirable.

In this chapter, we have described a novel approach, derived from ICA, for unsupervised classification of non-Gaussian classes in hyperspectral remote sensing imagery. This approach models class distributions with non-Gaussian densities, formulating the ICA mixture model (ICAMM). We demonstrated the successful application of the ICAMM algorithm in classifying hyperspectral remote sensing data. In particular, we employed the AVIRIS dataset for our experiment. Four feature extraction techniques – Principal Component Analysis (PCA), Segmented Principal Component Analysis (SPCA), Orthogonal Subspace Projection (OSP) and Projection Pursuit (PP) were employed as a preprocessing step to reduce the dimensionality of the hyperspectral data.

For the AVIRIS dataset preprocessed by each of the four feature extraction techniques, the ICAMM classification algorithm produced significantly higher accuracy as compared to the K-means algorithm. Similar results have been obtained for the data acquired by another hyperspectral sensor – HYDICE and have been reported in Shah (2003).

References

Bae U-M, Lee T-W, Lee S-Y (2000) Blind signal separation in teleconferencing using ICA mixture model. Electronic Letters 36: 680–682

Chang C-I, Chiang S-S, Smith JA, Ginsberg IW (2002) Linear spectral random mixture analysis for hyperspectral imagery. IEEE Transactions on Geoscience and Remote Sensing 40(2): 375–392.

Common P (1994) Independent component analysis, a new concept?. Signal Processing 36: 287–314.

Duda RO, Hart PE, Stork DG (2000) Pattern classification, 2nd edition. John Wiley and Sons, New York

Harsanyi JC, Chang C-I (1994) Hyperspectral image classification and dimensionality reduction: An orthogonal subspace projection approach. IEEE Transactions on Geoscience and Remote Sensing 32: 779–785

Hughes GF (1968) On the mean accuracy of statistical pattern recognition. IEEE Transactions on Information Theory 14: 55–63

Hyvärinen A, Oja E (2000) Independent component analysis: algorithms and applications. Neural Networks 13: 411–430

Ifarraguerri A, Chang C-I (2000) Unsupervised hyperspectral image analysis with projection pursuit. IEEE Transactions on Geoscience and Remote Sensing 38: 2529–2538.

Landgrebe D (2002) Hyperspectral image data analysis. IEEE Signal Processing Magazine 19: 17–28

Lee T-W, Girolami M, Sejnowski TJ (1999) Independent component analysis using an extended infomax algorithm for mixed sub-Gaussian and super-Gaussian sources. Neural Computation 11: 417–441

Lee T-W, Lewicki MS, Sejnowski TJ (2000) ICA mixture models for unsupervised classification of non-Gaussian classes and automatic context switching in blind signal separation. IEEE Transactions on Pattern Analysis and Machine Intelligence 22: 1078–1089

Lee T-W, Lewicki MS (2002) Unsupervised image classification, segmentation, and enhancement using ICA mixture models. IEEE Transactions on Image Processing 11: 270–279

Papoulis A (1991) Probability, random variables, and stochastic processes, 3rd edition. McGraw Hill, New York

Richards JA, Jia X (1999a) Remote sensing digital image analysis: an introduction. Springer-Verlag, Berlin

Richards JA, Jia X (1999b) Segmented principal components transformation for efficient hyperspectral remote sensing image display and classification. IEEE Transactions on Geoscience and Remote Sensing 37: 538–542

Robila SA, (2002) Independent component analysis feature extraction for hyperspectral images, PhD Thesis, Syracuse University, Syracuse, NY

Robila SA, Varshney PK (2002a) Target detection in hyperspectral images based on independent component analysis. Proceedings of SPIE Automatic Target Recognition XII, 4726, pp 173–182

Robila SA, Varshney PK (2002b) A fast source separation algorithm for hyperspectral image processing. Proceedings of IEEE International Geoscience and Remote Sensing Symposium (IGARSS), pp 3516–3518

Shah CA (2003) ICA mixture model algorithm for unsupervised classification of multi/hyperspectral imagery. M.S. Thesis, Syracuse University, Syracuse, NY

Shah CA, Arora MK, Varshney PK (2004) Unsupervised classification of hyperspectral data: an ICA mixture model based approach. International Journal of Remote Sensing 25: 481–487

Shaw G, Manolakis D (2002) Signal processing for hyperspectral image exploitation. IEEE Signal Processing Magazine 19: 12–16

Stark H, Woods JW (1994) Probability, random processes, and estimation theory for engineers, 2nd edition. Prentice Hall, Upper Saddle River, NJ

Tadjudin S, Landgrebe DA (2001) Robust parameter estimation for mixture model. IEEE Transactions on Geoscience and Remote Sensing 38: 439–445

CHAPTER 10

Support Vector Machines for Classification of Multi- and Hyperspectral Data

Pakorn Watanachaturaporn, Manoj K. Arora

10.1
Introduction

As discussed in Chap. 5, support vector machines (SVMs) have originated from statistical learning theory for classification and regression problems. Unlike the popular neural network classifiers, SVMs do not minimize the empirical training error (Byun and Lee 2003). Instead, they aim to maximize the margin between two classes of interest by placing a linear separating hyperplane between them. While doing so, the upper bound on the generalization error is minimized. Thus, SVM based classifiers are expected to have more generalization capability than neural networks. Other advantages of SVMs are their ability to adapt their learning characteristic via a kernel function and to adequately classify data on a high-dimensional feature space with a limited number of training data sets thereby overcoming the *Hughes Phenomenon*. The theoretical background on SVMs has been presented in Chap. 5. This chapter builds on this theoretical knowledge to apply SVMs for classification of multi and hyperspectral remote sensing data.

SVMs are good candidates for remote sensing image classification for a number of reasons. Firstly, an SVM can work well with a small training data set. Selection of a sufficient number of pure training pixels (i.e. pixels belonging to only one class) has always been a problem even though a remote sensing image contains as many as hundreds of thousands of pixels. The problem becomes severe in coarse resolution images where mixed pixels, i.e. pixels containing more than one class, may be abundant. Secondly, SVMs have been found to perform with high classification accuracy for data having hundreds of dimensions in applications such as the classification of personal income using 123 parameters in the data set (Keerthi 2002). In remote sensing these days, the hyperspectral data are available in hundreds of bands. The processing of such high-dimensional data to extract quality information will always remain a challenge. Conventional statistical classifiers fail to process high dimensional data due to the requirement of large number of training samples. SVMs have the potential to produce accurate classifications from high dimensional data with limited number of training samples. Thirdly, unlike neural networks, SVMs are robust to the overfitting problem as they rely on margin maximization rather than finding a decision boundary directly from the training samples. Fourthly, the structure of an SVM is less complex even with the high dimensional data. Neural networks, for example, have a very complex structure for processing high dimensional data. Finally, as compared to another recent nonparametric classifier namely the decision tree classifier, an SVM does not require the generation of rules that heavily depend on the knowledge from experts. This is crucial to achieve high classification accuracy.

Classification of remote sensing data using SVMs has been introduced recently. Gualtieri and Cromp (1998) and Gualtieri et al. (1999) used SVMs to classify an AVIRIS image. The results were compared with those obtained from 'the extraction and classification of homogeneous objects' (ECHO) classifier and the Euclidean classifier (Tadjudin and Landgrebe, 1998). Accuracy of 87.3%, 82.9%, and 48.2% were obtained from SVM, ECHO, and Euclidean classifiers respectively. In 2002, Zhu and Blumberg (2002) classified an ASTER image using SVMs. They reported the results of a classification experiment with an SVM using polynomial kernels of different degrees and the radial basis function (RBF) kernel. The overall accuracies were obtained in the range of 87% to 91%. Melgani and Bruzzone (2002) used an SVM to classify an AVIRIS image and compared that with K-Nearest Neighbors (K-NN), and RBF neural network (RBF-NN) classifiers, and obtained accuracies of 93.42%, 83.94%, and 86.99% from SVM, K-NN, and RBF classifiers respectively. Huang et al. (2002) intensively investigated the accuracy obtained from SVM classifiers on a Landsat TM image. They reported an accuracy of 75.62% from the SVM, 74.02% from the back propagation neural network, 73.31% from the decision tree, and 71.76% from the maximum likelihood classifiers. Shah et al. (2003) reported the accuracies attained by SVM, MLC, and back-propagation neural network classifiers when applied on an AVIRIS image with and without the use of supervised feature extraction techniques. They used the LSVM (see Chap. 5) with a linear kernel, polynomial kernel of different degrees, a RBF kernel, and a sigmoid kernel. They showed that accuracies of 90.9% to 97.2% were obtained from SVM classifiers while the maximum likelihood and neural network classifiers could achieve accuracies of 59.4% and 61.9% respectively on the full dimensionality dataset. They also showed that the classification results from the full dimensionality dataset were better than the classification of features obtained using the discriminant analysis feature extraction (DAFE) and the decision boundary feature extraction (DBFE) techniques (Lee and Landgrebe 1993).

From these limited studies, it can be seen that in all the cases SVM derived classifications of both multi- or hyperspectral datasets produced the highest accuracy. However, there are many issues, which need to be further investigated before SVM classifiers can be implemented at the operational level.

The aim of this chapter is to understand the application of SVMs and the behavior of associated parameters for the classification of multi and hyperspectral remote sensing data. In the next section, the details of parameters considered are provided. Section 10.3 describes the remote sensing data used to produce SVM classification. Experimental set up, results and their analyses is presented in Sect. 10.4. A summary is provided in Sect. 10.5.

10.2
Parameters Affecting SVM Based Classification

Application of SVMs to any classification and regression problem requires the determination of several design parameters and the resolution of several issues. Some of the parameters and issues considered here are:

1. Determination of the penalty value (C value) to regulate the error term (see Sect. 5.3.2)
2. Choice of an appropriate kernel and its parameters (see Sect. 5.3.3).
3. Selection of a suitable multiclass method (see Sect. 5.4).
4. Decision on the type of optimizer to be used (see Sect. 5.5).

In practice, the training data do not comprise of only pure pixels. Some of them are mixed pixels due to either the noise present in the training data or the mixture of classes during the selection of training data due to mislabeling thereby introducing classification errors. A penalty parameter is introduced to handle this error. The penalty parameter is also called the *C value* and depends on the training data and the type of kernel. In general, it is found by trial and error.

A kernel function determines the characteristic of an SVM. When the classes in the input data are separable using a linear hyperplane, a linear kernel is used. Similarly, a nonlinear kernel function may be used to separate data when the classes are nonlinearly separable. However, it is very difficult to determine which non-linear kernel function is suitable for a given application. In practice, a user will apply standard kernel functions such as a linear, polynomial, RBF (also known as the Gaussian kernel), or sigmoid kernel. If these standard kernel functions are unable to perform adequately, a user may have to design a problem-specific kernel function. Moreover, the choice of the kernel further brings in another issue that a user has to consider, which is to determine the values of parameters for a given kernel function. These parameters may be called *hyperparameters*. For example, in case of an RBF kernel, the hyperparameter that needs to be determined is σ (see Sect. 5.3.3).

Thus, finding the parameters or hyperparameters that are able to accurately predict the unknown data for a given application requires a method of model selection or parameter search. The unknown data, in this case, refers to testing samples or the image to be classified during the allocation stage. It is also a fact that a model that achieves high training accuracy may not guarantee high accuracy for the unknown data. Therefore, to establish that the model works well on the whole dataset (i.e. the model has high generalization capacity), a common strategy is to split the data into two parts – *a training set*, and *a validation set* or *a testing set*. An SVM is trained on the training set and is evaluated for its accuracy on the testing set. Another model selection method is called k-fold cross-validation. In this method, the training data are divided into k subsets of equal size. An SVM is trained on the $k-1$ subsets of data and is tested on the remaining one subset. Training and testing are performed on the input data k times with different $k-1$ subsets for training and the remaining one subset for testing. In other words, the testing subset is used to assess the accuracy of the model trained by the $k-1$ subsets. After repeating the above process k times, the overall accuracy measured by the cross-validation method is the percentage of the number of correctly classified data for all k testing subsets. The model that produces the highest overall accuracy is used to classify data during the allocation stage.

A straightforward way to find a judicious combination of the penalty value and the hyperparameters of kernel functions is to use a grid-search method. In the grid-search method, various combinations of the penalty value and the hyperparameters are formed and their effect on the accuracy is assessed by using either a validation set or k-fold cross-validation. The model that gives the best accuracy is selected. Applying the grid-search method is very time consuming because it acts on every combination of the penalty value and hyperparameters. To reduce the model selection time, one may perform a coarse grid-search first. For example, the penalty value may take values 10^2 apart such as $10^{-5}, 10^{-3}, \ldots, 10^5$. Then, after identifying the best classification accuracy region obtained from the coarse grid-search, a finer grid-search may be conducted over that region only. For instance, the best accuracy from the coarse grid search might be between the penalty values of 10^1 to 10^3. The finer grid search might be performed over penalty values with finer resolution (e. g. $10^1, 10^{1.1}, \ldots, 10^3$). The final model is trained using the whole training set with the hyperparameters identified by the grid-search method.

Another issue related to SVM classifiers is the selection of appropriate multi-class method (see Sect. 5.4). SVMs were initially developed to perform binary classification; though, applications of binary classification are very limited. More practical applications involve multiclass classification. For example, in remote sensing, land cover classification is generally a multiclass problem. A number of methods have been proposed to employ SVMs to produce multiclass classification. Most of the methods generate multiclass classifications from a number of binary SVM classifiers. All the methods have their own merits and demerits. For example, the one against the rest approach gives an advantage in terms of simplicity and requires fewer binary classifiers. However, many have argued that in this method the number of training samples of each class for each binary classifier become significantly imbalanced. To solve this problem, methods such as a re-sampling method have been proposed. In the re-sampling method, pseudo training data are created by copying existing training data and adding/subtracting noise. In other words, a new training area is identified for a class and a number of training data are selected randomly within that area. The latter approach may help in improving the accuracy in some cases. However, it results in increased training time since the number of training data is increased due to the use of the pseudo training data. Both the approaches are simpler than the pairwise and directed acyclic graph (DAG) approaches. However, experiments show that, training time of the one against the rest both with and without balancing training data is longer and percent accuracy is lower. On the contrary, the pairwise and DAG approaches require more binary classifiers, but they take less training time and give more accurate results.

Finally, an important issue in the implementation of SVMs is the selection of the type of optimizer to be used to find the support vectors. Different methods for this purpose have been proposed (see Sect. 5.5). An optimizer based on QP or LP methods is a natural choice. However, these optimizers may not be able to efficiently handle large data volume such as the one found in remote sensing images. In such cases, more efficient optimizers must be employed. A more appropriate choice for an optimizer, for instance, can be a subset selection or an iterative method. In the subset selection method, such as the *chunking-decomposition* method, a large problem is solved by breaking it into sub-problems. Each sub-problem is solved separately by either a QP or an LP optimizer. Examples of optimizers based on the subset selection method may be found in Chang and Lin (2002) and Joachims (2002). Alternatively, an iterative method, such as the LSVM (Mangasarian and Musicant 2000), which is a fast and simple algorithm, may also be used.

Many of these issues will be investigated in this chapter through a number of experiments on classification of multi and hyperspectral remote sensing images, which are described in the next section.

10.3
Remote Sensing Images

10.3.1
Multispectral Image

A UTM rectified Landsat 7 ETM+ multispectral image acquired in 8 bands has been used. The image was acquired in 1999 and covers an urban region of Syracuse, NY. Only 7 bands excluding the panchromatic band, due to its finer spatial resolution than the remaining bands were considered.

Approximately 10% of the area in the image selected is covered by water whereas the remaining 90% covers a combination of built up area, small vegetation, and trees. The size of the image is 445 × 595 pixels, which has been resampled to 25 m at the source (see Fig. 10.1).

A large number of pure pixels were extracted from the Landsat ETM+ image for the six classes of interest, namely water, highways/runways, grassland,

Table 10.1. Total pure pixels, training, and testing pixels from the Landsat ETM+ image

Class Name	Number of pure pixels identified
Water	2100
Highway/Runway	529
Grass	951
Commercial	1626
Tree	1378
Residential	3004
Total	9588

Fig. 10.1. Color composite from the Landsat 7 ETM+ multispectral image. For a colored version of this figure, see the end of the book

commercial areas, trees and residential areas. UTM rectified IKONOS MSS and PAN images at 4 m and 1 m fine spatial resolution and the previous knowledge about the study area were used as reference data (ground data) to determine the class identity of the pure pixels. From the database of pure pixels, the training and testing datasets consisting of randomly selected 200 pixels of each class were generated (Table 10.1).

10.3.2
Hyperspectral Image

Hyperspectral image used here comes from the AVIRIS sensor (see Sect. 4.4.3 of Chap. 4 for more details on this dataset).

In our experiments, similar to the other studies, twenty water absorption bands numbered [104 – 108], [150 – 163], and 220 were removed from the original image. In addition, fifteen noisy bands [1 – 3], 103, [109 – 112], [148 – 149], [164 – 165], and [217 – 219], as observed from visual inspection, were also discarded. A pure pixel database was created from which a number of training and testing pixels for each class were randomly selected (Table 10.2).

Table 10.2. Total number of pure pixels, training, and testing pixels from the AVIRIS image

Class Name	Number of pure pixels	Number of training pixels	Number of testing pixels
Alfalfa	13	7	6
Corn-notill	234	117	117
Corn-min	140	70	70
Corn	60	30	30
Grass/pasture	94	47	47
Grass/trees	116	58	58
Grass/pasture-mowed	18	9	9
Hay-windrowed	160	80	80
Oats	26	13	13
Soy-notill	231	115	116
Soy-mintill	291	146	145
Soy-clean	91	45	46
Wheat	49	25	24
Woods	208	104	104
Bldg-grass-trees-drives	81	40	41
Stone-steel towers	19	10	9
Total	1831	916	915

10.4
SVM Based Classification Experiments

This section illustrates the use of SVMs to produce land cover classification from multi and hyperspectral remote sensing data described in the previous section. The effect of a number of factors on the accuracy of classification produced by the SVM classifier has been investigated. These factors are – selection of the multiclass method, choice of the optimizer and type of the kernel function. As a result, a number of SVM classifications have been performed. The accuracy of all the classifications has been assessed using the most widely used measure namely overall accuracy obtained from the error matrix (see Sect. 2.6.5 of Chap. 2).

10.4.1
Multiclass Classification

To perform multiclass classification, four methods have been used to examine the effect of each on classification accuracy. These methods are one against the rest, one against the rest with a balanced number of training data, pairwise classification, and directed acyclic graph (DAG). A number of classifications using the Lagrangian Support Vector Machine and the linear kernel for different values of C have been produced. The variations in classification accuracy obtained by applying each multiclass method are analyzed with the help of plots shown in Fig. 10.2 and Fig. 10.3 for multi and hyperspectral datasets respectively.

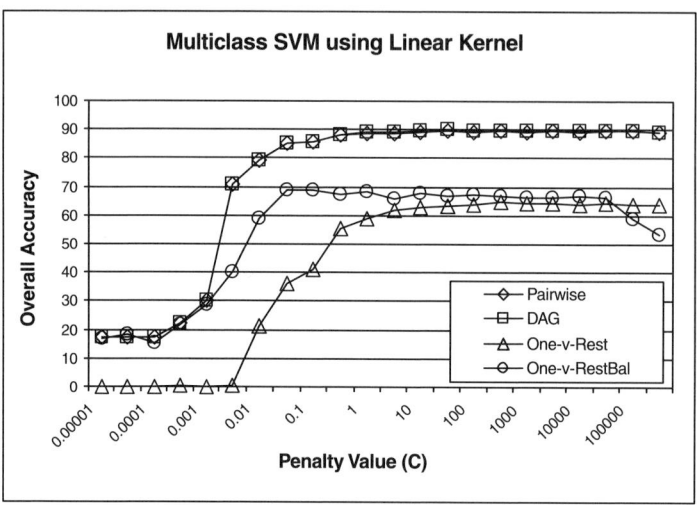

Fig. 10.2. Overall accuracy of multiclass approaches applied on the multispectral image

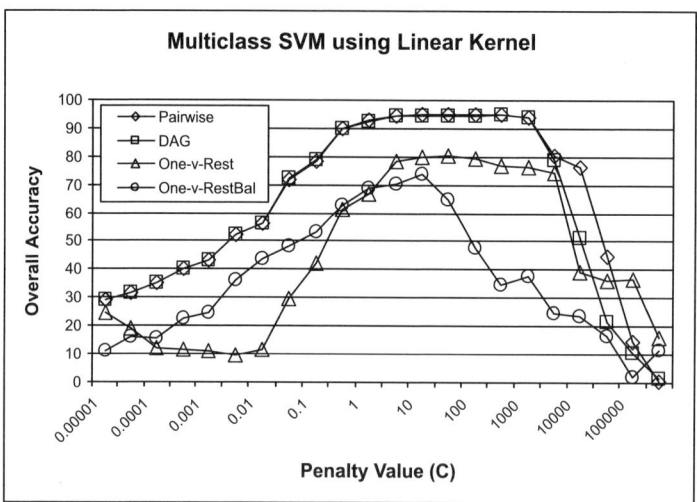

Fig. 10.3. Overall accuracy of multiclass approaches applied on the hyperspectral image

It can be seen that for both the datasets, the pairwise and DAG methods result in significantly higher accuracies than the other two methods. Both these methods show a similar trend with accuracy reaching 95%, even for a small value of C. The other two methods could attain an accuracy of only 70%. In all the methods, as the value of C is increased, the accuracy increases but gets saturated at a certain value of C in case of multispectral data. However, in the hyperspectral dataset, the accuracy drops after attaining a maximum value for each method. This shows that there is an optimum value of C irrespective of the multiclass method used for the classification of a certain dataset. Thus, the choice of C is highly data dependent.

10.4.2
Choice of Optimizer

The selection of an optimization method is of paramount importance to implement SVM classification in an efficient manner. Though, there are a number of optimization methods (see Sect. 5.5), only two have been considered here: the chunking-decomposition (CD) method (Vapnik 1982), which has been widely used, and the recently developed LSVM method (Mangasarian and Musicant 2000). Several classifications using the pairwise classification approach and the linear kernel for different C values are performed.

Looking at the classification accuracy of multi-spectral data (Fig. 10.4), it can be seen that both the optimizers reach an accuracy of 90% for C values greater than 0.5. The training time of both the optimizers is also the same and very low (i. e. just one second) when C varies from 10^{-5} to 5 (Fig. 10.5). With any increase in C value beyond 5, both the optimizers take a longer training time but without any appreciable increase in accuracy. Further, there appears to be no significant difference in testing time of the two optimizers (Fig. 10.6). Thus, both the optimization methods show similar performance in classifying the multi-spectral dataset.

In case of hyperspectral data, the LSVM performs marginally better than the CD optimizer for C values up to 1, after which both the methods achieve the same accuracy with the best value of 95% occurring when C is between 5

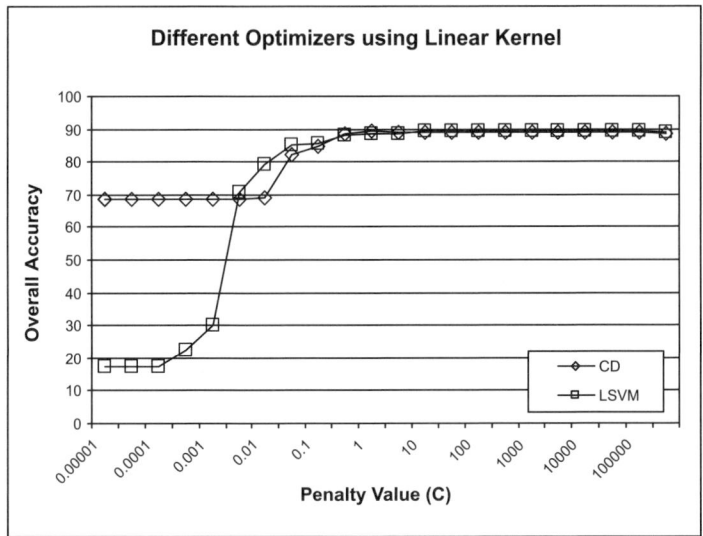

Fig. 10.4. Overall accuracy resulting from different optimizers for the multispectral image

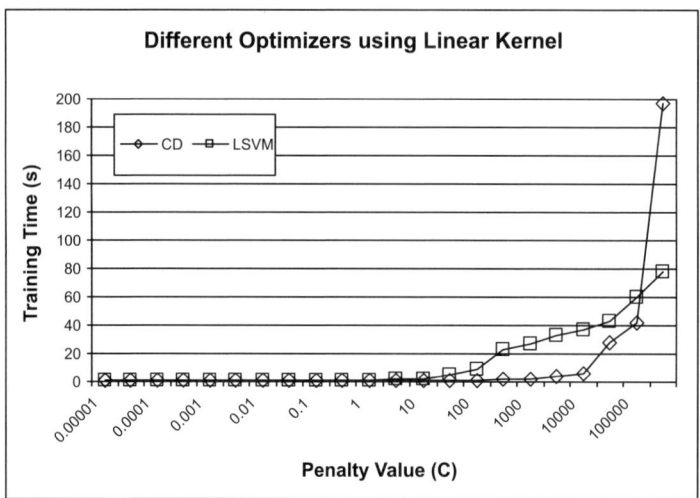

Fig. 10.5. Training time when using different optimizers on the multispectral image

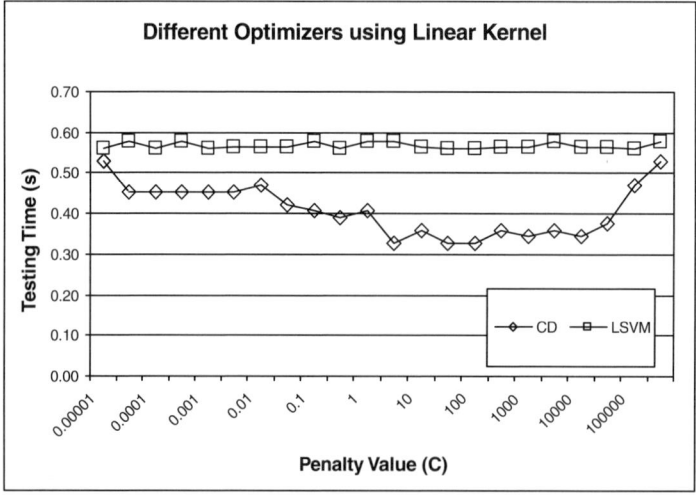

Fig. 10.6. Testing time when using different optimizers on the multispectral image

and 10^3 (Fig. 10.7). For larger values of C, while the accuracy of the CD method remains the same, it drops for LSVM.

In terms of training and testing times (Fig. 10.8 and Fig. 10.9), both the optimizers take very little time and are very close to each other up to a C value of 5, where the accuracy reaches its maximum. For values of C greater than 5, the LSVM takes longer time than the CD method. But this becomes immaterial since the maximum value of accuracy has been achieved at $C = 5$.

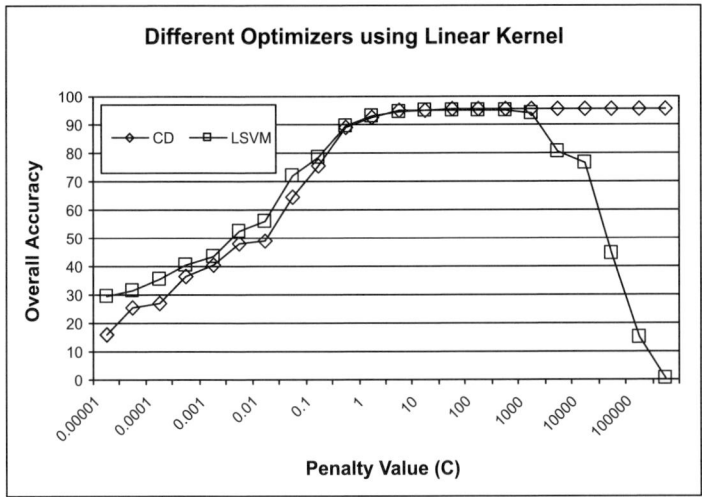

Fig. 10.7. Overall accuracy of different optimizers for the hyperspectral image

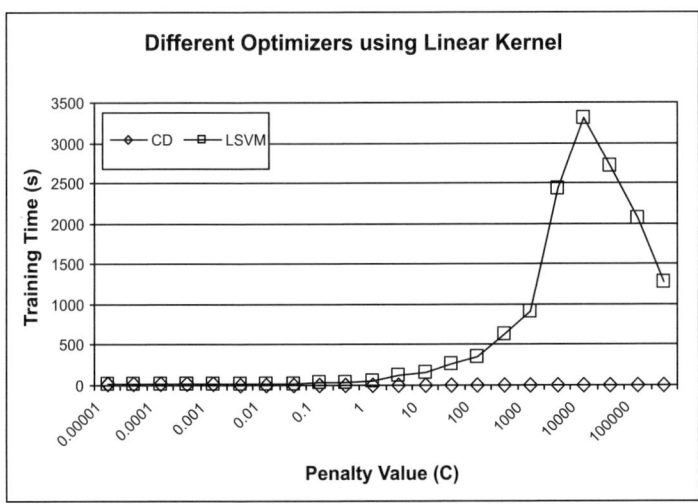

Fig. 10.8. Training time when using different optimizers on the hyperspectral image

Thus, there appears to be no appreciable difference between the performances of the two optimization methods for the classification of both multi and hyperspectral data. Either of these can be employed to perform SVM based classification.

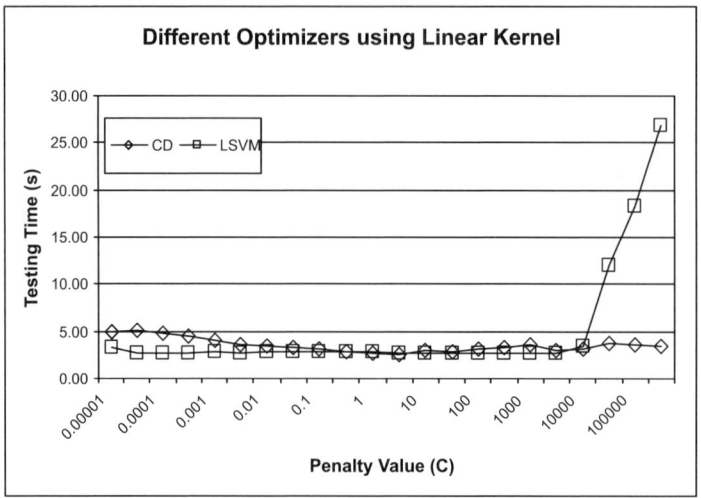

Fig. 10.9. Testing time when using different optimizers on the hyperspectral image

10.4.3
Effect of Kernel Functions

The choice of a proper kernel function also plays an important role in SVM based classification. Four types of kernels, namely the linear kernel, the polynomial kernel, the RBF kernel, and the sigmoid kernel have been investigated. Five polynomial kernels with degrees varying from 2 to 7 have been used. Thus, the effect of 9 kernel functions on the accuracy of classification implemented using the LSVM optimizer and the pairwise multiclass method for different values of the penalty parameter has been assessed. Classification of both multi- and hyperspectral datasets has been performed. The variation in the overall accuracy of multispectral classification over different C values is presented in the form of plots drawn for different types of kernel functions (Fig. 10.10 to Fig. 10.14).

It can be seen from these plots that the use of all the kernels, except the sigmoid function, results in an accuracy of more than 90%. The sigmoid kernel could attain a maximum accuracy of 79% at $C = 0.5$, which gradually dropped with any increase in C. This shows a relatively poor performance of this kernel with respect to others. In contrast, all other kernels showed an increase in accuracy as C increased and reached their maximum after which accuracies remained constant in general. This shows that there is an optimum C value for a particular kernel. Focusing on the performance of only polynomial kernels, it can be seen that the polynomials with degrees 2 to 4 produced very low accuracies (of the order of 20%) when C was very small. But the accuracy shoots up with a small increase in C. On the other hand, the initial accuracy obtained from polynomials with higher degrees (from 5 to 7) is high (of the order of 70% to 80%) at small C values. Also,

Support Vector Machines for Classification of Multi- and Hyperspectral Data

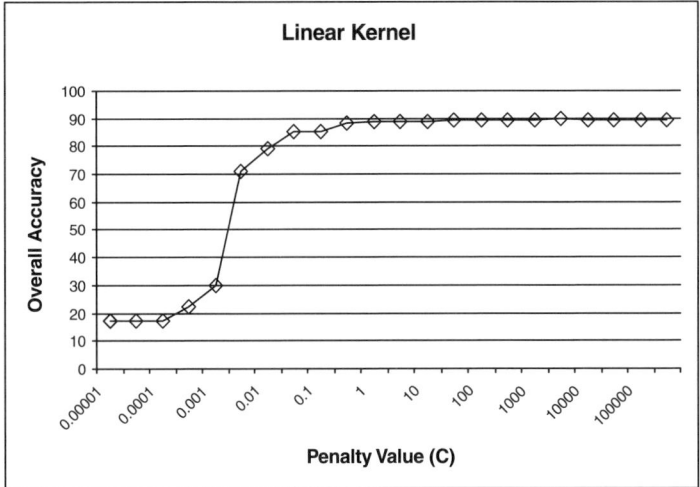

Fig. 10.10. SVM performance using the linear kernel applied on the multispectral image

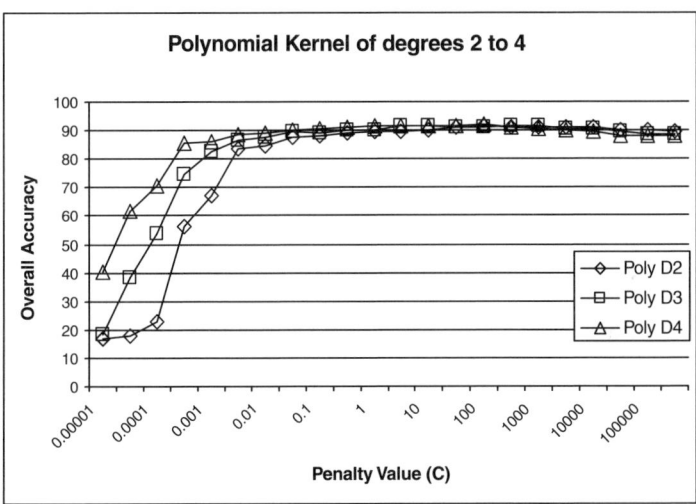

Fig. 10.11. SVM performance using the polynomial kernels of degrees 2 to 4 applied on the multispectral image

there is a gradual increase in accuracy for small increases in C values. This demonstrates that if a polynomial with a higher degree is used as the kernel, higher accuracy can be achieved at very low C values. In fact for polynomials with higher degrees, the accuracy may decrease after attaining the maximum at a certain C value unlike the polynomials with lower degrees, where the accuracy gets saturated. In case of the RBF kernel also, the accuracy increases as the penalty value increases but drops marginally after attaining

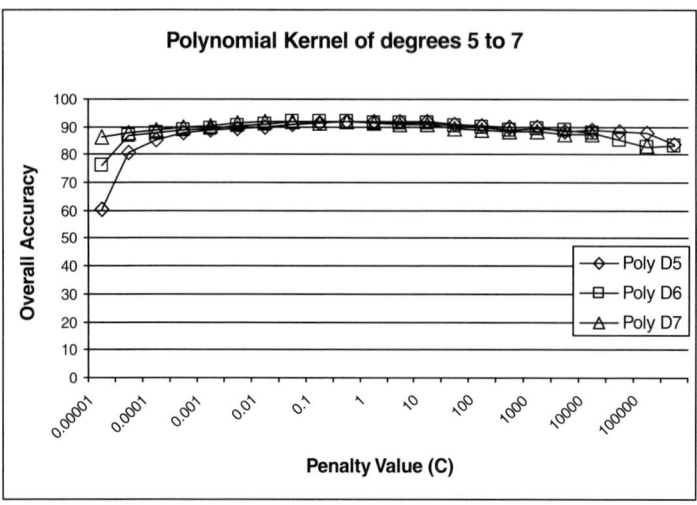

Fig. 10.12. SVM performance using the polynomial kernels of degrees 5 to 7 applied on the multispectral image

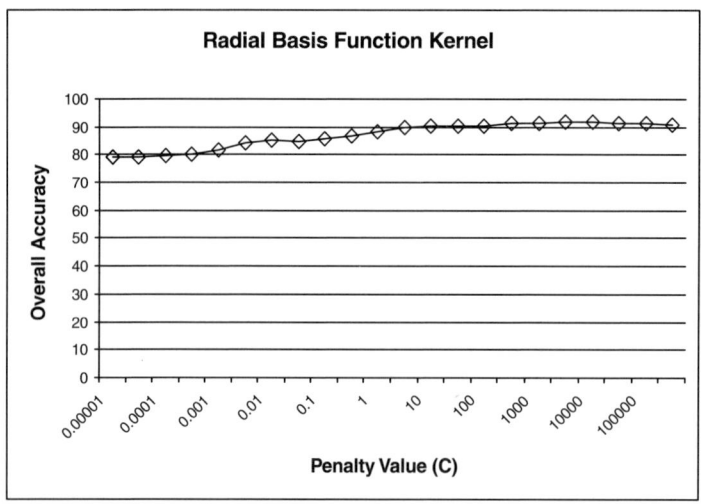

Fig. 10.13. SVM performance using the Radial Basis Function kernel applied on the multispectral image

the maximum. This shows that there is a limiting C value for all the kernel functions.

In the case of hyperspectral data (see Fig. 10.15 to Fig. 10.19), generally, a similar trend has been observed except for a couple of observations worth mentioning. Although, all the kernels achieved a maximum accuracy of more than 90%, this maximum occurs at different C values for each kernel. For

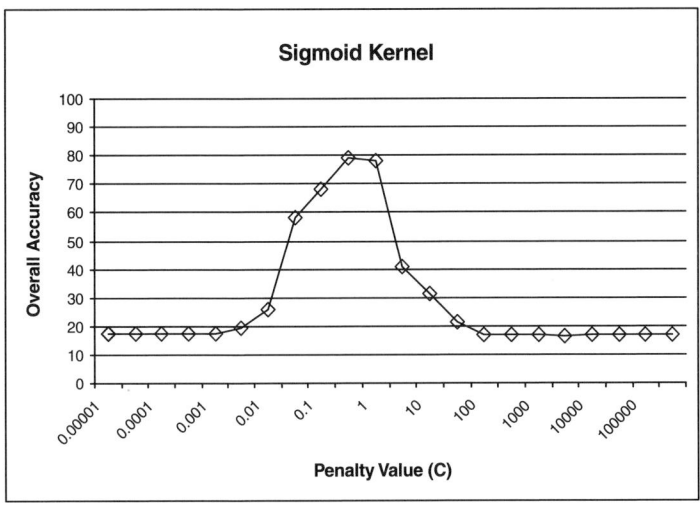

Fig. 10.14. SVM performance using the sigmoid kernel applied on the multispectral image

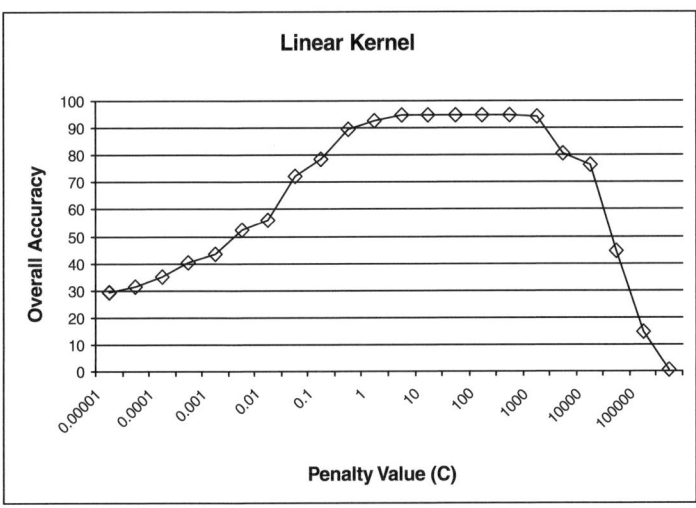

Fig. 10.15. SVM performance using the linear kernel applied on the hyperspectral image

example, for the polynomial kernel with degree 5, an accuracy of 92% is achieved with a C value as low as 10^{-5}. This shows that if the degree of the polynomial is high, a lower C value can be adopted. However, polynomials with lower degrees (i.e. 2 to 4) achieved higher accuracy than those obtained from polynomials with higher degrees (i.e. 5 to 7). Hence, a trade-off between the C value and the degree of polynomial may be maintained to get maximum classification accuracy for hyperspectral data. The linear kernel, although it attains a maximum accuracy of 95% at a certain C value, it drops dramatically

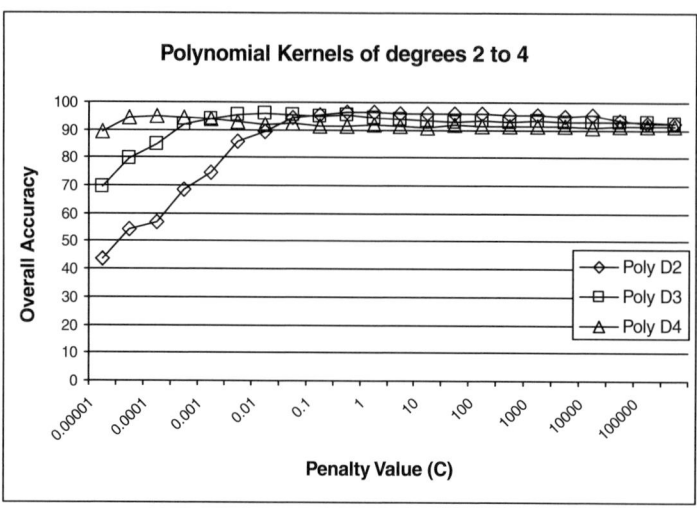

Fig. 10.16. SVM performance using the polynomial kernels of degrees 2 to 4 applied on the hyperspectral image

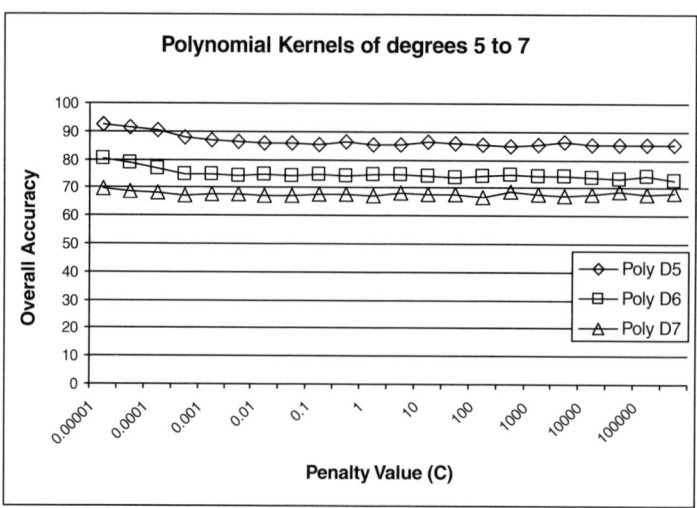

Fig. 10.17. SVM performance using the polynomial kernels of degrees 5 to 7 applied on the hyperspectral image

with any further increase in the C value unlike the accuracy achieved with multispectral data which gets stabilized after the maximum has been achieved. For the RBF and sigmoid kernels, the maximum occurs at very high values of C. But, since an increase in C is directly proportional to the training time required for LSVM optimization methods (as observed in the previous section), any kernel functions, which can produce the highest accuracy at lower C values

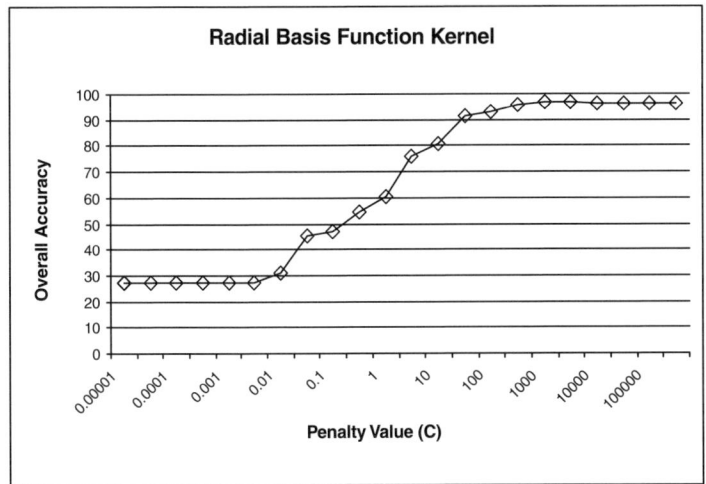

Fig. 10.18. SVM performance using the RBF kernel applied on the hyperspectral image

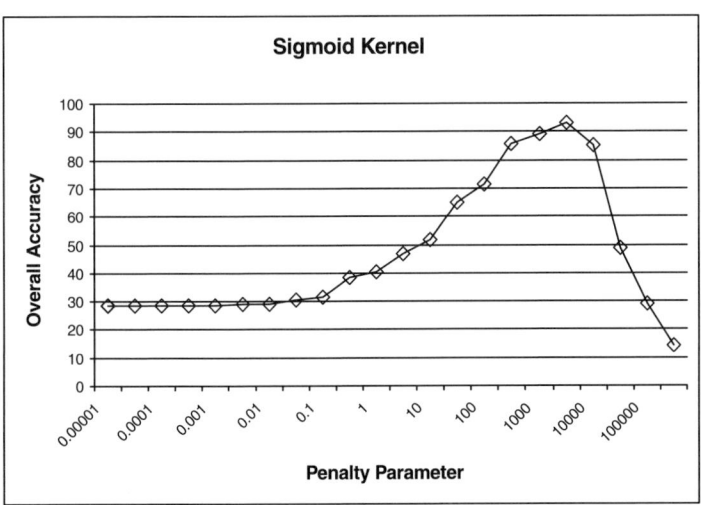

Fig. 10.19. SVM performance using the sigmoid kernel applied on the hyperspectral image

are preferred to perform efficient classification. This is particularly important in case of high dimensional data such as hyperspectral remote sensing images.

Thus, for the hyperspectral image, the SVM with the RBF and polynomial degree 2 kernels performed the best with an accuracy of 97%. Other kernels performed their best with an accuracy ranging from 70% to 96%. These results illustrate that the choice of kernel function affects the performance of SVM based classification.

10.5
Summary

In this chapter, we have considered the application of SVMs for the classification of multi and hyperspectral remote sensing datasets. We also investigated the effect of various parameters on the accuracy of SVM based classification. It is clear from the results that with an appropriate selection of the multiclass method, optimizer and the kernel function, accuracy of the order of 95% can be obtained for the classification of multi and hyperspectral image classification. The penalty value (the C value) has an important bearing on the performance of any SVM classifier. However, this may have to be selected by trial and error for a given dataset.

References

Byun H, Lee SW (2003) A survey on pattern recognition applications of support vector machines. International Journal of Pattern Recognition and Artificial Intelligence 17: 459–486

Chang CC, Lin CJ (2002) LIBSVM: a library for support vector machines (http://www.csie.ntu.edu.tw/~cjlin/libsvm)

Gualtieri JA, Cromp RF (1998) Support vector machines for hyperspectral remote sensing classification. In: Merisko RJ (ed), Proceeding of the SPIE, 27th AIPR Workshop, Advances in Computer Assisted Recognition, 3584, pp 221–232

Gualtieri JA, Chettri SR, Cromp RF, Johnson LF (1999) Support vector machine classifiers as applied to AVIRIS data. Proceedings of Summaries of the Eighth JPL Airborne Earth Science Workshop (ftp://popo.jpl.nasa.gov/pub/docs/workshops/99_docs/31.pdf)

Huang C, Davis LS, Townshend JR (2002) An assessment of support vector machines for land cover classification. International Journal of Remote Sensing 23(4): 725–249

Joachims T (2002) The SVM^{light} package (http://svmlight.joachims.org or ftp://ftp-ai.cs.uni-dortmund.de)

Keerthi SS (2002) Efficient tuning of SVM hyperparameters using radius/margin bound and iterative algorithms. IEEE Transactions on Neural Networks 13: 1225–1229

Lee C, Landgrebe DA (1993) Feature Extraction and Classification Algorithms for High Dimensional Data. Technical Report, TR-EE 93-1, School of Electrical Engineering, Purdue University

Mangasarian OL, Musicant D (2000) Lagrangian support vector machines, Technical Report (0006), Data Mining Institute, Computer Science Department, University of Wisconsin, Madison, Wisconsin (ftp://ftp.cs.wisc.edu/pub/dmi/tech-report/0006.ps)

Melgani F, Bruzzone L (2002) Support vector machines for classification of hyperspectral remote-sensing images. International Geoscience and Remote Sensing Symposium, IGARSS'02, CD.

Shah CA, Watanachaturaporn P, Arora MK, Varshney PK (2003) Some recent results on hyperspectral image classification. Proceedings of IEEE Workshop on Advances in Techniques for Analysis of Remotely Sensed Data, NASA Goddard Space Flight Center, Greenbelt, MD, CD

Tadjudin S, Landgrebe DA (1998) Classification of High Dimensional Data with Limited Training Samples. PhD thesis, School of Electrical Engineering and Computer Science, Purdue University.

Vapnik VN (1982) Estimation of dependences based on empirical data. Springer Verlag, New York

Zhu G, Blumberg DG (2002) Classification using ASTER data and SVM algorithms; the case study of Beer Sheva, Israel. Remote Sensing of Environment 80: 233–240

CHAPTER 11

An MRF Model Based Approach for Sub-pixel Mapping from Hyperspectral Data

Teerasit Kasetkasem, Manoj K. Arora, Pramod K. Varshney

11.1
Introduction

Image classification is a key task in many remote sensing applications. As discussed in Sect. 2.6 of Chap. 2, the objective of classification is to allocate each pixel of a remote sensing image into only one class (i.e. hard or per-pixel classification) or to associate the pixel with many classes (i.e. soft, sub-pixel or fuzzy classification). A number of hard classifiers are in vogue based on approaches such as statistical (Mather 1999), neural networks (Foody 2000a) and decision tree (Hansen et al. 2001).

However, in general and particularly in coarse spatial resolution images such as those obtained from hyperspectral MODIS and multispectral AVHRR sensors that provide data at spatial resolutions ranging from 250 m to 1.1 km, a majority of pixels are mixed at the scale of measurement. Even where the spatial resolution is medium (e.g. 30 m Landsat ETM+ sensor) or fine (e.g. 4 m IKONOS multi-spectral sensor), the high spatial frequency of some classes may result in a large number of mixed pixels (Aplin and Atkinson 2001). The causes for the occurrence of mixed pixels are well known (Foody and Cox 1994; Foody and Arora 1996) and their presence is a recurring problem for extracting accurate information from remote sensing images. Therefore, sub-pixel classification may be more appropriate than per-pixel classification.

In sub-pixel classification, a pixel is decomposed into a number of classes to reflect their proportions contained within the mixed pixel. Some of the prevalent techniques used for sub-pixel classification are fuzzy c-means clustering (see Sect. 2.6.4 of Chap. 2) (Bezdek et al. 1984), linear mixture modeling (Settle and Drake 1993) and artificial neural networks (Foody 1996; Binaghi et al. 1999). Recently, support vector machines (Brown et al. 2000) and possibilistic c-means clustering (Foody 2000b; Ibrahim et al. 2003) have also been used to unmix the class proportions in a pixel. The result after sub-pixel classification is a set of fraction images equal to the number of classes; a fraction image describes the proportion of a particular class within each pixel. While, under most circumstances, classification at the sub-pixel level is meaningful and informative, it fails to account for the spatial distribution of class proportions within the pixel (Verhoeye and Wulf 2002). A limited amount of work has been reported in the literature (e.g. Atkinson 1997; Foody 1998; Tatem et al. 2001) that considers this spatial distribution within and between pixels in order to produce maps at sub-pixel scale. This process is called super-resolution mapping (Tatem et al. 2002) or sub-pixel mapping (Verhoeye and Wulf 2002) to distinguish it from sub-pixel classification. Thus, a sub-pixel map is a map that is derived at an improved spatial resolution finer than the size of the pixel of the coarse resolution image being classified. Tatem et al. (2002) provide an excellent review on this subject. A range of algorithms based on knowledge-based procedures (Schneider, 1993), Hopfield neural networks (Tatem et al. 2002) and linear optimization methods (Verhoeye and Wulf 2002) have been proposed for sub-pixel mapping. Knowledge based procedures depend on the accurate identification of boundary features that divide the mixed pixels into pure components at improved resolution. The drawbacks of this technique are:

1. The information on accurate boundary features may not be readily available and

2. It does not consider the spatial dependence within and between pixels.

Similarly, for Hopfield neural network and linear optimization based methods, the availability of an accurate sub-pixel classification derived from some other techniques is a pre-requisite. Thus, the accuracy of the resulting sub-pixel map is limited by the accuracy of the sub-pixel classification technique used. Moreover, in these algorithms, the spatial dependence within and between pixels is incorporated only after the fraction images from a sub-pixel classification technique are obtained. In contrast, the Markov random field (MRF) model based algorithm, proposed here, neither relies on the availability of accurate boundary features nor on sub-pixel classification produced from other techniques. Under an MRF model, the intensity values of pixels in a particular spatial structure (i.e. neighborhood) are allowed to have higher probability (i.e. weight) than others. For instance, in a remotely sensed land cover classification, the spatial structure is usually in the form of homogenous regions of land cover classes. As a result, an MRF model assigns higher weights to these regions than to the isolated pixels thereby accounting for spatial dependence in the dataset.

The aim of this chapter is to introduce an MRF model based approach for obtaining a sub-pixel map from hyperspectral images. The approach is based on an optimization algorithm whereby raw coarse resolution images are first used to generate an initial sub-pixel classification, which is then iteratively refined to accurately characterize the spatial dependence between the class proportions of the neighboring pixels. Thus, spatial relations within and between pixels are considered throughout the process of generating the sub-pixel map. Therefore, the proposed approach may be more suitable for sub-pixel mapping as the MRF models can describe the spatial dependence in a more accurate manner than algorithms proposed in Verhoeye and Wulf (2002) and Tatem et al. (2002).

This chapter is organized as follows. In Sect. 11.2, the theoretical concept of MRF models for sub-pixel mapping is presented. The details of an MRF based optimum sub-pixel mapping approach are provided in Sect. 11.3. Experimental results are discussed in Sect. 11.4 followed by a summary of the chapter in Sect. 11.5.

11.2
MRF Model for Sub-pixel Mapping

Let Y be the observed coarse spatial resolution image having $M \times N$ pixels and X be the fine resolution sub-pixel map (SPM) having $aM \times aN$ pixels where a is the scale factor of the SPM. This means that a particular pixel in the observed coarse resolution image contains a^2 pixels of the SPM. We assume that the pixels in the fine resolution image are pure and that mixed pixels can only occur in the observed coarse resolution image. Thus, more than one class can occupy a pixel in a coarse resolution image. In general, a can be any positive real number, but, for simplicity, here it is assumed to be a positive integer number. Also, let $\mathcal{S}$ and $\mathcal{T}$ denote the sets of all sites (i.e. pixels) belonging to the observed image and the SPM, respectively. Thus, the number of sites belonging to $\mathcal{T}$ will be a^2 times the number of sites in the set $\mathcal{S}$. Let $\mathcal{T}^j = \{t_1^j, \ldots, t_{a^2}^j\}$ represent the set of all the pixels in $\mathcal{T}$ that correspond to the same area as the pixel s_j in $\mathcal{S}$. The observed coarse resolution multi or hyperspectral image is usually represented in vector form so that, $y(s_j) \in \mathbb{R}^K$ for the pixel s_j where $\mathbb{R}$ denotes the set of real numbers (e.g. intensity values) and K is the number of spectral bands. As stated earlier, each pixel in the SPM is assumed pure. That is, its configuration $x(t)$ (i.e. attribute) denotes one and only one class. Hence, $x(t) \in \{1, \ldots, L\}$ can only take an integer value corresponding to the class at a pixel t in the actual scene, where L is the number of classes. There can be $L^{a^2 MN}$ dissimilar sub-pixel maps $X(\mathcal{T}) \in \{1, \ldots, L\}^{\mathcal{T}}$ each having a different class allocation in at least one pixel.

It is further assumed that the SPM has the MRF property, i.e., the conditional probability of a configuration (i.e. intensity value) of a pixel given the configurations of the entire image excluding the pixel of interest is equal to the conditional probability of the configuration of that pixel given the configurations of its neighboring pixels. This can mathematically be represented as

$$\Pr\left(x(t) \,|\, X\left(\mathcal{T} - \{t\}\right)\right) = \Pr\left(x(t) \,|\, X(N_t)\right), \qquad (11.1)$$

where $\mathcal{T} - \{t\}$ is the set of all the pixels in $\mathcal{T}$ excluding the pixel t, and N_t is the set of pixels in the neighborhood of pixel t. For example, in the context of classification of remotely sensed images, this property implies that the same class is more likely to occur in connected regions than at isolated pixels. Hence, the conditional probability density functions (pdfs) in (11.1) have a higher value if the configuration of a pixel t is similar to the configurations of its neighboring pixels than the cases when it is not. From Winkler (1995) and Bermaud (1999), the marginal PDF of X takes the form of Gibbs distribution, i.e.,

$$\Pr(X) = \frac{1}{Z} \exp\left[-\sum_{C \subset \mathcal{T}} V_C(X)\right] \qquad (11.2)$$

where Z is a normalizing constant, C is a clique, and $V_C(X)$ is a Gibbs potential function. The value of the Gibbs potential function depends on the configurations of the entire SPM and the clique. A useful example of potential functions is the Ising model (see Chap. 6), given by,

$$V_{\{r,s\}}(X) = \begin{cases} -\beta & \text{if } x(r) = x(s) \text{ and } r \in N_s \\ +\beta & \text{if } x(r) \neq x(s) \text{ and } r \in N_s \\ 0 & r \notin N_s \end{cases} \qquad (11.3)$$

This model is also applied here to describe the SPM since, in general, the distribution of classes is similar to the phenomenon described above (i.e. classes occupying neighboring pixels are likely to be the same).

Assume a fine resolution remote sensing image with spatial resolution equal to the SPM, such that each pixel of the image corresponds to only one class, and that each class has normal distribution. The pdfs with mean vector μ_l and covariance matrix Σ_l for each class are given by,

$$\Pr(z|x = i) = \frac{1}{(2\pi)^{K/2}\sqrt{|\det(\Sigma_l)|}}$$
$$\times \exp\left[-\frac{1}{2}(z - \mu_l)'\Sigma_l^{-1}(z - \mu_l)\right], \quad l = 1,\ldots,L, \qquad (11.4)$$

where z is the observed vector of the fine resolution image and x' denotes the transpose matrix of x.

As discussed earlier, a^2 pixels in the SPM correspond to one pixel in the observed coarse resolution image. Hence, the pdf of an observed vector $y(s)$ in the coarse resolution image is assumed normally distributed with the mean vector and covariance matrix given by,

$$\mu(s) = \sum_{l=1}^{L} b_l(s)\mu_l \qquad (11.5)$$

and

$$\Sigma(s) = \sum_{l=1}^{L} b_l(s)\Sigma_l, \qquad (11.6)$$

respectively, where μ_l and Σ_l are the mean vector and covariance matrix for each class, $b_l(s)$ is the proportion of class l present in $\mathcal{T}^s$ such that $\sum_{l=1}^{L} b_l(s) = 1$. Note here that both the mean and the variance of the observed image are functions of pixels s. Hence, the conditional PDF of the observed coarse resolution image can be written as

$$\Pr(Y|X) = \prod_{s \in \mathcal{S}} \Pr(y(s)|b(s))$$

$$= \prod_{s \in \mathcal{S}} \frac{1}{(2\pi)^{K/2} \sqrt{|\det(\Sigma_s)|}}$$

$$\times \exp\left[-\frac{1}{2}(y(s) - \mu(s))' \Sigma^{-1}(s)(y(s) - \mu(s))\right], \qquad (11.7)$$

where $b(s) = [b_1(s), \cdots, b_L(s)]^T$.

Direct maximization of (11.7) yields the maximum likelihood estimate (MLE) of b. However, the MLE of b does not utilize the connectivity property of the SPM given in (11.2). To utilize this property, sub-pixel mapping is treated as an M-ary hypothesis testing problem (Trees 1968; Varshney 1997) where each hypothesis corresponds to a different SPM. In other words, for each SPM (hypothesis), the observation space is divided into several homogeneous segments each corresponding to one type of class attribute. The complete algorithm is described in the next section.

11.3
Optimum Sub-pixel Mapping Classifier

The algorithm based on the maximum *a posteriori* probability (MAP) criterion selects the most likely SPM among all possible SPMs given the observed image. The MAP criterion is expressed as (Trees 1968; Varshney 1997),

$$X^{\text{opt}} = \arg\left\{\max_X [\Pr(X|Y)]\right\}, \qquad (11.8)$$

where $\Pr(X|Y)$ is the posterior probability of the SPM when the coarse resolution observed image is available. By using the definition of the conditional pdf, (11.8) can be rewritten as

$$X^{\text{opt}} = \arg\left\{\max_X \left[\frac{\Pr(Y|X)\Pr(X)}{\Pr(Y)}\right]\right\}. \qquad (11.9)$$

Since $\Pr(Y)$ is independent of X, (11.9) reduces to

$$X^{\text{opt}} = \arg\left\{\max_X [\Pr(Y|X)\Pr(X)]\right\}. \qquad (11.10)$$

Substituting (11.2) and (11.7) into (11.10) yields

$$X^{\text{opt}} = \arg\left\{\max_X [C \exp(-E_{\text{post}}(X|Y))]\right\}, \qquad (11.11)$$

where $C = \frac{1}{Z(2\pi)^{K/2}}$, and

$$E_{\text{post}}(X|Y) = \sum_{C \subset \mathcal{T}} V_C(X) + \frac{1}{2} \sum_{s \in \mathcal{S}} (y(s) - \mu_s)^T \Sigma_s^{-1} (y(s) - \mu_s)$$
$$+ \frac{1}{2} \sum_{s \in \mathcal{S}} \log |\det(\Sigma_s)|. \qquad (11.12)$$

Since the exponential function is monotonic and C is a constant that is independent of X, (11.11) can be further reduced to

$$X^{\text{opt}} = \arg \left\{ \min_X \left[E_{\text{post}}(X|Y) \right] \right\}. \qquad (11.13)$$

In general, $E_{\text{post}}(X|Y)$ is a non-convex function and the number of possible SPMs is extremely large, therefore, the simulated annealing (SA) algorithm is employed here for optimization. From Chap. 6, we know that the SA algorithm generates a sequence of $\{X_p\}_{p=1,2,3,...}$ using a random number generator where the subscript p denotes the iteration number. The PDF of a proposed state X_{p+1} depends on the current state X_p, the observed data Y, and a temperature parameter T. The SA algorithm uses the temperature parameter to manipulate the randomness of X_p. A high temperature value results in high randomness of X_p while a low temperature value corresponds to low randomness of X_p. Here, the SA algorithm allows the randomness of a number generator to decrease in such a way that the optimum solution of (11.13) can be obtained as the iteration number p approaches infinity.

One major drawback of the SA algorithm is that convergence occurs when the number of iterations approaches infinity. Due to practical considerations, we need to stop the SA algorithm after a certain finite number of iterations and, hence, the accuracy of this algorithm depends largely on the initial estimate of the SPM. As a result, the optimization algorithm proposed here to solve (11.13) consists of two major parts. In the first part, initialization is performed to determine a fairly good initial estimate of the SPM whereas, in the second part, iterations take place to determine the optimum SPM.

Before we go into further details of the algorithm, it is worth mentioning that due to a large number of bands in hyperspectral images, a feature extraction technique such as PCA may also have to be applied to increase the efficiency of the algorithm. We employ the PCA algorithm for feature extraction and assume that the observed image consists of the best features extracted by PCA. Thus, in the initialization phase (shown in Fig. 11.1), the observed data in the form of first few principal components is selected. Then similar to a conventional supervised classification approach, the analyst selects sufficient number of pure training pixels (containing only one class) from this observed data. The training pixels are used to estimate unknown parameters used in (11.5) and (11.6) (i.e. mean vectors and variance-covariance matrices) for classes to be mapped. Due to its consistency, maximum likelihood estimation (MLE) is employed to estimate the parameters.

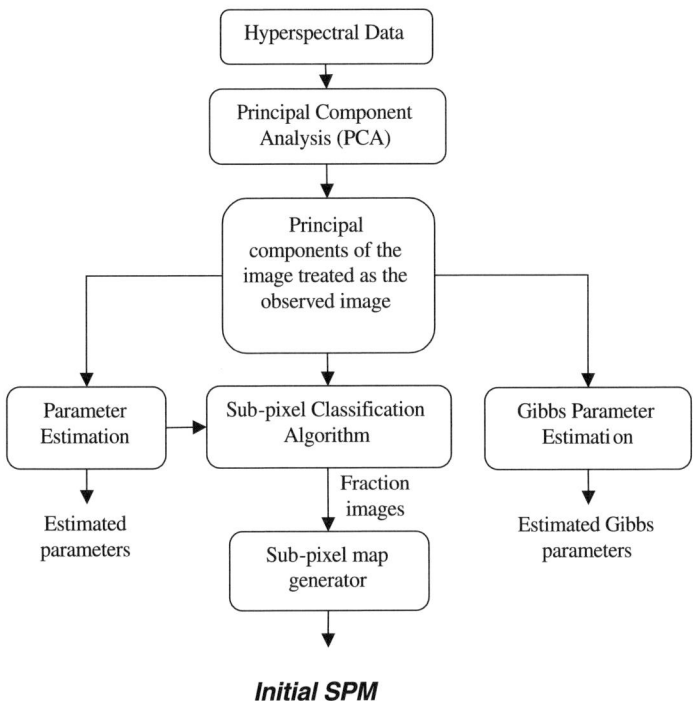

Fig. 11.1. Initialization phase

In addition, parameters related to Gibbs potential functions (Winkler 1995; Bremaud 1999) (e.g. β for the Ising model in (11.3)), required in the iteration phase, are also estimated, or they may be known beforehand (e.g. by reasonable approximation) in the initialization phase itself. MLE can again be used for the estimation of Gibbs potential functions, but here, the maximum pseudo likelihood estimation (MPLE) algorithm (Winkler 1995; Bremaud 1999) is used since it is computationally inexpensive.

The estimated parameters are used to perform sub-pixel classification from the observed data at coarse resolution. Any sub-pixel classification technique such as fuzzy c-means clustering, as discussed in Sect. 2.6.4 of Chap. 2, may be used. Here, MLE derived probabilities have been used to perform sub-pixel classification and generate fraction images. These fraction images are put through a sub-pixel map generator to produce an initial sub-pixel map at fine spatial resolution based on the following process.

Given the scale factor a for a pixel in the fraction image at coarse spatial resolution and the associated proportion of a class in that pixel, the corresponding number of pixels with the same class value may be found in the fine resolution SPM. For instance, if a fraction image of class A has a value of 0.5 in pixel s_j, there are $0.5a^2$ pixels out of a^2 in the set $\mathcal{T}^j$ belonging to class A in the

SPM. A similar procedure is adopted for each fraction image and the pixels in the set $\mathcal{T}^j$ are, then, randomly labeled with all the classes. This is referred to as the initial SPM. It is expected that the initial SPM will have a large number of isolated pixels since their neighboring pixels may belong to different classes. The initial SPM is used as the starting point in the iteration phase of the search process. An appropriate starting point (which is often difficult to obtain) may result in a faster convergence rate of the SA algorithm.

The initial SPM together with the observed data and its estimated parameters are the inputs to the iteration phase of the algorithm (Fig. 11.2). Since the SA algorithm is used for the optimization of (11.13), a visiting scheme to determine the order of pixels in the SPM whose configurations (i.e. class attributes) are updated, needs to be defined. As stated in Chap. 6, different visiting schemes may result in different rates of convergence (Winkler 1995). Here, a row wise visiting scheme is used. The temperature T, which controls the randomness of the optimization algorithm (Winkler 1995; Bremaud 1999) is set to a predetermined value T_0, and the counter, h, is set to zero. In the next step, the configurations of pixels are updated. Without loss of generality, let us assume that the pixel t_1 is currently being updated. Then, the value of the energy function, $E_{post}(X|Y)$, from (11.12), associated with all possible configurations at the pixel t_1 are determined as

$$E_{post}^{t_1}(X|Y) = \sum_{C \ni t_1} V_C(X) + \frac{1}{2}\left(y(s_{t_1}) - \mu(s_{t_1})\right)' \left(\Sigma(s_{t_1})\right)^{-1} \left(y(s_{t_1}) - \mu(s_{t_1})\right)$$
$$+ \frac{1}{2} \log \left|\det\left(\Sigma(s_{t_1})\right)\right| + r, \qquad (11.14)$$

1. Initial SPM
2. Observed Data
3. Estimated Parameters

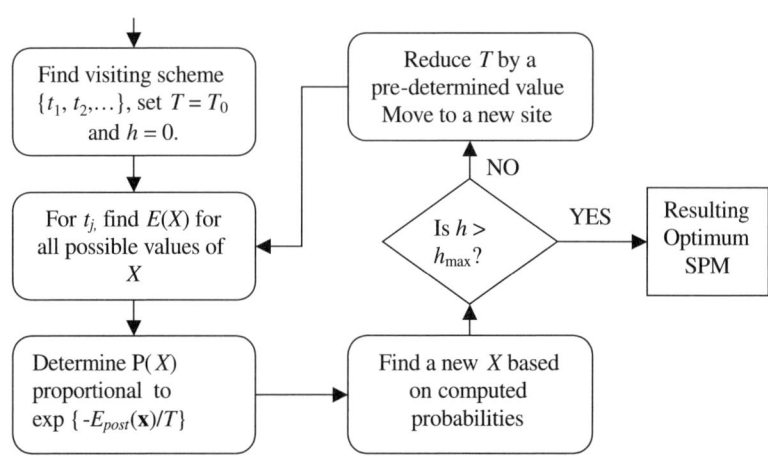

Fig. 11.2. Iterative phase of the optimization algorithm to generate the SPM

where $\sum_{C \ni t_1}$ denotes the summation over all the cliques to which t_1 belongs, s_{t_1} is the site in the observed image corresponding to t_1, and r is due to the remaining terms in (11.12) that are independent of the configuration of t_1. Note that depending upon the neighborhood system considered in a particular problem, different types of cliques could be employed. Since r is independent of $x(t_1)$, a list of probabilities associated with all possible configuration values at t_1 can be determined by ignoring r,

$$P[x(t_1)] = \frac{1}{Z_1} \exp \left(-\frac{1}{T} \left[\begin{array}{l} \sum_{C \ni t_1} V_C(X) \\ +\frac{1}{2} \left(y(s_{t_1}) - \boldsymbol{\mu}(s_{t_1})\right)' \\ \quad \times \left(\boldsymbol{\Sigma}(s_{t_1})\right)^{-1} \left(y(s_{t_1}) - \boldsymbol{\mu}(s_{t_1})\right) \\ +\frac{1}{2} \log \left|\det \left(\boldsymbol{\Sigma}(s_{t_1})\right)\right| \end{array} \right] \right) \quad (11.15)$$

where Z_1 is the normalizing constant such that summation of (11.15) over all possible values of $P[x(t_1)]$ is equal to one. A new configuration of pixel t_1 is, thus, generated based on (11.15). It is from this equation that the configuration corresponding to a lower energy value has higher likelihood of being generated than the one with a higher energy value. Next, we repeat the same process for each pixel until all pixels are updated to new configurations. Then, the algorithm increases the counter by one (i.e., $h = h + 1$) and determines a new temperature value such that the temperature decreases at the rate of $1/\log(n)$. In other words,

$$T = \frac{T_0}{\log(h+1)}. \quad (11.16)$$

The updating process is repeated with the new temperature value until the counter exceeds some predetermined value. Gradually, the number of isolated pixels in the SPM is reduced because the contextual information present in the Gibbs potential function V_C forces the SA algorithm to iteratively generate a new SPM that is closer to the solution of the MAP criterion in (11.13), which is the desired optimum SPM. From the resulting optimum SPM, a contextually refined version of fraction images at coarse resolution may also be generated as byproducts thereby producing sub-pixel classification with higher accuracy than that of the initial sub-pixel classification.

11.4
Experimental Results

In this section, we present two examples illustrating the use of the MRF model based algorithm to produce a sub-pixel map from multispectral and hyperspectral data respectively.

11.4.1
Experiment 1: Sub-pixel Mapping from Multispectral Data

A small dataset from multi-spectral (spatial resolution 4 m) and panchromatic (spatial resolution 1 m) sensors onboard the IKONOS satellite, acquired in 2001, has been used in this example. It consists of (26 × 60) pixels of 4 bands of the multi-spectral image in blue, green, red and near infrared (NIR) regions and (97 × 237) pixels of the single band PAN image, and covers a portion of the Syracuse University campus (Fig. 11.3a,b). The corresponding reference image showing six land cover classes generated from visual interpretation of the PAN image and verified in the field is shown in Fig. 11.4. The land cover classes are – grass, roof1, roof2, tree, shadow and road. Roof1 represents the top cover of a temporarily erected tent whereas roof2 denotes permanent roofs of buildings. Each pixel in this reference image has been assumed pure. Based on this reference image, fraction reference images for each land cover class at 4 m resolution are also generated for accuracy assessment purposes (Fig. 11.5). Note that black and white colors indicate either total absence or total presence of a class whereas intermediate gray shades represent proportion of a class within a pixel of the coarse resolution image. These reference images also

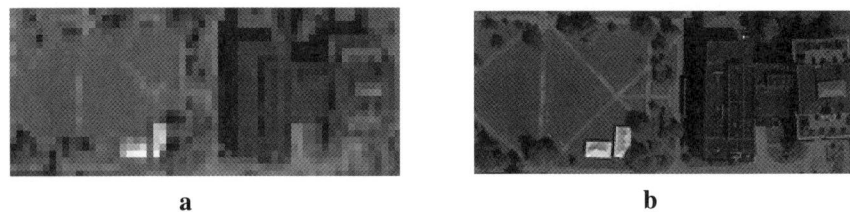

Fig. 11.3a,b. IKONOS images of a portion of the Syracuse University campus. **a** False color composite (Blue: 0.45 – 0.52 µm, Green: 0.52 – 0.60 µm, Red: 0.76 – 0.90 µm) of the multi-spectral image at 4 m resolution. **b** Panchronometric image at 1 m resolution. For a colored version of this figure, see the end of the book

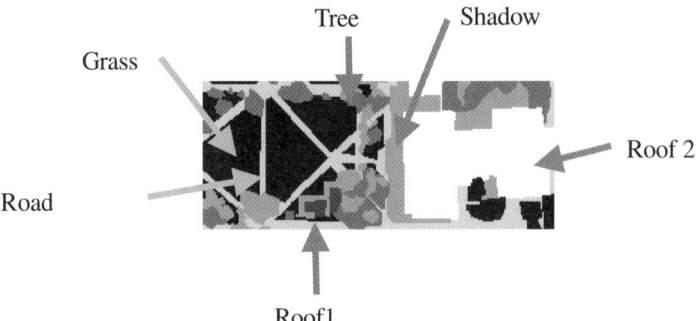

Fig. 11.4. Crisp reference image prepared from IKONOS PAN image at 1 m spatial resolution of the Syracuse University campus

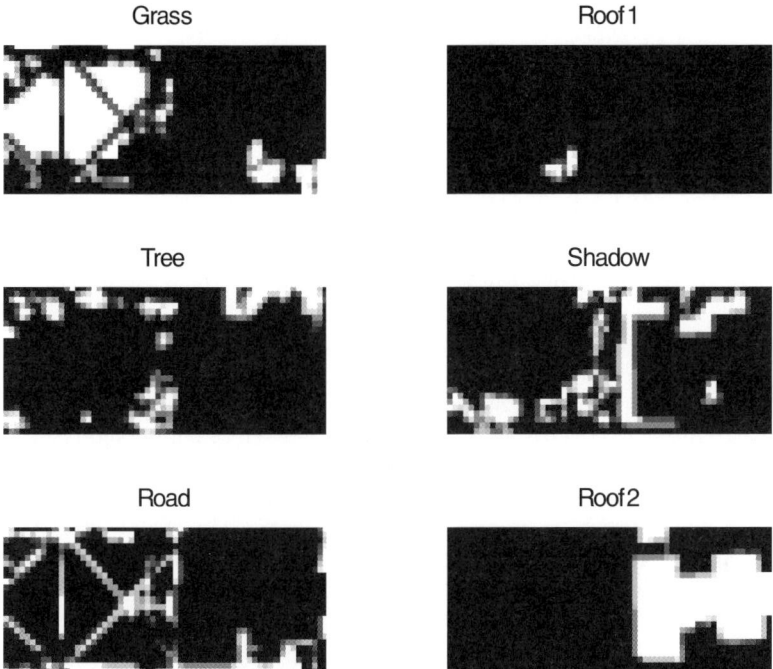

Fig. 11.5. Fraction reference images (4 m spatial resolution) of six land cover classes: grass, roof1, tree, shadow, road and roof2. These images have been generated from the crisp reference image shown in Fig. 11.4

assisted in identifying pure training pixels for initial sub-classification of the observed multi-spectral image.

The observed coarse resolution multi-spectral image (Fig. 11.3a) is submitted to the initialization phase of the algorithm to generate fraction images at 4 m resolution and to produce an initial SPM at 1 m resolution. The procedure begins with the estimation of mean vectors and covariance matrices associated with all the classes by selecting 100 pure training pixels for each class in the coarse resolution image. Purity of pixels has been examined from the fraction reference images – pixels with class proportions 100% are regarded as pure. The class roof1 had only 8 pure pixels in the dataset.

Recall that the parameters related to Gibbs potential function (Winkler 1995; Bremaud 1999) either need to be estimated or known beforehand. Here, it is assumed that the Gibbs potential functions depend only on four clique types C_2, C_3, C_4, and C_5, and follow the Ising model, i.e.,

$$V_{\{r,s\}}(X) = \begin{cases} -\beta_i & \text{if } x(r) = x(s) \text{ and } \{r,s\} \in C_i \\ +\beta_i & \text{if } x(r) \neq x(s) \text{ and } \{r,s\} \in C_i \\ 0 & \{r,s\} \notin C_i \end{cases} \quad . \tag{11.17}$$

Due to the availability of the crisp reference image of the SPM, the MPLE algorithm is employed for estimation of the parameters of the Gibbs potential function. The MPLE algorithm selects the sets of Gibbs parameters that maximize the product of all the local characteristic functions of $X(\mathcal{S})$, i.e.,

$$\beta_{MPLE} = \arg\left[\max_{\beta}\left(\prod_{s\in\mathcal{S}} \Pr\left(X(s)|X(N_s)\right)\right)\right]. \quad (11.18)$$

Here, for the four clique types, the associated parameters, $\beta_2, \beta_3, \beta_4$ and β_5 are obtained as 1.1, 1.1, 0.3 and 0.3, respectively. However, if the crisp reference image is not available, we can choose some appropriate values for β such as $\beta = \begin{bmatrix}1 & 1 & 1 & 1\end{bmatrix}$ to make the SPM smooth (or having more connected regions). It is obvious that this approach is quite heuristic, but it is based on the common knowledge that a land cover map is generally composed of connected regions rather than isolated points.

The estimated parameters have been used in the initialization phase to generate sub-pixel classification or fraction images (Fig. 11.6) from the coarse resolution image. The initial SPM at 1 m resolution is obtained from these fraction images (Fig. 11.7). A visual comparison of this SPM with the crisp reference map (Fig. 11.4), sufficiently demonstrates the poor quality of the MLE

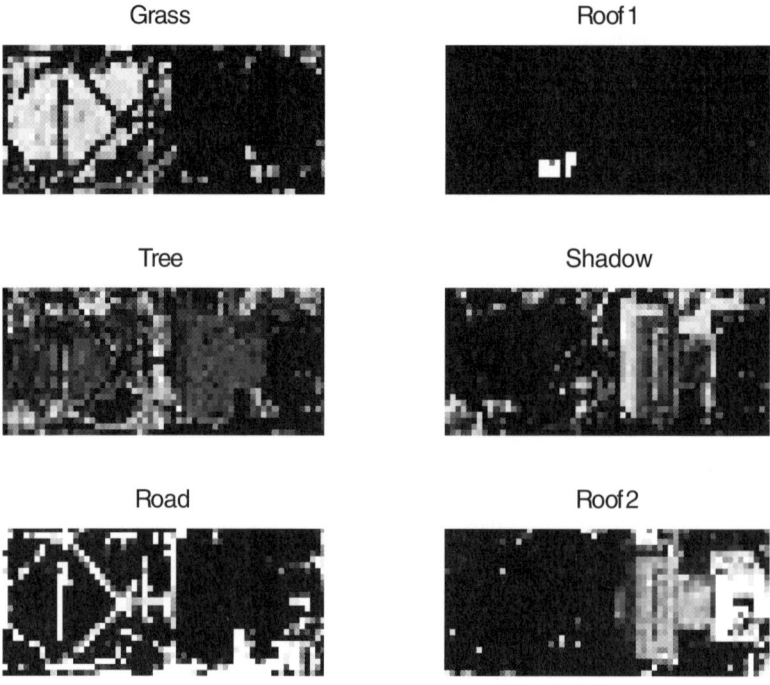

Fig. 11.6. Initial sub-pixel classification (fraction images) of six land cover classes: grass, roof1, tree, shadow, road and roof2

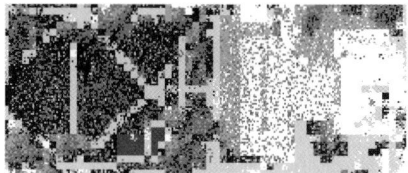

Fig. 11.7. Initial sub-pixel map derived from MLE

Fig. 11.8. Resulting sub-pixel map derived from the MRF model

derived product. A number of isolated pixels may be observed in this SPM. The initial SPM together with the estimated parameters and the observed data are submitted to the iterative phase of the optimization algorithm to obtain the resulting SPM (Fig. 11.8) after 100 iterations.

On visual comparison of Fig. 11.7 and Fig. 11.8, a significant improvement in the MRF based SPM over the initial SPM can be observed. Further, the optimum or the resulting SPM matches well with the reference image (Fig. 11.4). The accuracy of the SPM (both MLE derived initial SPM and resulting MRF model based SPM) has also been determined using the error matrix based Kappa coefficient (see Sect. 2.6.5 of Chap. 2 for details on accuracy assessment) with 2500 testing samples, and is provided in Table 11.1. The fine resolution (1 m) map (Fig. 11.4) prepared from visual interpretation has been used as reference data to generate the error matrix. The 95% confidence intervals for Kappa coefficient for both SPMs have also been computed. It can be seen that these intervals do not overlap, which shows that the resulting MRF based sub-pixel map has significantly higher accuracy than that of the initial SPM. The Student-t statistic is also used to confirm the previous statement. A larger value of the student t-statistic indicates more confidence that both classifiers perform differently – the classifier having higher Kappa value being the better one. This

Table 11.1. Classification accuracy of sub-pixel maps

Data type	Kappa	95% confidence interval of Kappa	Student-t Statistic
Initial MLE derived SPM	0.4562	(0.3907, 0.5217)	2.86
Resulting MRF model based SPM	0.6288	(0.5303, 0.7237)	

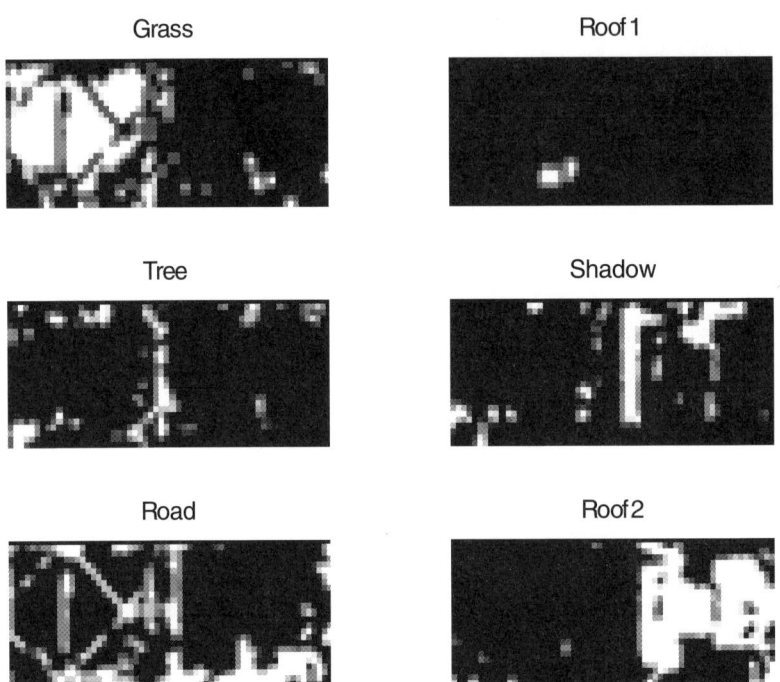

Fig. 11.9. Resulting MRF model based sub-pixel classification for grass, roof1, tree, shadow, road and roof2

Table 11.2. Overall accuracies of MLE and MRF derived sub-pixel classification

Data type	Overall accuracy
MLE derived sub-pixel classification	59.49 %
MRF derived sub-pixel classification	70.03 %

clearly demonstrates the excellent performance of the MRF based algorithm to produce the sub-pixel map from multispectral data.

Further, as a by-product of the SPM, a new set of sub-pixel classifications or fraction images has also been generated (Fig. 11.9). The improvement over the initial fraction images (Fig. 11.8) can easily be observed in the MRF derived fraction images. Many isolated pixels particularly in areas covered with grass and roof2 have been eliminated. Thus, the land cover class proportions have smoother appearance, which matches well with fraction reference images shown in Fig. 11.5. To quantitatively evaluate the performance of sub-pixel classification (i.e. fraction images), an overall accuracy measure based on the fuzzy error matrix (see Sect. 2.6.5 of Chap. 2) has been used. The significant increase in the overall accuracy (Table 11.2) of the resulting MRF derived sub-pixel classification over the initial MLC derived sub-pixel classification, illustrates the excellent performance of the proposed algorithm.

11.4.2
Experiment 2: Sub-pixel Mapping from Hyperspectral Data

In this experiment, we consider a hyperspectral image acquired in 126 bands in the wavelength region from 0.435 to 2.488 μm from the HyMap sensor (for specifications see Chap. 2) at spatial resolution 6.8 m. A portion consisting of (75 × 120) pixels has been selected (Fig. 11.10). Since the HyMap image contains a large number of bands, PCA has been employed as the feature extraction technique to reduce data dimensionality. First six principal components obtained from PCA contain 99.9% of the total energy in all the bands (Fig. 11.11) and, therefore, these have been used as observed data to produce the sub-pixel map.

Fine resolution (spatial resolution 0.15 m) digital aerial photographs from Kodak have been used for generating the reference image for accuracy assessment purposes. Here, we use visual image interpretation to assign one land cover class to each pixel. Thus, each pixel in this reference image has been assumed pure. The entire land cover map was produced based on this procedure at 0.15 m spatial resolution. The aerial photograph and the corresponding reference image prepared from visual interpretation are shown in Fig. 11.12 and Fig. 11.13, respectively. Six land cover classes – water, tree, bare soil, road, grass and roof are represented in blue, dark green, brown, yellow, light green, and white colors respectively. Based on the procedure given in Experiment 1, fraction images for each class at 6.8 m resolution of the HyMap image have been generated (Fig. 11.14a–f). From the knowledge of these fractions, pure pixels of each individual class are identified in the observed HyMap image (Table 11.3). A pixel is regarded as pure if it contains 100% proportion of a class. However, by this definition, the classes water and road contain very few pure pixels. As a result, for this class, all the pixels having more than 60% class proportion have been assumed pure. This may have an effect on the accuracy later during the analysis. The pure pixels are used as training data (Table 11.3) to estimate the mean vectors and covariance matrices of each class in the PCA derived reduced-size observed HyMap image.

Fig. 11.10. False color composite of the HyMap image (Blue: 0.6491 μm, Green: 0.5572 μm, Red: 0.4645 μm). For a colored version of this figure, see the end of the book

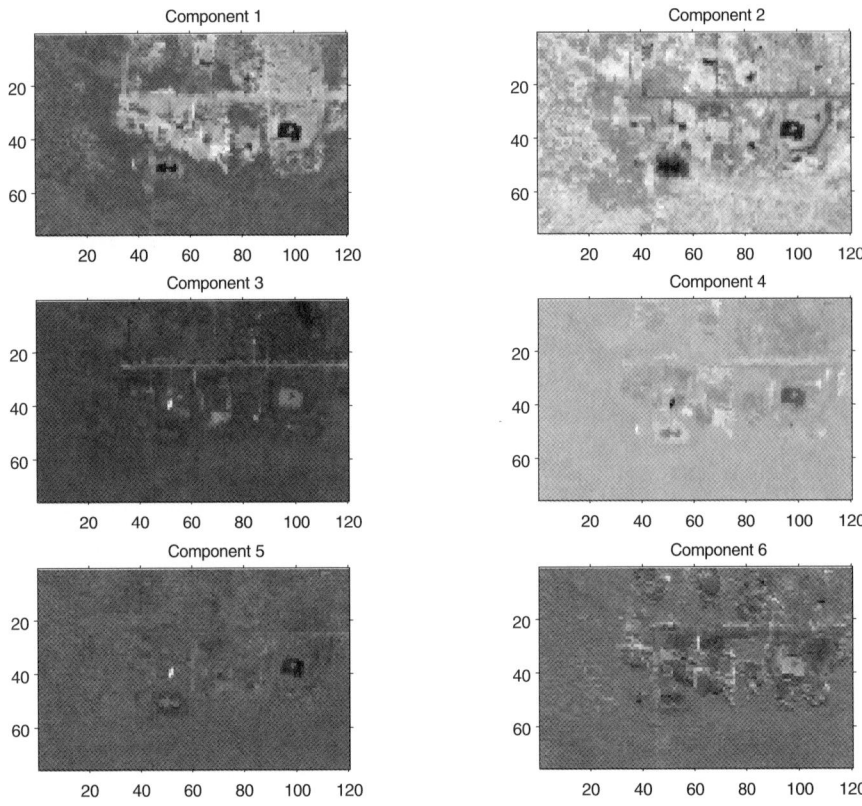

Fig. 11.11. First six principal components generated from 126 bands of the HyMap data

The MRF model is then used to produce a sub-pixel map at a spatial resolution of 0.85 m. Accordingly, the scale factor has been kept as 8 for the generation of a resampled reference map at 0.85 m spatial resolution (Fig. 11.15). Here, we define the resampled reference map as the one generated by degrading the highest resolution map (i. e. 0.15 m) to the corresponding spatial resolution according to the value of the scale factor. The class attribute of a pixel in the resampled reference map is determined by applying the majority rule to the corresponding area in the original fine resolution map. Since, the Gibbs potential functions represent the spatial dependence of class attributes for SPM at a scale factor, there is a unique set of parameters associated with Gibbs potential functions for a given scaled reference map. In other words, the parameters of Gibbs potential functions are estimated directly from the reference map. Again, the Ising model with clique types C_2, C_3, C_4 and C_5 given in (11.17) has been applied.

Similar to the previous experiment, the initial MLE-based SPM is determined in the initialization step of the algorithm and is shown in Fig. 11.16. The initial SPM then forms the input to the iteration phase of the proposed

An MRF Model Based Approach for Sub-pixel Mapping from Hyperspectral Data

Fig. 11.12. True composite of the digital aerial photograph at 0.15 m spatial resolution, used as the reference image. For a colored version of this figure, see the end of the book

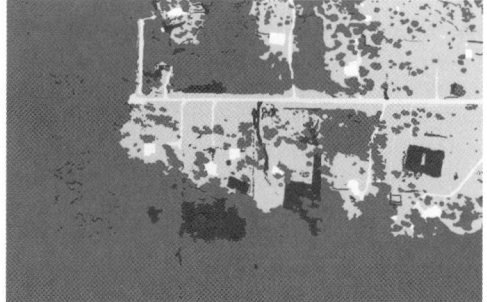

Fig. 11.13. The corresponding land cover map produced from the aerial photograph, used as the reference image. For a colored version of this figure, see the end of the book

Table 11.3. Number of pure pixels for different classes used as training data (Note that for the classes water and road, all the pixels containing at least 60% proportion of the respective class have been considered as pure)

Land Cover Class	Number of pure pixels	Total number of pixels
Water	10	47
Tree	4724	7054
Bare Soil	132	1349
Road	10	482
Grass	926	3319
Roof	26	275

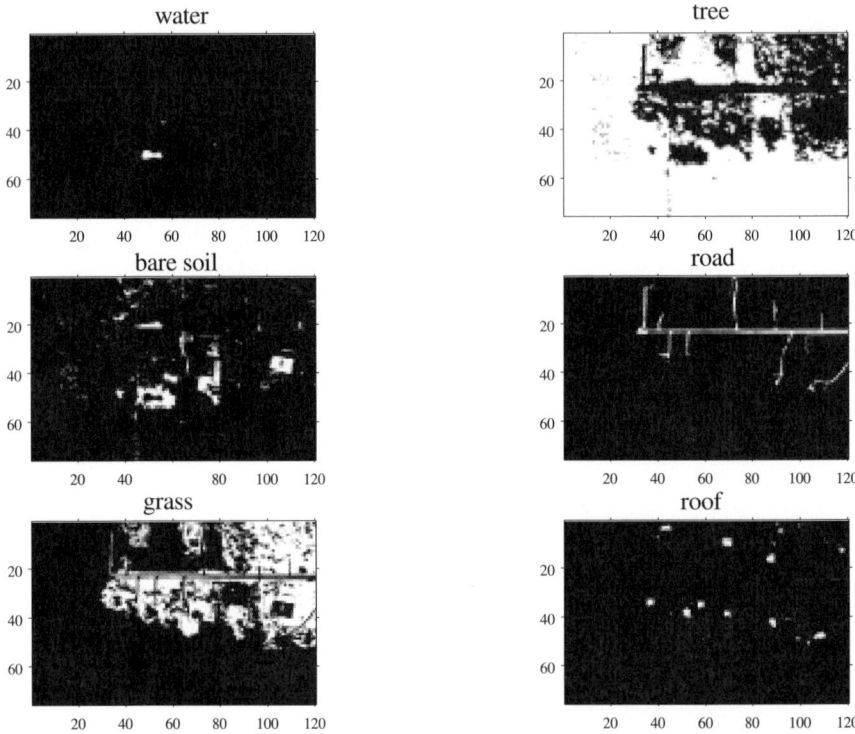

Fig. 11.14. Reference fraction images of the six land cover classes at 6.8 m resolution water; tree; bare soil; road; grass; and roof

Fig. 11.15. The resampled ground reference map at 0.85 m resolution. For a colored version of this figure, see the end of the book

algorithm to obtain the resulting SPM shown in Fig. 11.17. We observe that the MRF-based algorithm has produced a more connected map than the initial SPM. For instance, the class road in the initial SPM has a large number of speckle-like noise pixels whereas the resulting SPM represents this class as more connected and smooth, and compares well with the reference image. Similar observations can also be made for the classes grass and tree. The accuracy of sub pixel maps (both MLE derived and MRF derived SPMs) is also determined using the Kappa coefficient obtained from the error matrix generated from 10000 testing samples and is provided in Table 11.4. The 95% confidence intervals for Kappa coefficient for both initial and final SPMs are also given. Clearly, the intervals do not overlap indicating that the proposed MRF model based derived sub-pixel map is significantly better than the initial MLE derived sub-pixel map.

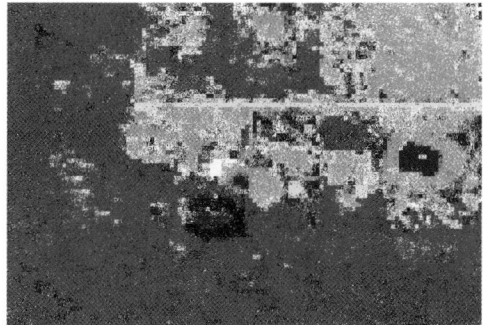

Fig. 11.16. Initial SPM at 0.85 m resolution derived from MLE. For a colored version of this figure, see the end of the book

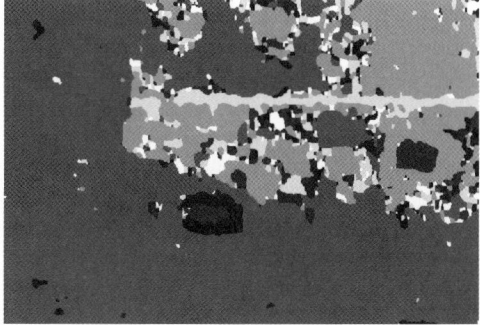

Fig. 11.17. Resulting SPM at 0.85 m resolution derived from MRF model. For a colored version of this figure, see the end section of the book

Table 11.4. Classification accuracy of sub-pixel maps

Initial SPM		Resulting SPM	
Kappa coefficient	95% Confidence Interval	Kappa coefficient	95% Confidence Interval
0.4126	(0.3728 0.4524)	0.5511	(0.4894 0.6128)

11.5
Summary

In this chapter, a new sub-pixel mapping algorithm based on the MRF model is proposed. It is assumed that a sub-pixel map has MRF properties, i.e., two adjacent pixels are more likely to belong to the same class than different classes. By employing this property of the model, the proposed MRF based algorithm is able to correctly classify a large number of misclassified pixels, which often appear as isolated pixels. In our experimental investigations, the efficacy of the proposed algorithm has been tested on both multi and hyperspectral datasets obtained from IKONOS and HyMap sensors respectively. The results show that for both the datasets, a significant improvement in the accuracy of the sub-pixel map over that produced from the conventional and most widely used maximum likelihood estimation algorithm can be achieved. The approach has been able to successfully reduce a significant number of isolated pixels thereby producing a more connected and smooth land cover map.

References

Aplin P, Atkinson PM (2001) Sub-pixel land cover mapping for per-field classification. International Journal of Remote Sensing 22(14): 2853–2858

Atkinson PM (1997) Mapping sub-pixel boundaries from remotely sensed images. Innovation in GIS 4: 166–180

Bezdek JC, Ehrlich R, Full W (1984) FCM: the fuzzy c-means clustering algorithm. Computers and Geosciences 10: 191–203

Binaghi E, Brivio PA, Ghezzi P, Rampini A (1999) A fuzzy set-based accuracy assessment of soft classification. Pattern Recognition Letters 20: 935–948

Bremaud P (1999) Markov chains Gibbs field, Monte Carlo simulation and queues. Springer Verlag, New York

Brown M, Lewis HG, Gunn SR (2000) Linear spectral mixture models and support vector machines remote sensing. IEEE Transactions on Geosciences and Remote Sensing 38: 2346–2360

Foody GM (1996) Approaches for the production and evaluation of fuzzy land cover classifications from remotely sensed data. International Journal of Remote Sensing 17: 1317–1340

Foody GM (1998) Sharpening fuzzy classification output to refine the representation of sub-pixel land cover distribution. International Journal of Remote Sensing 19(13): 2593–2599

Foody GM (2000a) Mapping land cover form remotely sensed data with a softened feedforward neural network. Journal of Intelligent and Robotic System 29: 433–449

References

Foody GM (2000b) Estimation of sub-pixel land cover composition in the presence of untrained classes. Computers and Geosciences 26: 469–478

Foody GM, Arora MK (1996) Incorporating mixed pixels in the training, allocation and testing stages of supervised classifications. Pattern Recognition Letters 17(13): 1389–1398

Foody GM, Cox DP (1994) Sub-pixel land cover composition estimation using a linear mixture model and fuzzy membership functions. International Journal of Remote Sensing 15: 619–631

Hansen M, Dubayah R, DeFries R (1996) Classification trees: an alternative to traditional land cover classifiers. International Journal of Remote Sensing 17: 1075–1081

Ibrahim MA, Arora MK, Ghosh SK (2003) A comparison of FCM and PCM for sub-pixel classification. Proceedings of 1st Indian International Conference on Artificial Intelligence, Hyderabad, India. CD

Mather PM (1999) Computer processing of remotely-sensed images: an introduction. Wiley, Chichester

Schneider W (1993) Land use mapping with subpixel accuracy from Landsat TM image data. Proceedings of 25th International Symposium on Remote Sensing and Global Environmental Change, Ann Arbor, MI, pp 155–161.

Settle JJ, Drake NA (1993) Linear mixing and the estimation of ground cover proportions. International Journal of Remote Sensing 14: 1159–1177

Tatem AJ, Hugh G, Atkinson PM, Nixon MS (2001) Super-resolution target identification from remotely sensed images using a Hopfield neural network. IEEE Transactions on Geosciences and Remote Sensing 39(4): 781–796

Tatem AJ, Hugh G, Atkinson PM, Nixon MS (2002) Super-resolution land cover pattern prediction using a Hopfield neural network. Remote Sensing of Environment 79: 1–14

Van Trees HL (1968) Detection, estimation and modulation theory. Wiley, New York

Varshney PK (1997) Distributed detection and data fusion. Springer Verlag, New York

Verhoeye J, Wulf RD (2002) Land cover mapping at sub-pixel scales using linear optimization techniques. Remote Sensing of Environment 79: 96–104

Winkler G (1995) Image analysis random fields and dynamic Monte Carlo methods. Springer Verlag, New York

CHAPTER 12

Image Change Detection and Fusion Using MRF Models

Teerasit Kasetkasem, Pramod K. Varshney

12.1
Introduction

The basic theory of Markov random fields was presented in Chap. 6. In this chapter, we employ this modeling paradigm for two image processing tasks applicable to remote sensing. The objectives of the two tasks are:

1. To investigate the use of Markov random field (MRF) models for image change detection applications
2. To develop an image fusion algorithm based on MRF models.

As mentioned in Chap. 2, image change detection is a basic image analysis tool frequently used in many remote sensing applications (such as environmental monitoring) to quantify temporal information. An MRF based image change detection algorithm is presented in the first part of this chapter. The image fusion algorithm is primarily intended to improve, enhance and highlight certain features of interest in remote sensing images for extracting useful information. In this chapter, an MRF based image fusion algorithm is applied to fuse the images from a hyperspectral sensor with a multispectral image thereby merging a fine spectral resolution image with a fine spatial resolution image. Examples are used to illustrate the performance of the proposed algorithms.

12.2
Image Change Detection using an MRF model

The ability to detect changes that quantify temporal effects using multitemporal imagery provides a fundamental image analysis tool in many diverse applications (see Sect. 2.7 of Chap. 2). Due to the large amount of available data and extensive computational requirements, there is a need to develop efficient change detection algorithms that automatically compare two images taken from the same area at different times to detect changes. Usually, in the comparison process (Singh 1989; Richards and Jia 1999; Lunneta and Elvigge 1999; Rignot and van Zyle 1993), differences between two corresponding pixels belonging to the same location for an image pair are determined on the basis of a quantitative measure. Then, a change is labeled, if this measure exceeds a predefined threshold, and no change is labeled, otherwise. Most of the comparison techniques described in Singh (1989) only consider information contained within a pixel even though intensity values of neighboring pixels of images are known to have significant correlation. Also, changes are more likely to occur in connected regions than at disjoint points. By using these facts, a more accurate change detection algorithm can be developed. To accomplish this, an MRF model for images is employed here so that the statistical correlation of intensity values among neighboring pixels can be exploited.

Bruzzone and Prieto (2000) and Wiemker (1997) have tried to employ the MRF model for image change detection (ICD). In Bruzzone and Prieto (2000), the authors subtracted one original image at one time, say t_1, from another original image at a different time, say t_2, to obtain a difference image. A change image was obtained directly from this difference image without using any knowledge from original images. Two separate algorithms were developed in their paper, namely distribution estimation and image change detection. The first algorithm employs the expectation maximization (EM) algorithm to estimate distributions of the difference image as a two-mode Gaussian mixture made up of the conditional densities of changed and unchanged pixels. In the second part, the MRF model was used to smoothen out the change image by using the iterative conditional mode (ICM) algorithm. A similar approach can be found in Wiemker (1997). This algorithm can be divided into two parts. In the first part, a pixel-based algorithm such as image differencing or image ratioing determines an initial change image that is further refined based on the MRF model in the second part. Some information is lost while obtaining the initial change image since the observed data is projected into a binary image whose intensity values represent change or no change. We observe that studies by Bruzzone and Prieto (2000) and Wiemker (1997) do not fully utilize all the information contained in images, and moreover, the preservation of MRF properties is not guaranteed. In Perez and Heitz (1996), the effect of image transformations on images that can be modeled by MRFs is studied. It has been shown that MRF properties may not hold after some transformations such as resizing of an image and subtraction of one image from another. For some specific transformations, MRF properties are preserved, but a new set of potential functions must be obtained. Since a difference image can be looked upon as a transformation, MRF modeling of a difference image in Bruzzone and Prieto (2000) and initial change image in Wiemker (1997) may not be valid. This provides the motivation for the development of an image change detection (ICD) algorithm that uses additional information available from the image and preserves MRF properties.

Here, we develop an ICD algorithm that consists of only one part. The observed images, modeled as MRFs, are directly processed by the MAP detector, which searches for the global optimum. The detector based on the MAP criterion chooses the most likely change image among all possible change images given the observed images. The resulting probability of error is minimum among all other detectors (Trees 1968; Varshney 1997). The structure of the MAP detector is based on statistical models. Therefore, the accuracy of the statistical models for the given images as well as for the change image is crucial. Here, we assume that the given images are obtained by the summation of noiseless images (NIM) and noises in the image. Both the NIMs and the change image are assumed to have MRF properties. Furthermore, we assume that configurations (intensity values) of the NIMs are the same for the unchanged sites. In addition, configurations of the changed sites from one NIM are independent of configurations of the changed sites from the other NIM when the configurations of the unchanged sites are given. Based on the above assumptions, the *a posteriori* probability is determined and the MAP criterion is used to select the optimum change image.

Due to the size of the search space, the solution of the MAP detection problem cannot be obtained directly. As a result, a stochastic search method such as the simulated annealing (SA) algorithm (see Sect. 6.4.1 of Chap. 6 and Geman and Geman 1984) is employed. Here, the SA algorithm generates a random sequence of change images in which a new configuration depends only on the previous change image and observed images by using the Gibbs sampling procedure (Geman and Geman 1984; Winkler 1995; Bremaud 1999; Begas 1986). The randomness of the new change image gradually decreases as the number of iterations increases. Eventually, this sequence of change images converges to the solution of the MAP detector.

12.2.1
Image Change Detection (ICD) Algorithm

The ICD algorithm uses the MAP detector whose structure is based on statistical knowledge. Therefore, statistical models for the given images as well as for the change image (CI) are required.

Noiseless Image and Change Image Models
Let $\mathcal{S}$ be a set of sites s, and $\Lambda = \{0, 1, \ldots, L-1\}$ be the phase space. Furthermore, we write $\{X_i(\mathcal{S})\}_{i \geq 0} \in \Lambda^\mathcal{S}$ as a sequence of image (configuration) vectors taken at time t_i, $i \in \{0, 1, \ldots, N-1\}$ where $t_0 < t_1 < \cdots < t_i < \cdots < t_{N-1}$. Here the value of N is 2. We assume that $X_i(\mathcal{S})$ satisfies MRF properties with a Gibbs potential $V_C(x_i)$. Note that $X_i(\mathcal{S})$ are noiseless image models (NIM) of the scene. These will be used extensively in our discussion. Due to noise, we cannot observe X_i directly. Instead, we observe noisy images (NI) $Y_i(\mathcal{S}) \in \Theta^\mathcal{S}$ given by

$$Y_i(\mathcal{S}) = X_i(\mathcal{S}) + W_i(\mathcal{S}) ; \quad i = 0, 1, \ldots, N , \qquad (12.1)$$

where $\Theta = \{q_0, q_1, \ldots q_{J-1}\}$ is the phase space corresponding to noisy observed images and $W_i(\mathcal{S})$ is a vector of additive Gaussian noises with mean zero and covariance matrix $\sigma^2 I$. Here, I is an identity matrix of size $M \times M$, where $M = |\mathcal{S}|$.

Obviously, there are a total of $K = 2^M$ possible change images (CIs) that may occur between any pair of NIMs (including the no-change event). Let

$H_k, k \in \{0, 1, \ldots, K-1\}$, correspond to the kth CI which is a binary image whose intensity values are either 0 or 1. Here, we use the notation $H_k(a) = 1$ to indicate a change at site a in the kth CI. When $x_i(a) = x_j(a)$, we have $H_k(a) = 0$ to indicate no change. We assume that all CIs satisfy MRF properties with the Gibbs potential $U_C(H_k)$ (defined in Chap. 6). We can write the marginal PDFs for X_i and H_k as

$$\pi_X(x_i) = \frac{1}{Z_X} \exp\left[-\sum_{C \subset \mathcal{S}} V_C(x_i)\right], \qquad (12.2)$$

and

$$\pi_H(H_k) = \frac{1}{Z_H} \exp\left[-\sum_{C \subset \mathcal{S}} U_C(H_k)\right], \qquad (12.3)$$

where $Z_X = \sum_{x \in \Lambda^\mathcal{S}} \exp\left[-\sum_{C \subset \mathcal{S}} V_C(x)\right]$ and $Z_H = \sum_{k=0}^{K-1} \exp\left[-\sum_{C \subset \mathcal{S}} U_C(H_k)\right]$, respectively. Since the elements of $W_i(\mathcal{S})$ are additive Gaussian noises which are independent and identically distributed, the conditional probability density of $Y_i(\mathcal{S})$ given $X_i(\mathcal{S})$ is given by

$$P(y_i|x_i) = \frac{1}{(2\pi\sigma^2)^{M/2}} \exp\left[-\frac{1}{2\sigma^2}(y_i - x_i)^T(y_i - x_i)\right]; \quad i = 0, 1, \cdots N.$$
$$(12.4)$$

Given H_k, we can partition a set $\mathcal{S}$ into two subsets $\mathcal{S}^{CH_k}$ and $\mathcal{S}^{NCH_k}$ such that $\mathcal{S}^{CH_k} \cap \mathcal{S}^{NCH_k} = \phi$ and $\mathcal{S}^{CH_k} \cup \mathcal{S}^{NCH_k} = \mathcal{S}$, where $\mathcal{S}^{CH_k}$ and $\mathcal{S}^{NCH_k}$ contain all changed and unchanged sites, respectively. We further assume that the configurations of changed sites between a pair of NIMs are independent given the configurations of unchanged sites, i.e.,

$$P\left\{X_i(\mathcal{S}^{CH_k}) = x_i^{CH_k}, X_j(\mathcal{S}^{CH_k}) = x_j^{CH_k} \middle| X_i(\mathcal{S}^{NCH_k}) = X_j(\mathcal{S}^{NCH_k}) = x_{ij}^{NCH_k}\right\}$$
$$= P\left\{X_i(\mathcal{S}^{CH_k}) = x_i^{CH_k} \middle| X_i(\mathcal{S}^{NCH_k}) = x_{ij}^{NCH_k}\right\}$$
$$\times P\left\{X_j(\mathcal{S}^{CH_k}) = x_j^{CH_k} \middle| X_j(\mathcal{S}^{NCH_k}) = x_{ij}^{NCH_k}\right\} \qquad (12.5)$$

where $X_i(\mathcal{S}^{CH_k})$ and $X_i(\mathcal{S}^{NCH_k})$ are the configurations of changed sites and unchanged sites of X_i, respectively. Here, to keep the analysis tractable, we have made a simplifying assumption that the pixel intensities of the changed sites in a pair of images are statistically independent given the intensities of the unchanged sites. A justification for this assumption is that changes are not predictable based on the current knowledge. Note that for unchanged sites, configurations of the same site in a pair of images must be equal. This equal configuration value is denoted by $x_{ij}^{NCH_k}$. For notational convenience, let us denote $X_i(s^{CH_k})$ and $X_i(s^{NCH_k})$ by $X_i^{CH_k}$ and $X_i^{NCH_k}$, respectively. The joint PDF of the pair is given by

$$P(x_i, x_j | H_k) \stackrel{\Delta}{=} P \begin{Bmatrix} X_i^{CH_k} = x_i^{CH_k}, X_i^{NCH_k} = x_{ij}^{NCH_k}, \\ X_j^{CH_k} = x_j^{CH_k}, X_j^{NCH_k} = x_{ij}^{NCH_k} \end{Bmatrix}$$
$$= P \left\{ X_i^{CH_k} = x_i^{CH_k} \middle| X_i^{NCH_k} = x_{ij}^{NCH_k} \right\}$$
$$\times P \left\{ X_j^{CH_k} = x_j^{CH_k} \middle| X_j^{NCH_k} = x_{ij}^{NCH_k} \right\}$$
$$\times P \left\{ X_i^{NCH_k} = X_j^{NCH_k} = x_{ij}^{NCH_k} \right\} . \tag{12.6}$$

Since X_i and X_j correspond to the same scene, we assume that they are statistically identical. By summing over all possible configurations of the changed sites, the joint PDF can be expressed as,

$$P(x_i, x_j | H_k) = P \left\{ X_i^{CH_k} = x_i^{CH_k} \middle| X_i^{NCH_k} = x_{ij}^{NCH_k} \right\}$$
$$\times P \left\{ X_j^{CH_k} = x_j^{CH_k} \middle| X_j^{NCH_k} = x_{ij}^{NCH_k} \right\}$$
$$\times \sum_{\lambda \in \Lambda^{sCH}} P(X_i^{CH_k} = \lambda, X_i^{NCH_k} = x_{ij}^{NCH_k}) . \tag{12.7}$$

Substituting the Gibbs potential V_C and using the properties of conditional probability, (12.7) can be written as

$$P(x_i, x_j | H_k) = \frac{1}{Z_X} \frac{\exp \left[- \sum_{C \subset s} V_C(x_i) - \sum_{C \subset s} V_C(x_j) \right]}{\sum_{\lambda \in \Lambda^{sCH_k}} \exp \left[- \sum_{C \subset s} V_C \left(\lambda, x_{ij}^{NCH_k} \right) \right]} . \tag{12.8}$$

By using the notation $\sum_{C \ni s}$ to denote the sum over the sets C that contain site s, the term $\sum_{C \subset s} V_C(x_i)$ can be decomposed as

$$\sum_{C \subset s} V_C(x_i) = \sum_{C \ni s^{CH_k}} V_C \left(x_i^{CH_k}, \partial x_{ij}^{CH_k} \right) + \sum_{C \subset s^{NCH_k}} V_C \left(x_{ij}^{NCH_k} \right), \tag{12.9}$$

where the boundaries of the changed region, denoted by ∂x_{ij}^{CH} are contained within the changed region. Substituting (12.9) into (12.8), we obtain

$$P(x_i, x_j | H_k) = \frac{\exp \left[-E^k(x_i, x_j) \right]}{Z_X}, \tag{12.10}$$

where

$$E^k(x_i, x_j) = \sum_{C \ni \delta^{CH_k}} V_C\left(x_i^{CH_k}, \partial x_{ij}^{CH_k}\right) + \sum_{C \ni \delta^{CH_k}} V_C\left(x_j^{CH_k}, \partial x_{ij}^{CH_k}\right)$$
$$+ 2 \sum_{C \subset \delta^{NCH_k}} V_C\left(x_{ij}^{NCH_k}\right) + \ln(Z_k) \qquad (12.11)$$

and

$$Z_k = \left\{ \sum_{\lambda \in \Lambda^{\delta^{CH_k}}} \exp\left[-\sum_{C \subset \delta} V_C\left(\lambda, x_{ij}^{NCH_k}\right)\right]\right\} \qquad (12.12)$$

are the joint image energy functions associated with H_k and the normalizing constant, respectively. Based on these assumptions, we design the optimum detector in the next section. Here, we have considered the case of discrete-valued MRFs. The derivation in the case of continuous-valued fields (used in the examples) is analogous.

12.2.2
Optimum Detector

We formulate the change detection problem as an M-ary hypothesis testing problem where each hypothesis corresponds to a different change image or a no change image. The maximum *a posteriori* (MAP) criterion (Trees 1968; Varshney 1997) is used for detecting changed sites in our work. This criterion is expressed as

$$H_k = \arg\left\{\max_{H_l}\left[P(H_l | Y_i = y_i, Y_j = y_j)\right]\right\}. \qquad (12.13)$$

From Bayes' rule, (12.13) can be rewritten as

$$H_k = \arg\left\{\max_{H_l}\left[\frac{P(Y_i = y_i, Y_j = y_j | H_l)P(H_l)}{P(Y_i = y_i, Y_j = y_j)}\right]\right\}.$$

Since $P(Y_i = y_i, Y_j = y_j)$ is independent of H_k and the two noises are independent of each other, the above equation reduces to

$$H_k = \arg\left\{\max_{H_l}\left[P(Y_i = y_i, Y_j = y_j | H_l)P(H_l)\right]\right\}$$
$$= \arg\left\{\max_{H_l}\left[\sum_{x_i, x_j \in \Lambda^\delta} P(y_i | x_i)P(y_j | x_j)P(x_i, x_j | H_l)P(H_l)\right]\right\}, \qquad (12.14)$$

where $P(y|x)$ and $P(x_i, x_j|H_l)$ denote $P(Y=y|X=x)$, and $P(X_i=x_i, X_j=x_j|H_l)$, respectively.

Substituting (12.3), (12.4) and (12.10) into (12.14) and taking the constant term out, we obtain

$$H_k = \arg\left\{\max_{H_l}\left[\exp\left(-\sum_{C\subset\mathcal{S}} U_C(H_l) + C^l(y_i, y_j)\right)\right]\right\}$$

$$= \arg\left\{\max_{H_l}\left[\exp\left(-E_{\text{post}}(H_l)\right)\right]\right\}, \tag{12.15}$$

where

$$C^l(y_i, y_j) = \ln\left\{\sum_{x_i, x_j \in \Lambda^\mathcal{S}} \exp\begin{pmatrix} -\frac{1}{2\sigma^2}(y_i - x_i)^T(y_i - x_i) \\ -\frac{1}{2\sigma^2}(y_j - x_j)^T(y_j - x_j) \\ -E^l(x_i, x_j) \end{pmatrix}\right\} \tag{12.16}$$

and

$$E_{\text{post}}(H_l) = \sum_{C\subset\mathcal{S}} U_C(H_l) - C^l(y_i, y_j). \tag{12.17}$$

From (12.15), we can observe that maximization of (12.14) is equivalent to the minimization of $E_{\text{post}}(H_l)$. Therefore, our optimal ICD algorithm can be expressed as

$$H_k = \arg\left\{\min_{H_l}\left[E_{\text{post}}(H_l)\right]\right\}. \tag{12.18}$$

In practice, the solution of (12.18) cannot be obtained directly due to the large number of possibilities of change images. To cope with this problem, we employ SA algorithm, which is a part of Monte Carlo procedures to search for solutions of (12.18). In the next section, some examples are provided for illustration. Both a simulated dataset and an actual dataset are used.

12.3
Illustrative Examples of Image Change Detection

Simulated data and real multispectral remote sensing images are used to illustrate and evaluate the performance of the image change detection algorithm. In these examples, we consider only five clique types, C_1, C_2, C_3, C_4, C_5 associated with the singleton, vertical pairs, horizontal pairs, left-diagonal pairs, and right-diagonal pairs, respectively. Furthermore, we assume that image potential functions are translation invariant. The Gibbs energy function of images is assumed to be given by

$$\sum_{C\subset\mathcal{S}} V_C(x) = \frac{1}{2}\beta^T J, \tag{12.19}$$

where $\boldsymbol{\beta} = [\beta_1, \beta_2, \cdots, \beta_5]^T$ is the NIM parameter vector, and $\boldsymbol{J}$ is assumed to be

$$J = \left[\sum_{s \in \mathcal{S}} x^2(s), \sum_{(s,t) \in C_2} (x(s) - x(t))^2, \cdots, \sum_{(s,t) \in C_5} (x(s) - x(t))^2 \right]^T,$$

which is the NIM potential vector associated with clique types, $C_1, \ldots, C_5$, respectively. We observe that the NIM potential vector is in a quadratic form. The quadratic assumption has widely been used to model images in numerous problems (e.g. Hazel 2000), and is called the Gaussian MRF (GMRF) model. GMRF models are suitable for describing smooth images with a large number of intensity values. Furthermore, using the GMRF model, (12.19) can be solved more easily due to the fact that the summation over configuration space in (12.16) changes to infinite integration in the product space. Hence, the analytical solution for (12.18) can be derived.

Using the GMRF model, we can rewrite (12.19) as (Hazel 2000),

$$\sum_{C \subset \mathcal{S}} V_C(\boldsymbol{x}) = \frac{1}{2} \boldsymbol{x}(\mathcal{S})^T \left[\Sigma^{-1} \right] \boldsymbol{x}(\mathcal{S}), \tag{12.20}$$

where the element (s_a, s_b) of the $M \times M$ matrix $\left[\Sigma^{-1}\right]$ is given by

$$\left[\Sigma^{-1}\right](s_a, s_b) = \begin{cases} \beta_1 + \sum_{i=2}^{5} 2\beta_i, & \text{if } s_a = s_b \\ -\beta_i, & \text{if } \{s_a, s_b\} \in C_i \\ 0, & \text{otherwise} \end{cases} . \tag{12.21}$$

Similarly, we define the Gibbs energy function of a CI over the clique system $[C_2 \, C_3 \, C_4 \, C_5]$ as

$$\sum_{C \subset \mathcal{S}} U_C(H_k) = \boldsymbol{\alpha}^T \boldsymbol{L}_k, \tag{12.22}$$

where $\boldsymbol{\alpha} = [\alpha_2, \cdots, \alpha_5]^T$ is the CI parameter vector, and

$$\boldsymbol{L}_k = \left[\sum_{(s,t) \in C_2} I[H_k(s), H_k(t)], \cdots, \sum_{(s,t) \in C_5} I[H_k(s), H_k(t)] \right]^T \tag{12.23}$$

is the CI potential vector associated with clique types as mentioned above and $I(a, b) = -1$ if $a = b$, and $I(a, b) = 1$, otherwise. The optimum detector has been derived in Kasetkasem (2002).

In order to obtain the result in a reasonable time, NIMs can be divided into subimages of much smaller size, (7×7) pixels in our case, due to the intensive computation required for the inversion of large matrices (e.g. the matrix size is 4096 × 4096 for an image of size (64 × 64) pixels). The suboptimum approach of considering one subimage at a time sacrifices optimality for computational efficiency. However, the statistical correlation among sites is concentrated in regions that are only a few sites apart. Therefore, our approximation is reasonable and yields satisfactory results in the examples.

The optimal algorithm used here involves matrix inversion and multiplication, both of which have the computational complexity $O(n^3)$ for an image of size $n \times n$. During each iteration, L_k, given in (12.23), is computed at least $2n^2$ times. As a result, the total complexity of our optimal algorithm is $O(n^5)$. However, when we employ the suboptimum algorithm, the number of operations for each subimage is fixed. Consequently, the overall complexity reduces to $O(n^2)$.

12.3.1
Example 1: Synthetic Data

Two noiseless simulated images of size (128 × 128) pixels are shown in Fig. 12.1. The corresponding CI is shown in Fig. 12.2, where black and white regions denote change and no change respectively. We observe that changes occur in the square region from pixel coordinates (60,60) to (120,120). To test our proposed ICD algorithm, these simulated images are disturbed with an additive Gaussian noise with zero mean and unit variance. An SA algorithm, with initial temperature $T_0 = 2$, is employed for optimization. For the difference image, the average image intensity power to noise power is 3.4 dB. For both the noiseless images, the average signal power is about 3.9 dB. Next, our proposed ICD algorithm is employed, and the results are shown in Fig. 12.3a–d after 0, 28, 63 and 140 sweeps, respectively. At 0-sweep, an image differencing technique is used, and the resulting CI is extremely poor. The situation improves as more sweeps are completed. Significant improvement can be seen when we compare

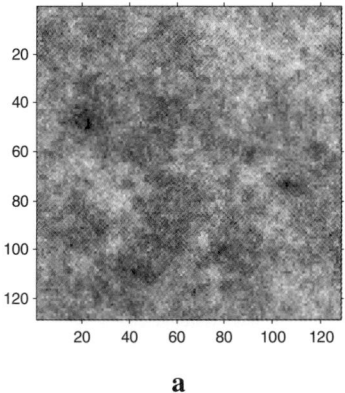
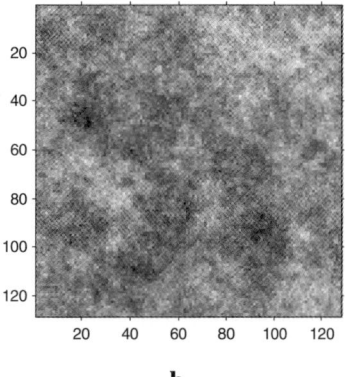

Fig. 12.1a,b. Two noiseless simulated images

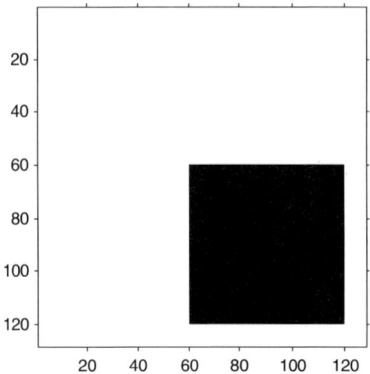

Fig. 12.2. Ideal change image (used as ground truth) produced from two noiseless images in Fig. 12.1

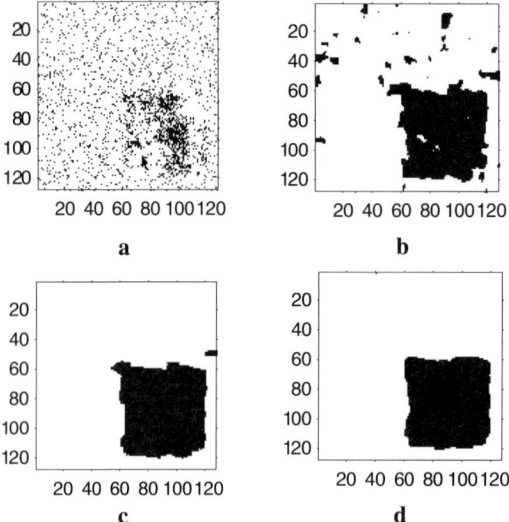

Fig. 12.3a–d. Results of MRF based algorithm after **a** 0 sweep, **b** 28 sweeps, **c** 63 sweeps, and **d**. 140 sweeps

Fig. 12.3a,b and Fig. 12.3b,c. However, very little improvement is seen in Figure 12.3d when we compare it with Fig. 12.3c. In order to evaluate the accuracy of our algorithm, we introduce three performance measures, average detection, false alarm and error rates. The average detection rate is defined as the total number of detected changed sites divided by the total number of changed sites. The average false alarm rate is defined as the total number of unchanged sites that are declared changed sites divided by the total number of unchanged sites. Similarly, the average error rate is given by the total number of misclassified sites (changed sites that are declared unchanged and vice versa.) divided by the total number of sites in the image. We plot the average detection, false alarm and error rates in Fig. 12.4 for a quantitative comparison. We observe rapid improvement (in accuracy) from 0 to 60 sweeps, with no further improvement after around 80 sweeps. The detection rate increases from about 60 percent to more than 90 percent and the average error rate decreases from around 35 percent to less than 5 percent.

We also compare the performance of our algorithm, with a variation of Bruzzone and Prieto's algorithm (abbreviated here as VBPA) described in Bruzzone and Prieto (2000) and apply it to the same dataset. In their algorithm, the two original images are replaced by their difference image as the input to the image change detection algorithm. Similar to our approach, the spatial information is characterized through an MRF model and the objective of their algorithm is to determine the most likely change image given the difference image of images at two different times. However, in Bruzzone and Prieto (2000), the expectation maximization (EM) algorithm is used for the estimation of means and variances associated with changed and unchanged regions, respectively. However, in this experiment, since the ground truth is available to us, we calculate these statistics beforehand. Furthermore, we substitute the ICM algorithm with the SA algorithm as the optimization algorithm since the SA algorithm is optimum.

The resulting change image after 500 iterations from VBPA is shown in Fig. 12.5a while Fig. 12.5b illustrates the change image presented in Fig. 12.3d for comparison purposes. The error rate of VBPA is 7% while our algorithm achieves the accuracy less than 1.5%. This is because VBPA uses the differ-

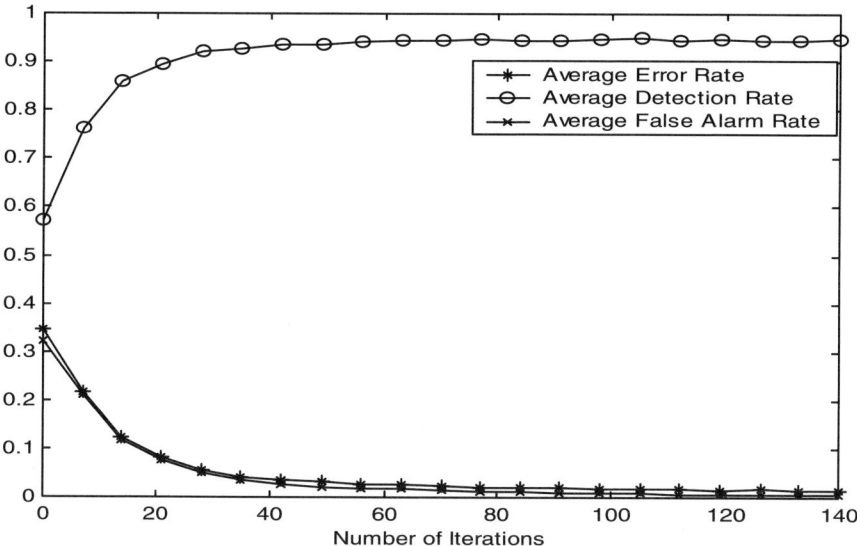

Fig. 12.4. Performance of the MRF based ICD algorithm in terms of average detection rate, average false alarm rate and average error rate

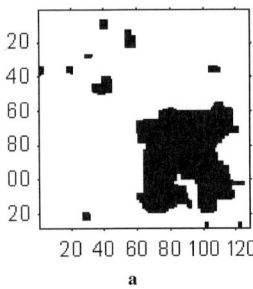 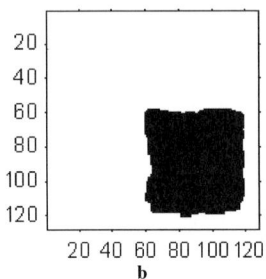

Fig. 12.5a,b. resulting change images: **a** VBPA, **b** our algorithm

ence image instead of original images. Some critical information may be lost when two images are transformed into one image through the subtraction operation. In particular, the noise power in the difference image is roughly the sum of the noise power levels of individual images. Hence, the SNR of the difference image is much lower than the original images. Our algorithm, on the other hand, depends mainly on the observed images rather than the difference image, which makes it more robust in a low SNR environment. The number of floating point operations for each iteration associated with VBPA is 4.6×10^5 when using MATLAB while for our algorithm, it is 3.1×10^{10}. VBPA has the complexity $O(n^2)$ because the number of instructions per iteration is independent of the image size. Hence, our algorithm consumes more computational resources to achieve performance gain and is, therefore, more suitable for off-line applications than real-time ones.

12.3.2
Example 2: Multispectral Remote Sensing Data

In this example, we apply our ICD algorithm to two real Landsat TM images of San Francisco Bay taken on April 6th, 1983 and April 11th 1988 (Fig. 12.6a,b). These images represent the false color composites generated from the images acquired in band 4, 5 and 6, and are given at http://sfbay.wr.usgs.gov/access/change_detect/Satellite_Images_1.html. For simplicity, we will apply the ICD algorithm only on images of the same bands acquired at two times to determine changes. In other words, we will separately find CIs for red, green, and blue band spectra, respectively.

Since we do not have any prior knowledge about the Gibbs potential and its parameters for both NIMs and CI, assumptions must be made and a parameter estimation method must be employed. Since the intensity values of NIMs can range from 0 to 255, we assume that they can be modeled as Gaussian MRF as in Hazel (2000), i.e., the Gibbs potential is in a quadratic form. Furthermore, we choose the maximum pseudo likelihood estimation (MPLE) method described in Lakshmanan and Derin (1989) to estimate the Gibbs parameters. Here, we estimate unknown parameters after every ten complete updates of the CI.

Image Change Detection and Fusion Using MRF Models

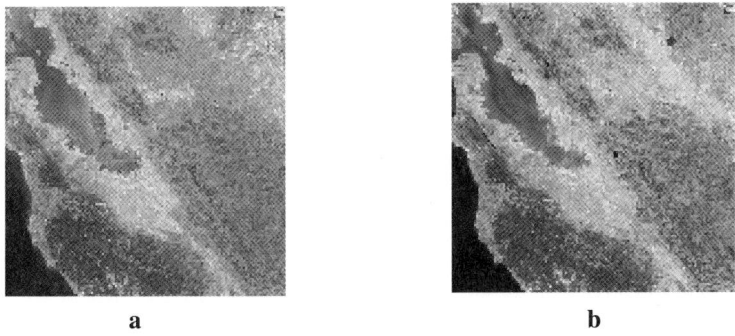

Fig. 12.6a,b. Landsat TM false color composites of San Francisco Bay acquired in **a** April 6th, 1983; **b** April 11th 1988. For a colored version of this figure, see the end of the book

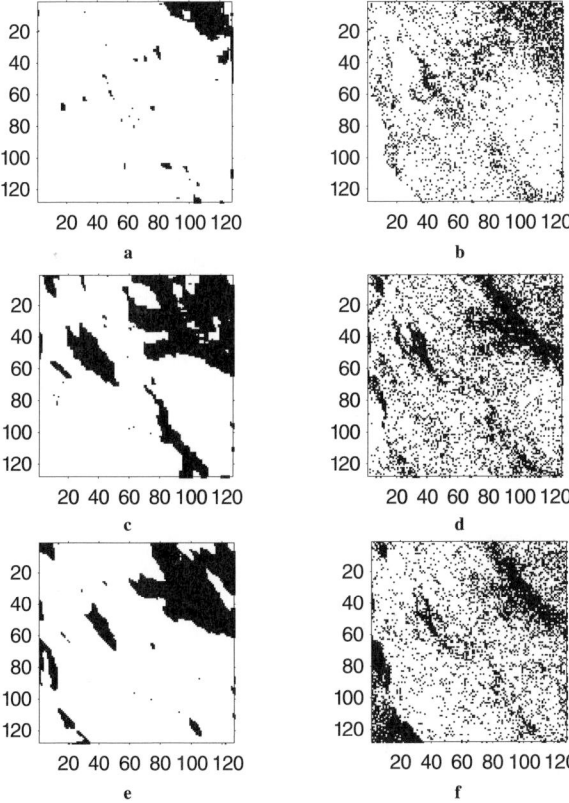

Fig. 12.7a–f. Change Images: MRF based ICD algorithm on the left and image differencing on the right: **a,b** red spectrum; **c,d** green spectrum; and **e,f** blue spectrum

The results of changed sites for individual spectra are displayed in Fig. 12.7a,b, 12.7c,d and 12.7e,f, respectively. Figure 12.7a,c and e are determined by the MRF based ICD algorithm while Fig. 12.7b,d and f result from image differencing. By carefully comparing results within each color spectrum, we conclude that our ICD algorithm detects changes that are more connected than those found using image differencing.

12.4
Image Fusion using an MRF model

The idea of combining information from different sensors both of the same type and of different types to either extract or emphasize certain features has been intensively investigated for the past two decades (Varshney 1997). However, most of the investigations have been limited to radar, sonar and signal detection applications (Chair and Varshney 1986). Image fusion has gained quite a bit of popularity among researchers and practitioners involved in image processing and data fusion. The increasing use of multiple imaging sensors in the fields of remote sensing (Daniel and Willsky 1997), medical imaging (Hill et al. 1994) and automated machine vision (Reed and Hurchinson 1996) has motivated numerous researchers to consider the idea of fusing image data. Consequently, image fusion techniques have emerged for combining multisensor and/or multiresolution images to form a single composite image. The final composite image is expected to provide more complete information content or better quality than the individual source image.

The simplest image fusion algorithm is the image averaging method (Petrovic and Xydeas 2000) in which a fused image is a pixel-by-pixel average of two or more raw images. This technique is very robust to image noise when raw images are taken from the same type of sensor under identical environments (Petrovic and Xydeas 2000). However, if images are taken from different types of sensors, the image averaging method may result in the loss of contrast information since the bright region in an image obtained by one type of sensor may correspond to a dark region in other images taken from different types of sensors. Furthermore, the image averaging method only utilizes information contained within pixels even though images are known to have high correlations among neighboring pixels. To utilize this information, Toet (1990) developed a hierarchical image fusion algorithm based on the pyramid transformation in which a source image is decomposed successively into a set of component patterns. Each component pattern corresponds to the representation of a source image at different levels of coarsenesses. Two or more images are fused through component patterns by some feature selection/fusion procedures to obtain the composite pyramid. The feature selection procedure must be designed in such a manner that it enhances the information of interest while removing irrelevant information from the fused image. Then, the fused image is regenerated back through the composite pyramid. Burt and Kolczynski (1993) have developed a similar algorithm with gradient pyramid transformation. Uner et al. (1997) have developed an image fusion algorithm specifically for concealed weapons detection by using wavelet decomposition. In their work, infrared (IR) and millimeter wave (MMW) imaging are used to capture images of a suspect. IR images have high resolution, but are not sensitive to concealed weapons because IR does not penetrate clothing. MMW images are highly sensitive to concealed weapons, but have very poor resolution. As a result, any concealed weapon detection algorithm that utilizes only either an IR or a MMW image is likely to perform poorly and will be unreliable. To increase reliability, IR and MMW images should be fused to a single image in such a manner that high resolution information of IR images and high contrast information of MMW images are retained.

The goal of resolution merging in remote sensing applications is slightly different from the concealed weapon detection problem. Unlike the concealed weapon detection problems where the only goal is to detect weapons by enhancing their shape, remote sensing applications are concerned more with the mixed-pixel problem (see Sect. 2.6.4 of Chap. 2) where the observed data corresponding to a pixel is a mixture of several materials such as grass and soil. In crisp image classification, the general assumption that a pixel can only be occupied by a single type of material is violated in mixed-pixel cases. As a result, these mixed pixels severely degrade performance of many hard image analysis tools including image classification. To avoid this mixed-pixel problem, multispectral images must be taken at very high resolution which may not be possible or too expensive to obtain with current photographic technologies. For instance, the IKONOS satellite provides the multispectral images in red, green, blue, and near IR spectra at 4×4 meter resolution and panchromatic image at 1×1 meter resolution. Figure 12.8a and b display red-color spectrum and panchromatic images of the Carrier Dome on the Syracuse University campus, respectively. Here, we observe finer details of the roof in the panchromatic image than in the red-spectrum image. As a result, the idea of combining the high resolution data of a panchromatic image with the multispectral images through image fusion is suitable in this case.

For resolution merge, we assume that there is a high resolution version of a multispectral image with the same resolution as a panchromatic im-

a

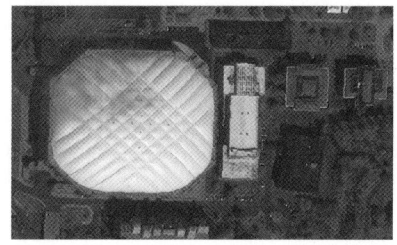
b

Fig. 12.8a,b. IKONOS images of the Carrier Dome (Syracuse University Campus): **a** red spectrum image and **b** panchromatic image

age (PI) and there is a pixel-to-pixel statistical relationship between a high resolution multispectral image (HRMI) and a PI. This assumption implies that observations of any two or more pixels in the PI are statistically independent when the HRMI is given. Next, we assume that the HRMI is distorted by a linear filter and then disturbed by noise before being captured by multispectral sensors. Furthermore, the HRMI is assumed to satisfy the Gaussian MRF properties since images can be very well described by the MRF model and the Gaussian MRF is suitable for images with a large number of intensity levels. The MAP criterion together with the Metropolis algorithm is used to search for the optimum HRMI image. We test our algorithm with two actual datasets. Visual inspection is used for performance evaluation.

12.4.1
Image Fusion Algorithm

In this section, we develop an image fusion algorithm based on an MRF model for spatial enhancement. The algorithm employs the MAP criterion to pick the "best" fused image from given observed images. The Metropolis optimization algorithm is used to search for the solution of the MAP equation.

12.4.1.1
Image Model

In this problem, we are interested in fusing two images coming from different sensing modalities and with different resolutions. The image from the first modality is assumed to have low resolution while the image from the second modality has high resolution. The goal is to obtain an enhanced image in the first modality at the same resolution as the image in the second modality (high resolution). Here, let $\mathcal{S}$ be a set of sites (pixels) s, and $\Lambda = \{0, 1, \ldots, L-1\}$ be the phase space. Note that L is the number of intensity values in the image (for example, 256 for an 8-bit gray-scaled image.) Furthermore, let $X(\mathcal{S}) \in \Lambda^{\mathcal{S}}$ denote the high-resolution image (HI) vector, or the enhanced image vector of the first modality (e.g. multispectral image) whose element $x(s) \in \Lambda$ is a configuration (intensity value) of a site (pixel) s in the HI. We assume that $X(\mathcal{S})$ satisfies the MRF properties with Gibbs potential function $V_C(x)$, i.e.,

$$\pi_X(x_i) = \frac{1}{Z_X} \exp\left[-\sum_{C \subset \mathcal{S}} V_C(x_i)\right], \qquad (12.24)$$

where $Z_X = \sum_{x \in \Lambda^{\mathcal{S}}} \exp\left[-\sum_{C \subset \mathcal{S}} V_C(x)\right]$ is the normalizing constant and C is a clique. $\sum_{C \subset \mathcal{S}} V_C(x)$ is called the Gibbs energy function. To make our algorithm computationally efficient, we only consider clique types composed of a single site and pairs of sites in an 8-neighborhood system that is C_1 to C_5 described in Chap. 6. Furthermore, the Gaussian MRF model (GMRF) is chosen to characterize the Gibbs potential function since natural images usually have a smooth texture. Hence, we have

$$V_{C_i}(X) = \begin{cases} \beta_i \left(x(s) - x(r)\right)^2 ; & \text{if } (s, r) \in C_i \\ 0 ; & \text{otherwise} \end{cases}, \quad (12.25)$$

where β_i is a Gibbs parameter associated with clique type C_i. Note here that, in general, these parameters may also depend on the location of a site s.

Let $Y_1(\mathcal{S}) \in \mathbb{R}^{\mathcal{S}}$ be the observed image of the first modality (i.e. coarser spatial resolution) where $\mathbb{R}$ denotes the set of real numbers. Here, we further assume that the HI is distorted by a linear filter, and then is disturbed by an additive noise before being captured by the sensor of the first modality. Thus, we can write the relationship between Y_1 and X as

$$Y_1(\mathcal{S}) = X(\mathcal{S}) \otimes F(\mathcal{S}) + N(\mathcal{S}), \quad (12.26)$$

where $A \otimes B$ denotes convolution of A and B, $F(\mathcal{S})$ is a linear filter of size $(k \times k)$ and $N(\mathcal{S})$ is a noise image. Since the observed image in the first modality is generally blurred (otherwise the vital information can be seen very clearly), a linear filter $F(\mathcal{S})$ is often a low pass filter. Here, we assume the linear filter $F(\mathcal{S})$ to be a finite impulse response (FIR) filter due to the complexity involved in convolving an infinite impulse response (IIR) filter with $X(\mathcal{S})$. Thus, a larger size of $F(\mathcal{S})$ is more suitable to describe the observed image of the first modality with a higher degree of blurriness. Furthermore, for simplicity, let us assume that the configurations at two different pixels of a noise image are statistically independent with an identical variance σ^2. For convenience, let $Z(\mathcal{S})$ denote the filtered version of the HI before being disturbed by the additive noise, i.e.,

$$Z = X \otimes F. \quad (12.27)$$

We do not explicitly show the dependence on $\mathcal{S}$ throughout the rest of this chapter for notational convenience. From (12.26) and (12.27), we have

$$P(Y_1 | X) = \prod_{s \in \mathcal{S}} P(y_1(s) | z(s), F). \quad (12.28)$$

Next, let $Y_2(\mathcal{S}) \in \Lambda^{\mathcal{S}}$ be the observed image in the second modality (e.g. high spatial resolution image). Its observations at sites s_i and s_j are statistically independent given the associated HI $X(\mathcal{S})$. Hence, we have

$$P(Y_2(\mathcal{S}) | X(\mathcal{S})) = \prod_{s \in \mathcal{S}} P(y_2(s) | X(\mathcal{S})). \quad (12.29)$$

Furthermore, we assume that the intensity value at a site s of Y_2 depends only on the intensity of the HI at the same site, i.e.,

$$P(Y_2(\mathcal{S}) | X(\mathcal{S})) = \prod_{s \in \mathcal{S}} P(y_2(s) | x(s)). \quad (12.30)$$

We formulate the image fusion problem as an M-ary hypothesis testing problem where each hypothesis corresponds to a different HI.

12.4.1.2
Optimum Image Fusion

The maximum *a posteriori* (MAP) criterion used for solving the above problem is expressed as

$$X_k = \arg\left\{\max_{X_j}\left[P(X_j|Y_1=y_1,Y_2=y_2)\right]\right\}. \tag{12.31}$$

From Bayes' rule and assuming conditional independence of Y_1 and Y_2 given X_j, (12.31) can be rewritten as

$$X_k = \arg\left\{\max_{X_j}\left[\frac{P(Y_1=y_1|X_j)P(Y_2=y_2|X_j)P(X_j)}{P(Y_1=y_1,Y_2=y_2)}\right]\right\}. \tag{12.32}$$

Since $P(Y_1=y_1,Y_2=y_2)$ is independent of X_k, it can be omitted and the above equation reduces to

$$\begin{aligned} X_k &= \arg\left\{\max_{X_j}\left[P(Y_1=y_1|X_j)P(Y_2=y_2|X_j)P(X_j)\right]\right\} \\ &= \arg\left\{\max_{X_j}\left[\left(\prod_{s\in\mathcal{S}}p(y_2(s)|x_j(s))\right)P(Y_1=y_1|X_j)P(X_j)\right]\right\}. \end{aligned} \tag{12.33}$$

Substituting (12.24), (12.28) and (12.30) into (12.33) we have

$$X_k = \arg\left\{\max_{X_j}\left[\begin{array}{c}\left(\prod_{s\in\mathcal{S}}p(y_2(s)|x_j(s))\right)\\ \times\left(\prod_{s\in\mathcal{S}}p(y_1(s)|z_j(s),F)\right)\\ \times\frac{1}{Z_X}\exp\left(-\sum_{C\subset\mathcal{S}}V_C(x_j)\right)\end{array}\right]\right\}. \tag{12.34}$$

Equation (12.34) can be rewritten as

$$X_k = \arg\left\{\max_{X_j}\left[\frac{1}{Z_X}\exp\left(\begin{array}{c}-\sum_{s\in\mathcal{S}}\left[E_2\left(y_2(s)|x_j(s)\right)\right.\\ \left.+E_1\left(y_1(s)|z_j(s),F\right)\right]\\ -\sum_{C\subset\mathcal{S}}V_C(x_j)\end{array}\right)\right]\right\}, \tag{12.35}$$

where

$$E_1\left(y_1(s)|z_j(s),F\right) = -\log\left[p\left(y_1(s)|z_j(s),F\right)\right]$$

and

$$E_2\left(y_2(s)|x_j(s)\right) = -\log\left[p\left(y_2(s)|x_j(s)\right)\right].$$

Finally, from (12.35), we obtain

$$X_k = \arg\left\{\min_{X_j}\left[E_{\text{post}}\left(Y_1, Y_2, X_j\right)\right]\right\}, \tag{12.36}$$

where

$$E_{\text{post}}\left(Y_1, Y_2, X_j\right) = \sum_{s \in \mathcal{S}}\left[E_1\left(y_1(s)\,|z_j(s), F\right) + E_2\left(y_2(s)\,|x_j(s)\right)\right] + \sum_{C \subset \mathcal{S}} V_C(x_j)$$

denotes the energy of X_j.

12.4.1.3
Proposed Algorithm

The main objective of this algorithm is to find the HI that is a spatially enhanced version of the image of the first modality. The solution of (12.36) yields the most likely HI for given images of the first and second modalities. However, a direct search for this solution is infeasible due to the enormous number of possibilities of HIs. In order to find solutions of (12.36) in a reasonable time, we adopt the Metropolis algorithm (described in Chap. 6) to expedite the search process. Unlike an exhaustive search technique where posterior probabilities associated with all possible HIs are calculated and the HI having the maximum posterior probability is selected, the Metropolis algorithm generates a new HI through a random number generator whose outcome depends on the current HI and observed images. Here, a new configuration is randomly proposed and is accepted with the probability given by

$$\alpha\left(x_{\text{old}}, x_{\text{new}}\right)$$
$$= \min\left\{1, \exp\left[-\frac{1}{T(n)}\left(E(y_1, y_2, x_{\text{new}}) - E(y_1, y_2, x_{\text{old}})\right)\right]\right\}, \tag{12.37}$$

where $\alpha\left(x_{\text{old}}, x_{\text{new}}\right)$ is the acceptance probability for the current stage x_{old} and the proposed stage x_{new}, $T(n)$ is the temperature, and n is the iteration number. From (12.37), if a proposed configuration corresponds to a lower energy, it will be accepted with probability one. However, if a proposed configuration corresponds to a higher energy, it will be accepted with probability associated with the difference between energies of x_{old} and x_{new}, and the temperature. In the early stages of the Metropolis algorithm, the temperature is set high to ensure that the resulting HI can escape from any local optimum points. However, in later stages of optimization, the temperature should be low so that a single solution is obtained. The rate at which this randomness decreases (i.e. decrease in T) must be carried out properly to ensure the convergence of the induced Markov chain. Chapter 6 discusses the properties of $T(n)$ that are restated here,

1) $\lim_{n \to \infty} T(n) = 0$,

2) $T(n) \geq \dfrac{\Delta \tau}{\log(n)}$, \hfill (12.38)

where Δ is the maximum value of the absolute energy change when a pixel is updated and τ is a constant defined in Chap. 6.

Theoretically speaking, regardless of the initial HI, the induced Markov chain eventually converges to the solution of (12.36) if necessary conditions given in (12.38) are satisfied. However, this convergence may require an infinite number of iterations, which are not feasible in practice. To actually implement this algorithm, the Metropolis algorithm must be terminated after a certain number of iterations. As a result, we must carefully select the initial HI that is relatively close to the desired solution so that the algorithm terminates in a reasonable time. Here, we pick the initial HI to be the average of intensity values of pixels in Y_1 and the maximum likelihood estimate (MLE) of HI based on Y_2, i.e.,

$$X_{\text{initial}} = \frac{1}{2}\left(Y_1 + \text{MLE}(Y_2)\right), \quad (12.39)$$

where MLE($\cdot$) denotes the MLE of HI based on observation Y_2. The main reason for employing this procedure is the fact that the MLE of HI for a given Y_2 usually has high resolution, but the important information cannot be clearly seen whereas Y_1 has high contrast and the vital information is clearly seen, but its resolution is very poor. By using the above procedure, we expect the initial HI to have high contrast making the vital information more visible while the background and texture information of the high-resolution image is also present. To use the MLE, the conditional probability of the HI given Y_2 must be estimated. Here, we employ the joint histogram of Y_2 and Y_1. Hence, we have

$$p_{Y_2|X}(a|b) = \frac{\text{number of pixels that have } Y_2 = a \text{ and } Y_1 = b}{\text{number of pixels that have } Y_1 = b}. \quad (12.40)$$

Furthermore, even with a reasonably good initial estimate given in (12.40), the total number of iterations required for the induced Markov chain to converge to a single value is still very large since we are dealing with gray-scaled images under the GMRF model. The Gibbs energy function under the GMRF model is in a quadratic form under which small changes (increase or decrease of one or two intensity values) of a configuration of a site has little effect on the Gibbs energy function unless the temperature is very low. This fact creates a noise-like effect on the fused image even with a fairly large number of iterations (500 in our examples.) To minimize this effect and to allow our algorithm to terminate in a reasonable time, we average the resulting HIs generated during the last few iterations. Obviously, this method violates the necessary conditions for the convergence of the Metropolis algorithm, but after a large number of iterations, the temperature change is very small due to the fact that the inverse of the log function (for example, $1/\log(400) = 0.167$ and $1/\log(500) = 0.161$) is used to determine temperature in (12.34). As a result, over a short period of time, an induced Markov chain under the Metropolis algorithm (when parameters are fixed) after a large number of iterations has similar properties as homogeneous Markov chains. Hence, the average of the resulting HIs should provide a reasonable estimate of the most likely HI for the given observations.

In addition, (12.36) depends on several unknown parameters such as the noise variance and the low-pass filter F as described earlier. Therefore, we couple parameter estimation with the Metropolis algorithm so that parameter estimation and the search for a solution of (12.36) can be accomplished simultaneously. Here also, the maximum likelihood estimate (MLE) is chosen because of its simplicity in implementation. The use of the MLE can sometimes severely affect the convergence of the induced Markov chain in (12.37) since, in many cases, the difference between parameters before and after estimation is too large and may cause the induced Markov chain to diverge. To deal with this problem, we limit the difference to the rate of n^{-1}. With this rate, parameters are permitted to change significantly in the early stages of the search whereas, in the later stages, these can only be changed by small amounts to allow the induced Markov chain to converge. More details of this algorithm have been presented in Chap. 6.

12.5
Illustrative Examples of Image Fusion

We test our algorithm on two sets of remote sensing data: multispectral and hyperspectral in the two examples respectively. Visual inspection is used for performance evaluation of the fused products in both the examples.

12.5.1
Example 1: Multispectral Image Fusion

The IKONOS data set for the Syracuse University campus is used to investigate the effectiveness of our image fusion algorithm for resolution merging, i.e., fused image resulting from images of two different resolutions. Four multispectral images of red, green, blue, and near infrared (NIR) color spectra and one panchronometric (PAN) image are used in this example. Figure 12.9 and Fig. 12.10 display false color composites of NIR, red and green spectra of the multispectral image, and the PAN image, respectively. Clearly, the PAN image has sharper feature boundaries than the multispectral image, specifically for road features.

We first apply principal component analysis (PCA) to the 4-band multispectral image to transform the multispectral image into four uncorrelated images. Then, the image corresponding to the highest eigenvalue (highest power) is

Fig. 12.9. False color composite of the IKONOS multispectral image of NIR, red and green color spectra of Syracuse University area. For a colored version of this figure, see the end of the book

Fig. 12.10. IKONOS PAN image of Syracuse University area

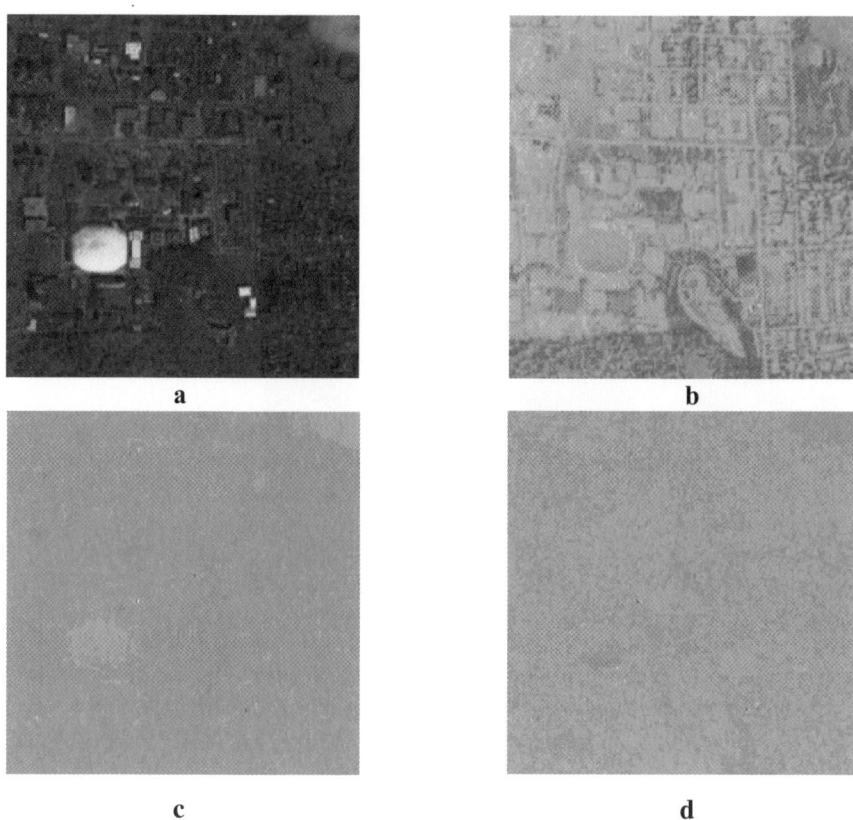

Fig. 12.11a–d. Image bands after PCA transformation corresponding to eigenvalues of a 2537, b 497, c 14, d 3.16

selected for fusion with the PAN image. The main reason for this procedure is to reduce the computational time since four times the amount of computation is needed to fuse all four color spectra to the PAN image. Furthermore, since the principal component corresponding to the highest eigenvalue contains most of the information of the four band multispectral image, we expect that very little improvement can be achieved by fusing the rest of the components with the PAN image. Figure 12.11a–d show the image bands after PCA transformation corresponding to eigenvalues 2537, 497, 14 and 3.16, respectively. As expected, higher variation in data can be observed in the component corresponding to higher eigenvalues while low variation in data corresponds to lower eigenvalues.

The resulting fused image after 500 iterations with a 5×5 pixel filter (F defined earlier) is shown in Fig. 12.12. Here, the same false color composite is used for display purposes. Clearly, the fused image appears to be sharper and clearer as compared to the original multispectral image. Furthermore, we also observe that more details in certain areas of the image are visible, e. g.,

Fig. 12.12. The false color composite of the resulting fused image assuming a filter size of 5 × 5 pixels. For a colored version of this figure, see the end of the book

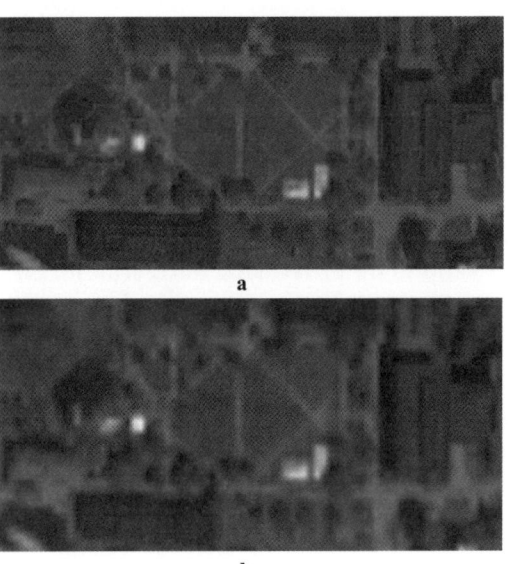

Fig. 12.13a,b. False color composite of the quad area of Syracuse University campus. **a** fused image, **b** original image. For a colored version of this figure, see the end of the book

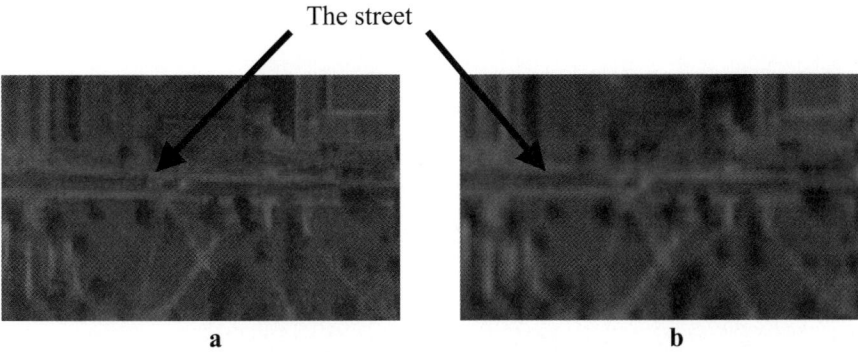

Fig. 12.14a,b. False color composite of the street next to Schine student center. **a** fused image, **b** original image. For a colored version of this figure, see the end of the book

textures on the roofs of the buildings and cars on the roads. Figure 12.13a,b and Fig. 12.14a,b display both the fused image and the original multispectral image of the quad area of the Syracuse University campus and of the street next to Schine student center, respectively, for comparison purposes where Fig. 12.13a and Fig. 12.14a display the fused image and Fig. 12.13b and Fig. 12.14b display the original multispectral image. We observe that small objects such as cars and roofs of the buildings in the fused image are more visible and separable than the original image.

12.5.2
Example 2: Hyperspectral Image Fusion

In this experiment, we consider a hyperspectral image acquired in 126 bands in the wavelength region from 0.435 to 2.4887 µm from the HyMap sensor (specifications given in Chap. 2) at spatial resolution 6.8 m. A portion consisting of (375 × 600) pixels has been selected for experimentation in this example. Figure 12.15 displays false color composite of images acquired in bands with mid-wavelengths of 2.04660 µm, 1.16690 µm, and 0.64910 µm. Fine resolution (0.15 m resolution) digital aerial photographs (shown in Fig. 12.16) obtained from Kodak are fused with the HyMap image.

Here again, we first apply principal component analysis (PCA) to all the 126 bands of the HyMap image to get uncorrelated transformed images. Figure 12.17a–d show the first four principal components after the PCA transformation with eigenvalues 4.40×10^6, 9.57×10^5, 4.35×10^4, 1.37×10^4, respectively. As expected, the higher variation in the data can be observed in the components with higher eigenvalues while the low variation in the data corresponds to lower eigenvalues.

The component corresponding to the highest eigenvalue (highest power) is fused with the digital aerial photograph. The false color composite of the resulting fused image after 500 iterations with a 5 × 5 pixel filter is shown in Fig. 12.18. Clearly, the fused image appears to be sharper and clearer when

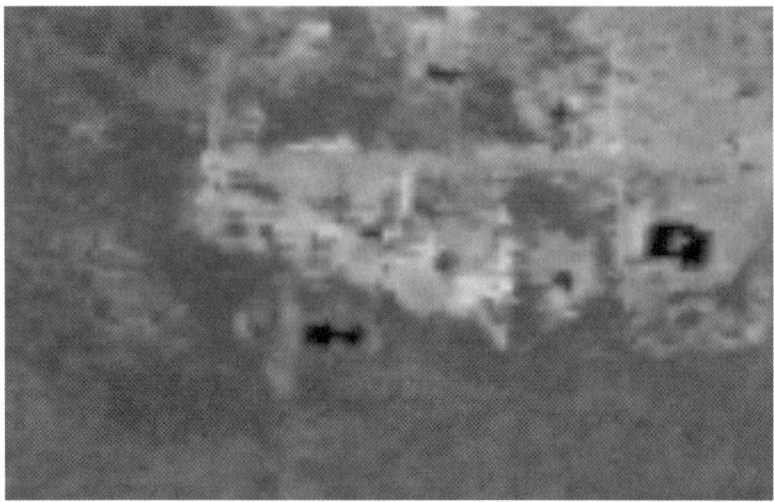

Fig. 12.15. Color composite of hyperspectral image (2.04660 µm, 1.16690 µm, and 0.64910 µm) acquired by the HyMap sensor. For a colored version of this figure, see the end of the book

Fig. 12.16. Digital aerial photograph of the same scene as shown in Fig. 12.15

compared with the original hyperspectral image (Fig. 12.15). Furthermore, we also observe that more details (e.g. roads and a lake) in certain areas of the image are visible. This shows that our algorithm can successfully be applied to fuse images at different resolutions.

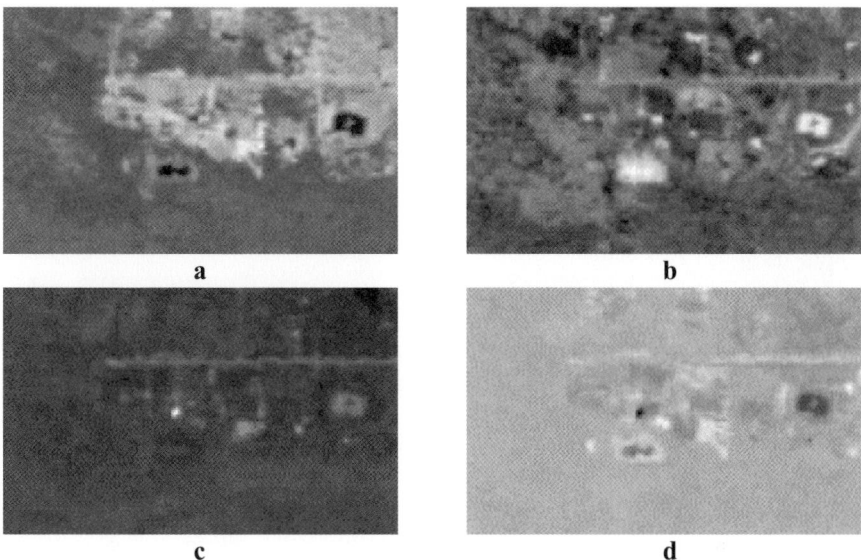

Fig. 12.17a–d. First four principal components corresponding to eigenvalues of **a** 4.40×10^6, **b** 9.57×10^5, **c** 4.35×10^4, **d** 1.37×10^4

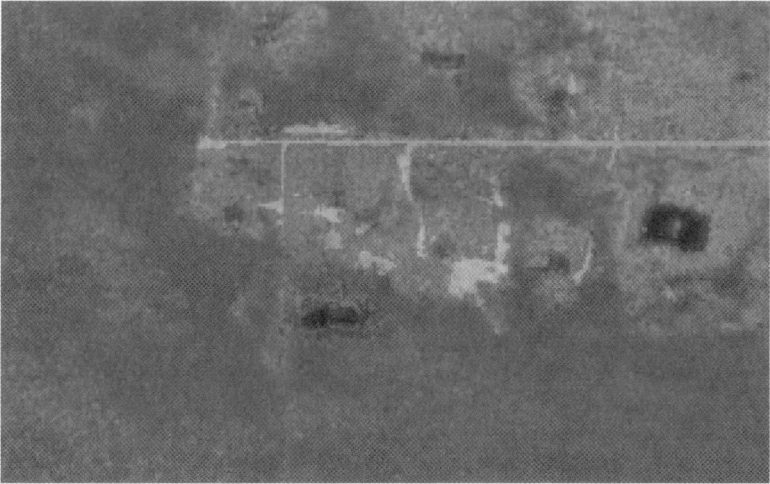

Fig. 12.18a–d. The false color composite of the resulting fused image assuming a filter size of 5×5. For a colored version of this figure, see the end of the book

12.6
Summary

In this chapter, the application of MRF models to image change detection and image fusion was demonstrated. The ultimate goal for both problems is the same that is to select the best solution for the problem under consideration under the statistical framework based on the observed data. Different optimization algorithms were employed depending upon the requirement of each application.

In image change detection, the simulated annealing algorithm was used to find the optimum solution. Here, we assumed that the change image and the noiseless image have MRF properties with different energy functions. Furthermore, we assumed that configurations of changed pixels from different images given configurations of unchanged pixels are statistically independent. Based on this model, the MAP equation was developed. The simulated annealing algorithm was employed to search for the optimum. The results showed that our algorithm performed fairly well in a noisy environment.

For the image fusion problem, the Gaussian MRF model was employed. The high-resolution remote sensing image was assumed to have pixel-to-pixel relation with the fused image, i.e., the observed configurations of any two pixels in the high resolution image were statistically independent when the fused image was given. Furthermore, the observed low resolution image was assumed to be the filtered version of the fused image. The Metropolis algorithm was employed to search for the optimum solution due to the large number of possible gray levels (256 in gray-scaled image.) The fused images obtained from our algorithm clearly showed improvement in the interpretation quality of the features to be mapped from both multispectral and hyperspectral remote sensing images.

References

Begas J (1986) On the statistical analysis of dirty pictures. Journal of Royal Statistical Society B 48(3): 259–302

Bremaud P (1999) Markov chains Gibbs field, Monte Carlo simulation and queues. Springer Verlag, New York

Bruzzone L, Prieto DF (2000) Automatic analysis of the difference image for unsupervised change detection. IEEE Transactions on Geoscience and Remote Sensing 38(3): 1171–1182

Burt PJ, Kolczynski RJ (1993) Enhance image capture through fusion. Proceedings of 4th International Conference on Computer Vision, 7(4): 593–600

Chair Z, Varshney PK (1986) Optimal data fusion in multiple sensor detection systems. IEEE Transactions on Aerospace and Electrical Systems AES-22: 98–101

Daniel MM, Willsky AS (1997) A multiresolution methodology for signal-level fusion and data assimilation with applications to remote sensing. Proceedings IEEE 85(1): 164–180

Geman S, Geman D (1984) Stochastic relaxation, Gibbs distributions and the Bayesian restoration of images. IEEE Transactions on Pattern Analysis and Machine Intelligence PAMI-6(6): 721–741

References

Hazel GG (2000) Multivariate Gaussian MRF for multispectral scene segmentation and anomaly detection. IEEE Transactions on Geoscience and Remote Sensing 39(3): 1199–1211

Hill D, Edwards P, Hawkes D (1994) Fusing medical images. Image Processing 6(2): 22–24

Kasetkasem T (2002) Image analysis methods based on Markov random field models. PhD Thesis, Syracuse University, Syracuse, NY

Lakshmanan S, Derin H (1989) Simultaneous parameter estimation and segmentation of Gibbs random fields using simulated annealing. IEEE Transactions on Pattern Analysis and Machine Intelligence 11(8): 799–813

Lunetta RS, Elvigge CD (eds) (1999) Remote sensing change Detection. Taylor and Francis, London, UK

Perez P, Heitz F (1996) Restriction of a Markov random field on a graph and multiresolution statistical image modeling. IEEE Transactions on Information Theory 42(1): 180–190

Petrovic VS, Xydeas CS (2000) Objective pixel-level image fusion performance measure. Proceedings of the SPIE, pp 89–98

Reed JM, Hutchinson S (1996) Image fusion and subpixel parameter estimation for automated optical inspection of electronic components. IEEE Transactions on Industrial Electronics 43: 346–354

Richards JA, Jia X (1999) Remote sensing digital image analysis: an introduction. Springer-Verlag, Berlin

Rignot EJM, van Zyle JJ (1993) Change detection techniques for ERS-1 SAR data. IEEE Transactions on Geoscience and Remote Sensing 31(4): 896–906

Singh A (1989) Digital change detection techniques using remotely sensed data. International Journal of Remote sensing 10(6): 989–1003

Toet A (1990) Hierarchical image fusion. Machine Vision and Applications 3(1): 1–11

Uner MK, Ramac LC, Varshney PK, Alford M (1997) Concealed weapon detection: an image fusion approach. Proceeding of SPIE 2942, pp 123–132

Van Trees HL (1968) Detection, estimation and modulation theory. Wiley, New York

Varshney PK (1997) Distributed detection and data fusion. Springer Verlag, New York

Wiemker R (1997) An iterative spectral-spatial Bayesian labeling approach for unsupervised robust change detection on remotely sensed multispectral imagery. Proceedings of the 7th International Conference on Computer Analysis of Images and Patterns: 263–270

Winkler G (1995) Image analysis random fields and dynamic Monte Carlo methods. Springer Verlag, New York

Color Plate I

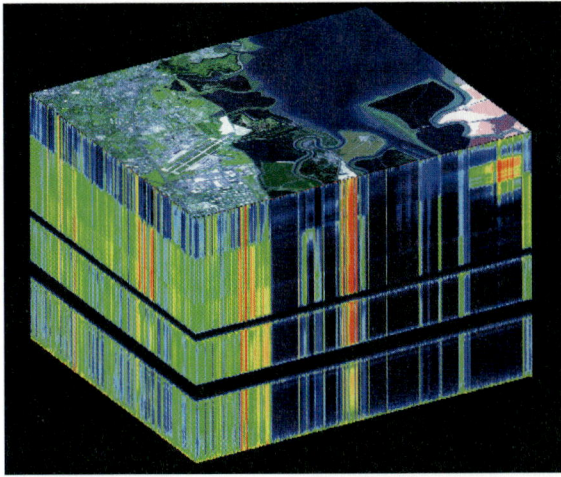

Fig. 1. The hyperspectral cube (source: http://aviris.jpl.nasa.gov/html/aviris.cube.html)

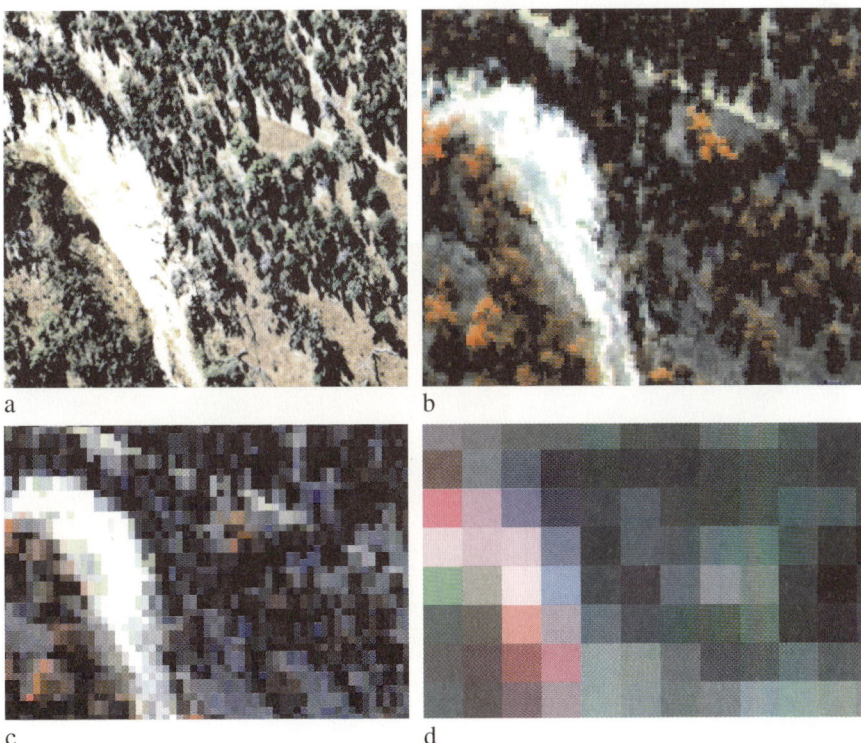

Fig. 1.1a–d. Observations of the same area of mixed species subtropical woodlands near Injune, central Queensland, Australia, observed using **a** stereo color aerial photography, **b** CASI, **c** HyMap, and **d** Hyperion data at spatial resolutions of < 1 m, 1 m, 2.8 m and 30 m respectively

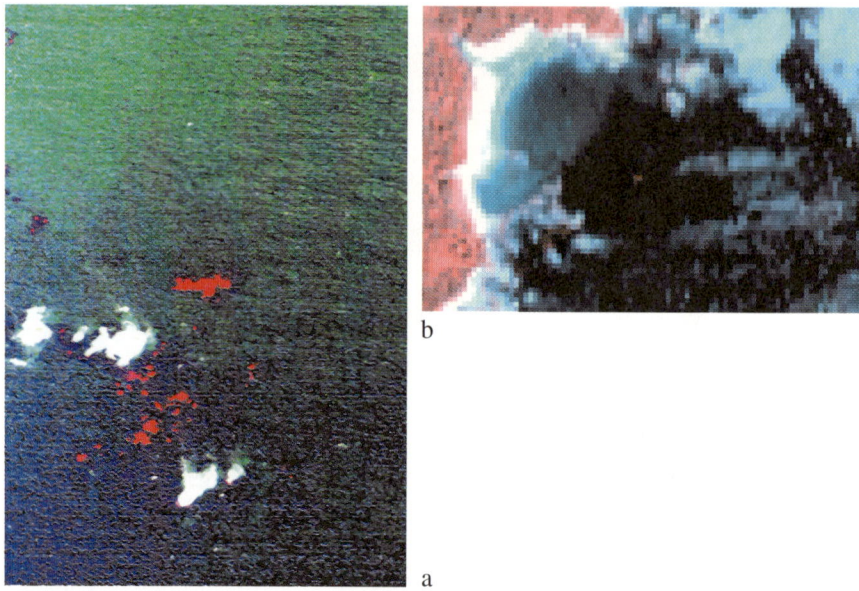

Fig. 1.2a,b. Example of detection of aquatic vegetation using airborne imaging spectrometers: partially submerged kelp beds (*bright red*) as observed using **a** 2 m CASI and **b** 20 m AVIRIS data

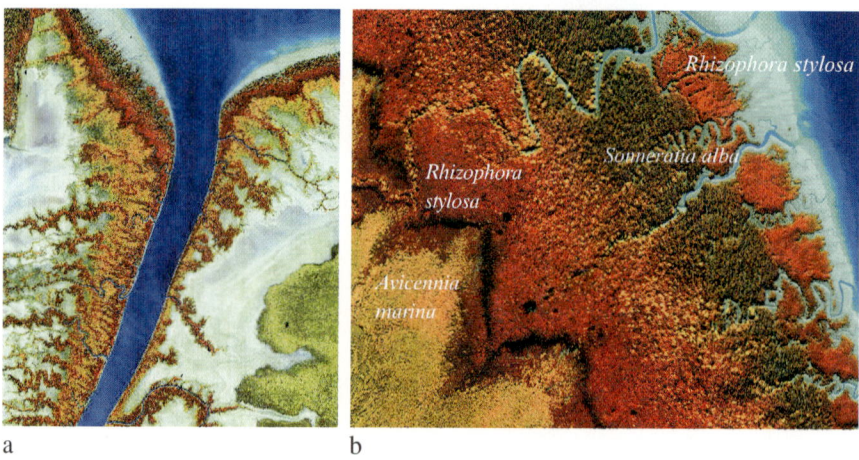

Fig. 1.3a,b. Color composite (837 nm, 713 nm and 446 nm in RGB) of fourteen 1 km × ~ 15 km strips of CASI data acquired over the West Alligator River mangroves, Kakadu National Park, Australia by BallAIMS (Adelaide). The data were acquired at 1 m spatial resolution in the visible and NIR wavelength (446 nm–838 nm) region. **b** Full resolution image of the west bank near the river mouth with main species communities indicated (Lucas et al. in preparation)

Color Plate III

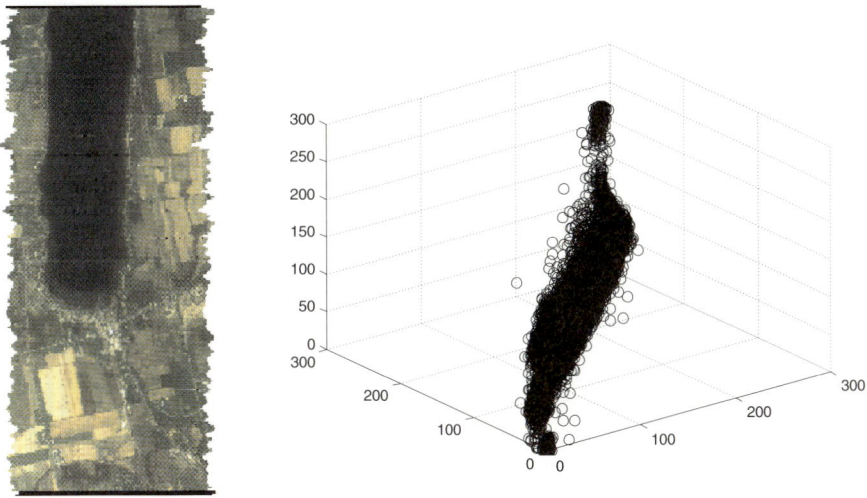

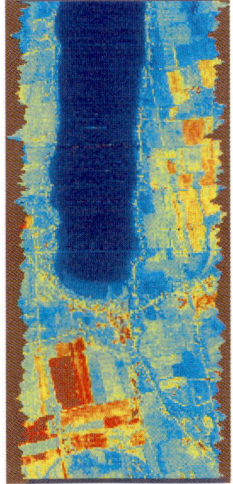

Fig. 2.10a–c. a Input color image. **b** 3D scatter plot of R, G and B components. **c** Pseudocolor map of unsupervised classification

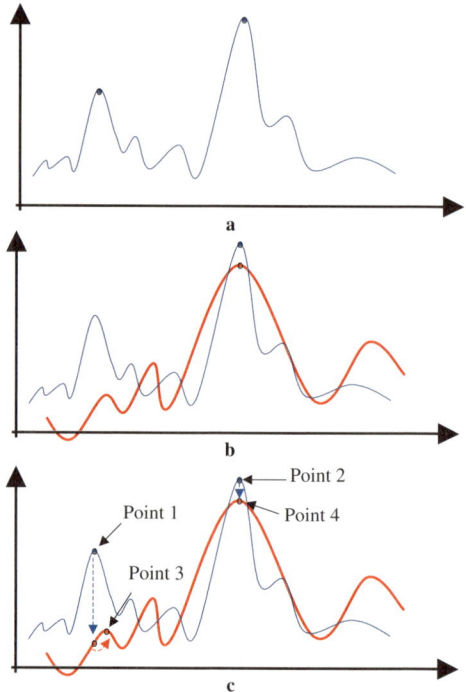

Fig. 3.16. An illustration of the heuristic test to identify the global maximum

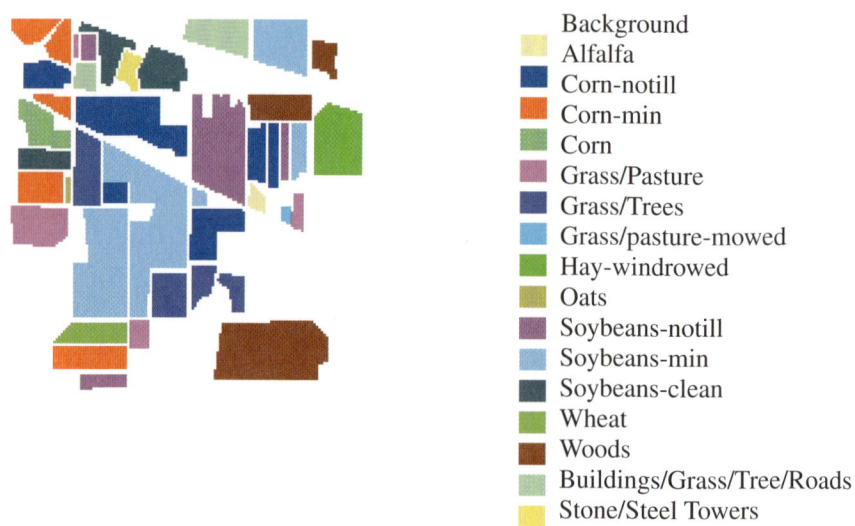

Fig. 4.8. Reference data (ground truth) corresponding to AVIRIS image in Fig. 4.7

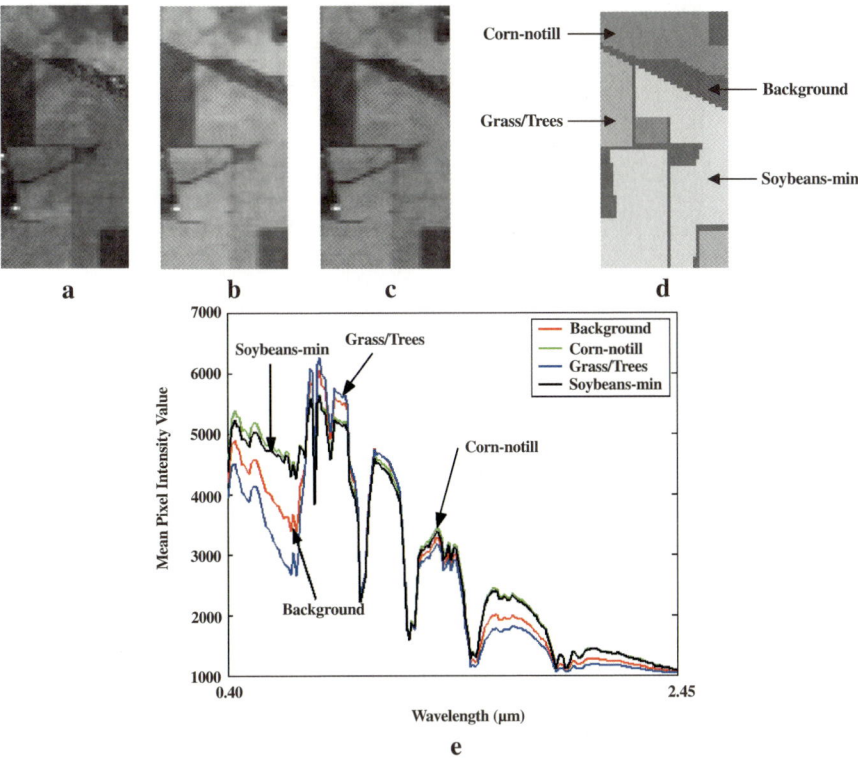

Fig. 9.2a–e. Three images (80 × 40 pixels) centered at **a** 0.48 μm, **b** 1.59 μm, and **c** 2.16 μm. These images are a subset of the data acquired by the AVIRIS sensor. **d** Corresponding reference map for this dataset consisting of four classes: background, corn-notill, grass/trees and soybeans-min. **e** Mean intensity value of each class plotted as a function of the wavelength in μm

Fig. 11.3a,b. IKONOS images of a portion of the Syracuse University campus. **a** False color composite (Blue: 0.45 – 0.52 μm, Green: 0.52 – 0.60 μm, Red: 0.76 – 0.90 μm) of the multispectral image at 4 m resolution. **b** Panchronometric image at 1 m resolution

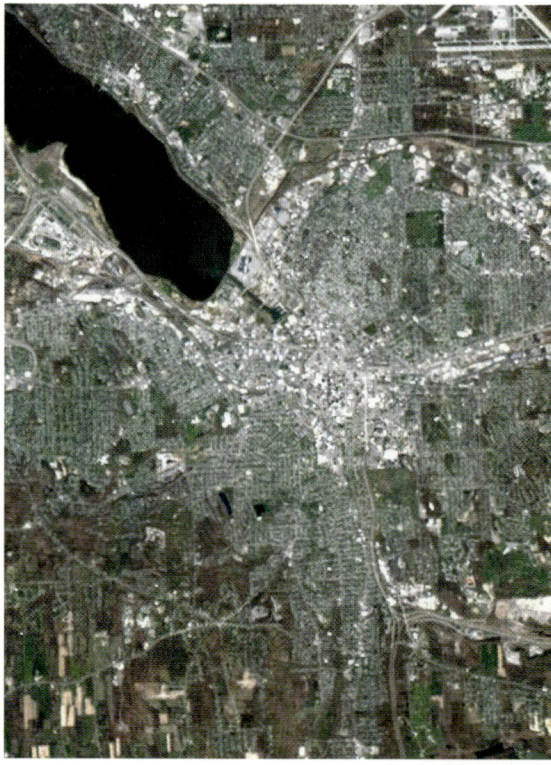

Fig. 10.1. Color composite from the Landsat 7 ETM+ multispectral image

Fig. 11.10. False color composite of the HyMap image (Blue: 0.6491 μm, Green: 0.5572 μm, Red: 0.4645 μm)

Fig. 11.12. True composite of the digital aerial photograph at 0.15 m spatial resolution, used as the reference image

Color Plate VII

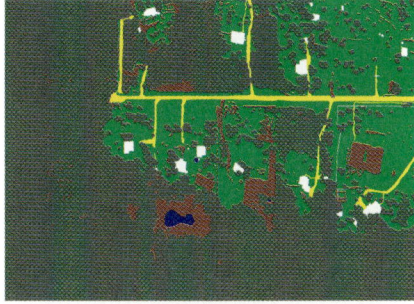

Fig. 11.13. The corresponding land cover map produced from the aerial photograph, used as the reference image

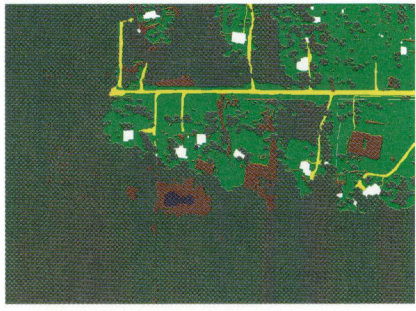

Fig. 11.15. The resampled ground reference map at 0.85 m resolution

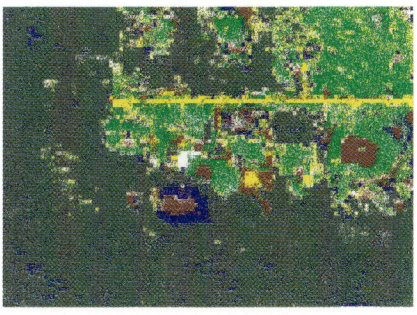

Fig. 11.16. Initial SPM at 0.85 m resolution derived from MLE

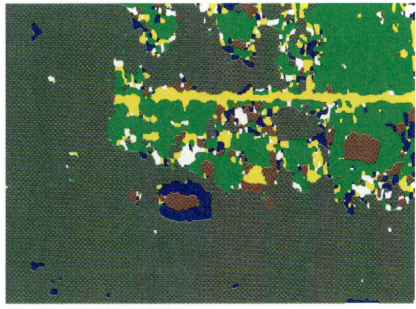

Fig. 11.17. Resulting SPM at 0.85 m resolution derived from MRF model

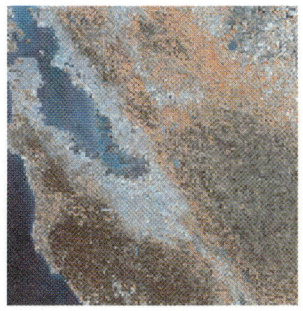

a b

Fig. 12.6a,b. Landsat TM false color composite of San Francisco Bay acquired in a April 6th, 1983; b April 11th 1988

Fig. 12.9. False color composite of the IKONOS multispectral image of NIR, red and green color spectra of Syracuse University area

Fig. 12.12. The false color composite of the resulting fused image assuming a filter size of 5 × 5 pixels

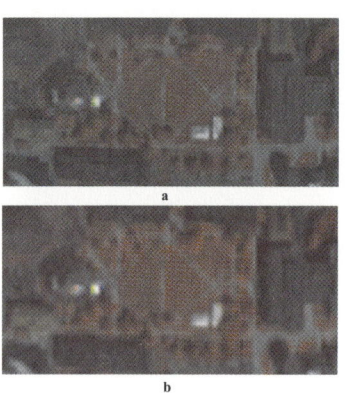
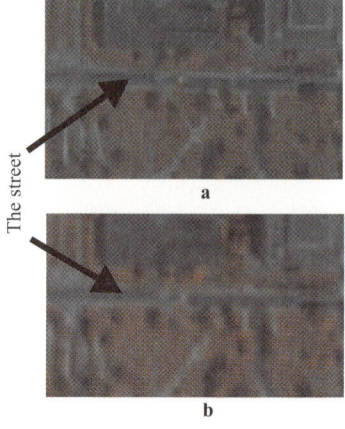

Fig. 12.13a,b. False color composite of the quad area of Syracuse University campus. **a** fused image, **b** original image

Fig. 12.14a,b. False color composite of the street next to Schine student center. **a** fused image, **b** original image

Fig. 12.15. Color composite of hyper spectral image (2.04660 µm, 1.16690 µm, and 0.64910 µm) acquired by the HyMap sensor

Fig. 12.18. The false color composite of the resulting fused image assuming a filter size of 5 × 5

Index

A

accuracy assessment   2, 76, 127, 266, 269, 271
Advanced Airborne Hyperspectral Imaging Sensor (AAHIS)   15
aerial photography   11, 20, 21, 31
Airborne Hyperspectral Imager (AHI)   15
Airborne Imaging Spectrometer (AIS)   14, 15
Airborne Prism Experiment (APEX)   16
Airborne Visible/Infrared Imaging Spectrometer (AVIRIS)   16, 19–22, 36–40, 127, 200, 210, 225, 230, 233–235, 238, 242
Analytical Spectral Devices (ASD)   27, 28
artifact   5, 54, 90, 96, 97, 99, 100, 102, 103, 107, 127, 182, 189, 192, 195–197, 202, 210
Advanced Solid-State Array Spectro-radiometer (ASAS)   15, 19, 20
Advanced Spaceborne Thermal Emission and Reflection (ASTER)   16, 21, 23–26, 238
at sensor radiance   3, 53
atmospheric absorption   20
Atmospheric Pre-Corrected Differential Absorption (APDA)   31
automatic target recognition (ATR)   81, 82
average detection rate   288
Advanced Very High Resolution Radiometer (AVHRR)   74

B

Bayes rule   284, 296
Bayes theorem   67, 68, 221
bi-directional reflectance distribution function (BRDF)   24, 25
binary classification   5, 67, 135, 137, 139, 148, 154, 240
blind source separation (BSS)   109, 110
B-spline   95, 100–102

C

change detection   2, 4–7, 33, 56, 79, 80, 165, 279–281, 284, 285, 289, 306
chunking method   153
chunking-decomposition method   241, 245
class-component density   221
clique   162–164, 260, 265–268, 272, 285, 286, 294, 295
Compact Airborne Spectrographic Imager (CASI)   16, 19, 20, 22, 31, 309
Compact High Resolution Imaging Spectrometer (CHRIS)   22, 24–27
conditional entropy   92
confusion matrix   76
contingency table   76
continuum interpolated band ratio (CIBR)   31
correlation coefficient   77, 78
CPLEX   153
crisp classification   71, 74–78
cross correlation   56, 107, 182, 190
curse of dimensionality   133

D

Daedalus   12
dark object subtraction method   53
data fusion   13, 292
datacube   14
decision boundary feature extraction (DBFE)   238

decision boundary   237, 238
Decision Directed Acyclic Graph (DDAG)   150
decision function   140, 142, 151
decision level fusion   81
decision surface   134
decision theory   67
decision tree   37, 72, 74, 134, 150, 151, 238, 257
decomposition method   153
de-striping   3
detection rate   289
Digital Airborne Imaging Spectrometer (DAIS)   16, 20
digital elevation model (DEM)   71
digital number (DN)   51
Directed Acyclic Graph (DAG)   240, 243, 244
discriminant analysis feature extraction (DAFE)   238
distortion
– radiometric   53, 54, 57
– geometric   54
– panoramic   54
divergence   77, 224
dual problem   140, 141

E

EASI-PACE   30
edge detection   64
edge intensity   65
edge operations   64
El Niño Southern Oscillation (ENSO)   33
electromagnetic radiation (EMR)   22
empirical line correction   53
empirical risk minimization (ERM)   134, 137
endmember   29, 125, 126, 129
Enhanced Thematic Mapper (ETM+)   13, 24, 27, 61, 241, 257
entropy   77, 78, 91–93, 97, 99, 115, 122, 199
ENVI   11, 30, 214
ENVISAT   23, 26
error matrix   76–78, 225, 243, 269, 275
Euclidean classifier   238
Euclidean distance   68, 69, 151
Euclidean norm   142
expectation maximization (EM)   280

expected risk   135–138, 143
extended information-maximization   218, 221
extraction and classification of homogeneous objects (ECHO)   238

F

feature based registration   56, 57, 181
feature extraction   2, 4–6, 25, 82, 114, 124, 129, 131, 199–202, 208, 215, 217, 218, 220, 222–225, 227, 229–231, 234, 235, 238, 262, 271
feature identification   181
feature level fusion   81
feature ranking   224
feature selection   201, 224, 225, 292
feature space   135, 146, 147, 154, 223, 237
field of view (FOV)   54, 55
Fluorescence Line Imager (FLI)   14
foliage projected cover (FPC)   36
foliar biochemistry   35
fraction of absorbed photosynthetically active radiation (fAPAR)   36
full-width-half-maximum (FWHM)   12, 14–16, 18, 20, 25, 26
fuzzy classification   71, 74, 75, 77, 78, 124, 257
fuzzy c-means (FCM)   75
fuzzy error matrix   78, 270

G

Gaussian mixture model   217, 219, 233
Gaussian MRF (GMRF)   286, 294, 306
generalization error   237
generalized least-square error   75
generalized partial volume estimation (GPVE)   90, 100–103, 182, 189, 193, 195–197
geocoding   3
geographical information system (GIS)   1, 2, 56, 71, 74
geometric calibration   3
geometric operations   57, 63
geometric rectification   5, 181
geometric transformations   96
geo-referenced   55

Index

Gibbs Distribution   161–164, 166, 168, 170, 171, 176, 177, 259, 298
Gibbs field   6, 160, 161, 164, 165, 167, 168, 171, 298
Gibbs potential   163, 165, 259, 260, 263, 265, 267, 268, 272, 281–283, 290, 294, 295
global positioning system (GPS)   55
graduated nonconvexity (GNC)   160
gross correction   53
ground control point (GCP)   55, 56, 181
ground spectroscopy   11, 27
ground truth   127, 289

H

Hammersley-Clifford theorem   165
Hermitian operation   166
hidden layer   72
High Resolution Visible (HRV)   13
High Resolution Visible Infra Red (HRVIR)   13, 26
histogram equalization   59, 60, 62
histogram   58–63, 68, 91, 112
Hughes phenomenon   133, 223, 237
HyMap   17, 19-23, 36, 39, 184–186, 188, 271, 276, 303
Hyperion   1, 21, 23–27, 36, 39
hyperparameters   239, 240
hyperplane   134, 135, 138–140, 142–144, 146, 147, 149, 154, 237, 239
Hyperspectral Digital Imagery Collection Experiment (HYDICE)   17, 215, 235
hyperspectral image cube   2, 124, 125
Hyperspectral Image Processing and Analysis System (HIPAS)   30
Hyperspectral product Generation System (HPGS)   30

I

IKONOS   13, 33, 258, 266, 276, 293, 299
image classification   5, 6, 66, 72, 81, 159, 165, 184, 254, 257, 293
image compression   159
image differencing   79, 190, 280, 287, 292
image distortion   53
image enhancement   5, 51, 57
image file formats
– band interleaved by pixel format (BIP)   52
– band interleaved by line format (BIL)   52
– band sequential format (BSQ)   52
– hierarchical data format (HDF)   52
image filtering   109
image fusion   4, 7, 56, 80, 81, 184, 279, 292–296, 299, 303, 306
image pyramid   185–188
image ratioing   79, 280
image registration   2, 4, 5, 56, 81, 89–91, 93–95, 103–105, 107, 181–183, 190, 192, 196
image restoration   160, 165
Imagine   1, 30
independent component analysis (ICA)   4–6, 109–111, 113–116, 119–121, 123–127, 129, 131, 199–204, 208, 211, 214, 215, 217–220, 234
Independent Component Analysis Feature Extraction (ICA-FE)   6, 7, 199, 200, 202–204, 208, 210–121, 214, 215, 218, 219
Independent Component Analysis Mixture Model (ICAMM)   6, 7, 217, 218, 220, 222, 224, 227, 229–235
Indian Remote Sensing (IRS) satellite   12, 13, 94, 105, 188–190, 192, 196
information maximization (infomax)   62, 120, 121, 219
infra red (IR)   293
instantaneous field of view (IFOV)   20
intensity based registration   5, 56, 79, 89, 103, 181, 182
intensity hue saturation (IHS)   81
International Geosphere-Biosphere Program (IGBP)   37
Ising model   159, 177, 263, 267, 272
Iterated Conditional Modes (ICM)   160, 167, 175–177, 280, 289

J

joint entropy   91, 92, 97, 99, 115, 116, 121
joint histogram   5, 89–97, 99–102, 107, 182, 188–190, 192, 195–197, 298
joint probability density function   283
joint probability mass function   90

K

kappa (κ)   77, 269, 275, 276
Karush-Kuhn-Tucker (KKT) theorem   142
Kernel Adatron (KA)   153
kernel functions   7, 100, 102, 135, 139, 147, 148, 237, 239, 240, 243, 248, 250, 252–254
kernel matrix   150
kernel trick   135
k-means algorithm   7, 69, 225, 229–232, 235
k-nearest neighbors (k-NN) algorithm   238
Kullback-Leibler distance   90
Kurtosis   111–113, 122–124, 126, 199, 221

L

L_1-distance   69, 77
L_2-distance   69
Lagrange multipliers   140–142, 145, 147, 153
Lagrangian SVM (LSVM)   154, 238, 241, 245, 246, 248, 252
LANDSAT   12, 13, 24, 27, 39, 61, 92, 96, 181, 191–193, 195, 238, 241, 257, 290
Land-Surface Processes and Interactions Mission (LSPIM)   24
Laplacian of Gaussian (LoG)   65
leaf area index (LAI)   36, 40
leptokurtic   111, 218
likelihood   34, 119, 219, 265
linear filters   63, 294, 295
linear interpolation   93, 94, 96, 97, 99, 189, 192, 194, 196, 197
linear mixture model/modeling (LMM)   75, 125, 126, 129
linear programming (LP)   152–154
linear separating hyperplane   134, 135, 138, 139, 143, 144, 146, 147, 237
linear spectral random mixture analysis (LSRMA)   218
log-likelihood   222
LOQO   153
low pass filter   295, 299

M

machine learning   5, 7, 133, 134
Mahalanobis distance   151
margin   134, 140, 142, 143, 149, 153, 154, 214, 237
marginal distribution   162, 167
marginal entropy   97
marginal probability mass functions   90, 91
Markov Chain   159, 168, 170, 175, 276, 297–299
Markov random field (MRF)   4, 6, 7, 75, 80, 81, 159–162, 164–167, 176, 177, 257–259, 265, 269, 270, 272, 275, 276, 279–282, 284, 286, 289, 292, 294, 306
M-ary hypothesis   261, 284, 295
maximum *a posteriori* (MAP)   67, 68, 160, 161, 167, 168, 170, 172, 175–177, 261, 265, 280, 281, 284, 294, 296, 306
maximum likelihood classifier (MLC)   72, 74, 75, 133, 134, 166, 238, 270
maximum likelihood estimate (MLE)   261–263, 269, 270, 272, 275, 298, 299
maximum pseudo likelihood estimation (MPLE)   263, 268, 290
mean square difference (MSD)   190–192, 194–197
Mercer's theorem   147
Medium Resolution Imaging Spectrometer (MERIS)   23–27, 74
Metropolis algorithm   160, 167, 173–175, 177, 294, 297–299, 306
Metropolis-Hasting (MH) algorithm   160, 173
millimeter wave (MMW) image   293
minimal mutual information (MMI)   116, 119, 120, 126, 204
MINOS   153
mixed pixel   7, 29, 31, 33, 71, 74, 75, 124, 237, 239, 357–259, 293
modified partial least squares (MPLS)   35
Moderate Resolution Imaging Spectrometer (MODIS)   17, 21, 23–27, 31, 37, 74, 257
Modular Airborne Imaging Spectrometer (MAIS)   17
Monte Carlo procedures   160, 285
multi-angle   12, 25, 40
Multi-angle Thermal/Visible Imaging System (AMTIS)   12
multi-camera   11

multi-class classification   148, 150, 152, 154, 240, 243
multi-sensor fusion   80
multi-sensor registration   182, 184, 188–190, 192, 197
Multispectral Infrared and Visible Spectrometer (MIVIS)   18, 38
multi-spectral scanning systems (MSS)   11–13, 22, 242
multi-temporal registration   6, 182, 183, 190, 197
multivariate data analysis   5, 109
mutual information (MI)   4, 5, 56, 89, 90, 92, 93, 99, 102, 103, 107, 113, 115–117, 120, 124, 184, 185, 189, 192, 200, 207, 212–214

N

NASA   2, 14, 17, 19, 23, 52
near infrared (NIR)   11, 13, 28, 34, 266, 299
nearest neighbor interpolation   56, 133, 134, 189, 190, 197
neural network   7, 35, 37, 72–75, 80, 81, 134, 135, 190, 237, 238, 257, 258
non-linear support vector machines   146
non-parametric classification   71
norm   140, 154, 205
normalized difference water band index (NDWBI)   36
normalizing constant   163, 260, 265, 284, 294

O

one against the rest   149, 150, 152, 240, 243
Operative Modular Airborne Imaging Spectrometer (OMIS)   18
optimization   6–7, 89, 90, 93, 103, 104, 107, 113, 116, 119, 135, 140–142, 144–147, 152–154, 161, 167, 175, 176, 182, 184–187, 189, 197, 245, 247, 252, 258, 262, 264, 269, 287, 289, 294, 297, 306
orthogonal subspace projection (OSP)   224, 225, 227–230, 232–234
overall accuracy (OA)   76–78

P

pairwise classification   149, 150, 243, 245
panchromatic (PAN)   13, 94, 105, 188–190, 192, 196, 242, 266, 299, 301
parametric classifier   67, 71
partial volume interpolation (PVI)   94–96, 99–103, 188–190, 193, 195–197
path radiance   53
penalty value   144, 146, 239, 240, 249, 254
platykurtic   111, 218
point operations   57, 58, 290
primal problem   140
principal component analysis (PCA)   5, 6, 81, 113–115, 119, 124–127, 129, 131, 199–204, 208–210, 214, 215, 217, 219, 224, 225, 227–231, 234, 262, 271, 299, 303
probability density function (PDF)   59, 60, 67, 68, 110, 111, 115, 118, 119–122, 160, 164–166, 174, 176, 204, 218, 219, 234, 259, 262, 283
producer's accuracy (PA)   76, 78
Project for On-Board Autonomy (PROBA)   24, 26
projection pursuit (PP)   224, 225, 229, 230, 234

Q

quadratic programming (QP)   140, 152–154, 241
quantization   13, 22, 51
Quickbird   13, 22, 52

R

radial basis function (RBF)   238, 239, 252, 253
RBF neural network   238
random field   161, 162, 164, 165
red edge inflection point (REIP)   34
Reflective Optics System Imaging Spectrometer (ROSIS)   18
registration consistency   6, 183, 184, 188, 190, 194–197
relative entropy   90
resolution
– radiometric   14, 22
– spatial   3, 6, 12, 13, 20, 21, 23–27, 29, 31, 34, 37, 39, 74, 80, 81, 127, 182, 184,

186–189, 192, 197, 241, 242, 257–260, 263, 266, 271, 272, 279, 295, 303
– spectral 2, 3, 20, 24, 28, 33, 39, 80, 279
– temporal 13, 22, 27, 79
resolution merge 293
RGB image 81

S

segmented PCA (SPCA) 224, 225, 227–230, 232–234
sequential minimal optimization (SMO) 153
shortwave infrared (SWIR) 13–16, 19, 20, 23–26, 34
Shortwave Infrared Airborne Spectrographic Sensor (SASI) 18–20
Short Wavelength Infrared Full Spectrum Imager (SFSI) 18
Shuttle Multi-spectral Infrared Radiometer (SMIRR) 14
signal-to-noise ratio (SNR) 14, 22, 23, 36, 82, 290
simulated annealing (SA) 160, 167, 168, 170–175, 262, 264, 265, 281, 285, 289
Sobel edge operator 65
spectral calibration 3
Spectral Image Processing System (SIPS) 30
spectral mixture analysis (SMA) 35, 36, 39
spectral reflectance 4, 21, 27, 33–36, 38, 71, 74
spectral response 2, 7, 74, 75, 79, 226, 227, 229
spectral sampling interval 220
spectral screening 60
spectral signature 14, 21, 25, 29, 33, 34, 37, 38, 40, 82, 218
SPOT 12, 13, 26, 39
statistical learning theory 5, 133–135, 145, 154, 237
structural risk minimization (SRM) 134, 137, 138, 142, 143
sub-Gaussian 111, 112, 120
sub-pixel classification 75, 257, 258, 263, 265, 268, 270
sub-pixel mapping (SPM) 259–265, 267–270, 272, 275, 276
successive over-relaxation (SOR) method 153
super-Gaussian 111, 112, 120
supervised classification 7, 67, 68, 71, 135, 200,
support vector machine (SVM) 4–7, 133–135, 138, 139, 142, 143, 146–154, 237–241, 243, 245, 247–254
Synthetic Aperture Radar (SAR) 94, 105, 188–190, 192, 195, 196

T

Thematic Mapper (TM) 13, 92, 181, 191, 192, 195, 238
thermal infrared (TIR) 14, 20, 24–26

U

Undercomplete ICA-FE algorithm (UICA-FE) 200, 203, 208–211, 214, 215
Universal Transverse Mercator (UTM) 181, 187, 241, 242
unsupervised classification 6, 7, 69, 131, 200, 214, 215, 217, 218, 220, 225, 234
user's accuracy (UA) 76–78

V

Variable Interference Filter Imaging Spectrometer (VIFIS) 18
VC-dimension 137, 138, 142, 143, 145
Vdata 52
Vegetation Mapping 30, 36, 37
Vgroup 52
Visible and Infrared Thermal Imaging Spectrometer (VIRTIS-H) 24
visible and NIR (VNIR) 13–16, 19, 23, 26, 28, 34
visiting scheme 170, 171, 177, 264
voxel 89

W

water band index (WBI) 36